Mathematics as a Second Language

Mathematics as a Second Language

SECOND EDITION

Frances Lake
The College of Staten Island

Joseph Newmark
The College of Staten Island and Brooklyn College

▲▼ **ADDISON-WESLEY PUBLISHING COMPANY**
Reading, Massachusetts · Menlo Park, California
London · Amsterdam · Don Mills, Ontario · Sydney

Second printing, May 1978

Copyright © 1977, 1974 by Addison-Wesley Publishing Company, Inc. Philippines copyright 1977, 1974 by Addison-Wesley Publishing Company, Inc.

ISBN 0-201-04099-9
ABCDEFGHIJK-HA-798

To my mother, Gertrude Weinstein Wilson

To Trudy, Sharon, and Rochelle

Preface

In writing this second edition of *Mathematics as a Second Language*, we have tried to reflect the changing approach to the teaching of liberal arts courses. To accomplish this, we have included more practical applications, both in the body of the text and in the exercises. Throughout the text we have expanded the introductory sections and included a substantial amount of historical material. We have also added a number of new sections, including discussions on infinite sets, the Babylonian number system, number patterns, abstract mathematical systems, random numbers, genetics, percentiles, and hypothesis testing. Numerous new exercises with interesting and wide-ranging applications have been added. We have also included a new chapter on computers.

For prospective elementary-school teachers, we have expanded our treatment of geometry and number bases. We have also included appendixes covering the metric system, some algebra review, and functions and graphs.

For the business and social science student, the chapters on probability and statistics have been considerably broadened. Furthermore, an entirely new chapter on computers has been added.

As in the first edition, the language is simple and clear. The writing is at a level which our experience has shown can be easily understood by students. As before, we have assumed no prior mathematical background beyond basic arithmetic.

We wish to thank the many teachers and students who received the first edition of this book so warmly. We are grateful to all those who sent us comments, suggestions, and corrections.

New York   F. L.
December 1976   J. N.

v

To the Instructor

This book was written for use in courses for liberal-arts students and for prospective elementary-school teachers. It can also be successfully used by business and social science students.

In recent years, as the admissions policies at many colleges broadened, we who were teaching these courses found that our student body had changed significantly. We no longer had a fairly homogeneous group of students. Rather, we were teaching (in the same course) heterogeneous groups whose backgrounds varied from four years of high-school mathematics to no high-school mathematics other than general math. Moreover, reading skills and motivation varied much more than previously. To overcome these problems and still provide an interesting and challenging course, we prepared a set of notes that were classroom tested over a period of two years, with both large groups and small recitation sections, at The College of Staten Island and Brooklyn College.

Our notes have proved successful because they are comprehensible to students of varying backgrounds and to those with little or no mathematical preparation. The mathematical content is complete and correct, but the language is elementary and understandable. We have purposely avoided words, phrases, and any modes of expression that might be difficult to understand. We carefully discuss, in frequent comments, points which our teaching experience has shown are often misunderstood or completely missed by students. These ideas are often taken for granted or slurred over in other texts. All our examples and exercises relate directly to the student's own experience. With the exception of a few examples, exercises, or explanations, no prior mathematical knowledge is assumed.

In selecting topics, we have allowed for a wide diversity of interest among students. We have found that many students can benefit from a solid survey in basic mathematical concepts, such as sets, logic, and real numbers, together with applications which make them more meaningful. The business- or social-science-oriented student will probably find that topics such as game theory, linear programming, probability, and statistics have more immediate relevance. For the prospective elementary-school teacher and the interested liberal-arts student, we have included material on geometry.

The instructor should find no difficulty in selecting material from these topics for a one- or two-semester course suitable for any group of students. Several suggested course outlines are appended.

To the Student

This book was written for people who do not like mathematics and for people to whom mathematics is a difficult and boring subject. We frequently meet students who tell us "Why should I study math? I am never going to use it! It has no real practical applications for me."

In this book, we hope to show how mathematics affects almost every aspect of our lives. The possible applications of mathematics are increasing every day. Did you know that an insurance company can use mathematical logic in writing its policies? and that a dress manufacturer can use advanced mathematics to decide how many size 6's and how many size 16's to produce? Every gambler (from the casual card player to the professional) can profit from a knowledge of probability. The use of statistics is so widespread that we need not comment on its importance. Regardless of your objectives in life, both career and recreational, a knowledge of mathematics is essential.

Perhaps you feel that your mathematical background is rather shaky. Don't worry about that! To read and understand this book, you need only a knowledge of elementary arithmetic. You do not need to know any high-school mathematics, such as algebra or geometry, but if you have this background, you will not find the material in this book to be a repetition. Instead, you will find how mathematics can be put to work in interesting and important ways. Occasionally a section, explanation, or exercise does require a little algebra. Such material has a † in front of it. If your background is weak, just omit it. It will not affect your understanding of what follows. Also, occasionally, a section or exercise is starred (*). This means that it is slightly more difficult, and may require more time and thought.

We hope that you will find reading and using this book an enjoyable experience, and that the mathematics in it will prove helpful to you in many of the things that you do.

Acknowledgments

We are grateful to our many colleagues and former students who gave us valuable suggestions and constructive criticisms.

Particular gratitude is expressed to our hard-working secretary, Renée Steinberg, who typed the original class notes and to Arlene Percoco who also typed parts of the manuscript. In addition, we wish to express our appreciation to the staff of Addison-Wesley for their enthusiastic interest in the project.

We also wish to thank the following people who reviewed the manuscript and made helpful suggestions for its improvement.

Robert D. Beckey, Towson State College
Joseph A. Betz, Tacoma Community College
Tom Davis, Sam Houston State University
Calvin A. Lathan, Monroe Community College
Carol L. Reagles, Western Kentucky University
Nelson G. Rich, Monroe Community College
Peter Shenkin, John Jay College of Criminal Justice

Finally, and most importantly, we wish to thank our families, Trudy, Sharon, Rochelle, and Mark, for their understanding and continued encouragement as they patiently endured the enormous strain associated with such an undertaking.

Suggested Course Outlines

	Course		Chapters
One-semester courses	Liberal arts students		1, 2, 3, 4, 5
			1, 2, 3, 8, 9
			1, 3, 6, 7
			1, 2, 3, 6, 12
			8, 9, 1, 3, 12
	Prospective elementary-school teachers		1, 2, 3, 4, 5, 7
			1, 2, 7, 8, 9
			1, 2, 3, 6, 7
	Business or social science students		8, 9, 10, 11, 13
			8, 9, 10, 12, 13
			4, 8, 9, 10, 12
Two-semester courses	Liberal arts students	Semester I	1, 2, 3, 4, 12
		Semester II	8, 9, 10, 11
	Prospective elementary-school teachers	Semester I	1, 2, 3, 4, 7
		Semester II	5, 6, 8, 9
	Business or social science students	Semester I	4, 8, 9, 12
		Semester II	10, 11, 13, 1

Background Needed for Each Chapter

Chapter	Prerequisite
1	None
2	None
3	None
4	Chapter 2 (or equivalent)
5	Chapter 4
6	None
7	None
8	Chapter 4 (or equivalent)
9	None
10	Chapter 4 (or equivalent)
11	Chapter 4 (or equivalent)
12	Chapters 3 or 4
13	Chapter 4 and some high-school algebra

Contents

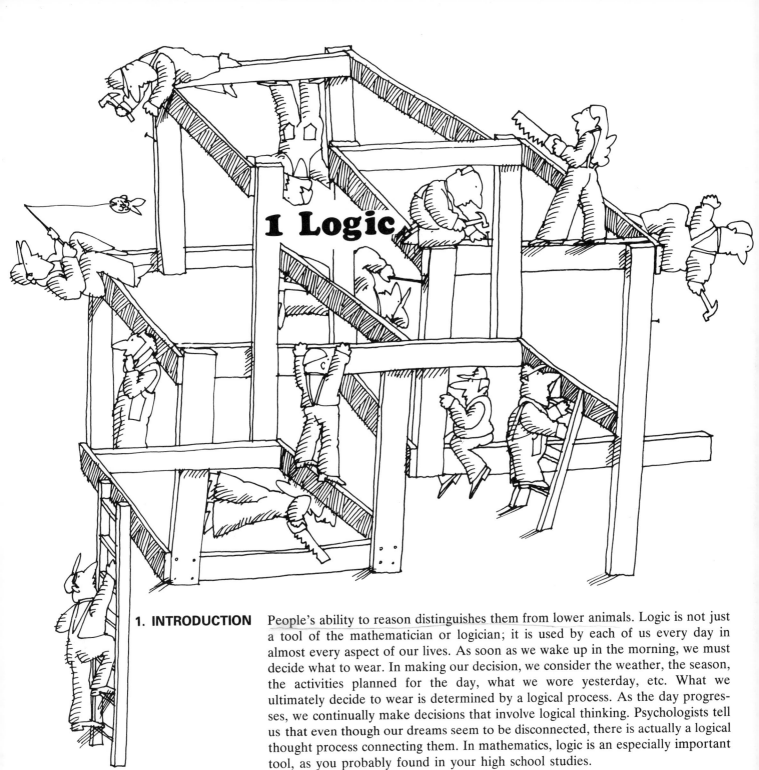

1 Logic

1. INTRODUCTION

People's ability to reason distinguishes them from lower animals. Logic is not just a tool of the mathematician or logician; it is used by each of us every day in almost every aspect of our lives. As soon as we wake up in the morning, we must decide what to wear. In making our decision, we consider the weather, the season, the activities planned for the day, what we wore yesterday, etc. What we ultimately decide to wear is determined by a logical process. As the day progresses, we continually make decisions that involve logical thinking. Psychologists tell us that even though our dreams seem to be disconnected, there is actually a logical thought process connecting them. In mathematics, logic is an especially important tool, as you probably found in your high school studies.

Aristotle

Historically, the study of logic can be traced back to the ancient Greeks, specifically to Aristotle (384–322 B.C.) He is generally considered to be the "father" of logic. The logic of Aristotle was based upon a formal kind of argument, called a *syllogism*. For example,

1. All men are mortal.

2. Socrates is a man.

Therefore, Socrates is mortal.

It can be shown that the statement "Socrates is mortal" follows logically from the first two sentences. Much of Aristotle's logic was devoted to a detailed study of such syllogisms. For many centuries, Aristotle was considered to be the ultimate authority on logic. In fact, it has been said that further developments in the study of logic were delayed for centuries because of the unquestioning faith that logicians had placed in Aristotle's work.

Aristotle's logic was concerned with statements expressed in ordinary language. Symbols were not used to any great extent. As we will see later, the use of symbols is very important in the mathematical applications of logic. The first serious attempt to use symbols was made by Gottfried Wilhelm Leibniz (1646–1716), who believed that all mathematical concepts could be derived from the principles of logic. No further progress in this area was made until the time of George Boole (1815–1865).

The English mathematician, Boole, is considered to be the founder of modern symbolic logic. He was born into a very poor family and was only able to attend third-rate schools. Nevertheless, by the age of 12 he had taught himself Latin and Greek. In *The Mathematical Analysis of Logic* and *The Law of Thought*, Boole applied the methods of algebra, using symbols, to logic. This was a major breakthrough in the theory of mathematical logic. Boole was a dedicated teacher and lecturer. In fact, he died at the age of 50 of pneumonia, which he caught when he went out in the rain to keep a lecture date.

In this chapter we will discuss logic and how it is applied.

2. INDUCTIVE AND DEDUCTIVE LOGIC

Dr. Smith recently announced that he has developed a serum for a certain disease. Over a period of ten years, he has administered his serum in varying dosages to 6,341 patients. All those receiving this medicine recovered shortly thereafter. He therefore claims that his serum is a cure for this dreadful disease. Do you believe his claim?

You would probably say yes. Let us analyze the claim more carefully. Suppose on the 6,342nd patient the serum fails. However, on the next 5,000 patients it works. Does this one failure constitute a disproof of Dr. Smith's claim? Not really. We would not expect his serum to work in every case. It should work in *almost* every case. This is how we interpret his claim. In other words, we expect that anyone who gets the serum will *probably* recover.

Now consider a traffic light at the intersection of Main Street and Broadway. A taxi pulls up just as the light turns red. Based on past experience, the driver knows that the light changes every 30 seconds; consequently, at the end of 30 seconds, he begins to move through the intersection. Since, in the past, the light has changed every 30 seconds, the driver assumes that the light will again change after 30 seconds, and that he can proceed safely. In most cases, this conclusion is correct. However, occasionally the light may be broken and will not change as anticipated. Although the driver's decision is probably justified, there still is a possibility that the light will not change.

Both of the above examples are illustrations of inductive reasoning.

Definition 2.1   *In inductive reasoning, we arrive at conclusions based on a number of observations of specific instances.* The conclusion is probably, but not necessarily, true.

Inductive reasoning is widely used in science. It is also the kind of reasoning we all use daily in making decisions. In inductive reasoning, we assume that the present or future will resemble the past and we act or reason accordingly. The following examples illustrate these ideas.

Example 1   Before leaving the house in the morning, Alice looks out the window. The skies are overcast. She has heard the weatherman forecast rain. She decides to take her umbrella. In making this decision, Alice is reasoning inductively. It *probably* will rain and Alice will need her umbrella. There is also a slim chance that it will clear up.

Example 2   Johnny is crying. His mother has just told him that they are going to the dentist. His past visits to the dentist were quite painful, so he concludes that the present visit will be painful. While his fears are *probably* justified, he may find that the visit will turn out otherwise.

Example 3   In its September 1975 issue, *Consumer Reports* published a report on scientific calculators. It concluded that the Texas Instrument Model SR-50 was a "clear first choice." *Consumer Reports* makes such conclusions on the basis of laboratory testing and surveys of users of the product. Their conclusion about the SR-50 was made inductively from the available information. It indicates that buyers of the SR-50 will *probably* be satisfied with the calculator, because the calculator will continue to perform in the future as it has in the past.

Example 4   In 1866, the Austrian monk Gregor Mendel published a major work in the theory of heredity. In experiments with garden peas, Mendel noticed that certain characteristics appeared in peas according to a recognizable pattern. For example, when he crossbred green and yellow peas, he found that out of every four peas produced, approximately three were green and one was yellow. Based upon these experiments, Mendel was able to state general "laws" of heredity, not only for plants, but also for humans. These enabled him to predict such things as eye color, hair color, etc.

Example 5   In recent years, doctors in the United States have been experimenting with the drug lithium in the treatment of mentally depressed people. Approximately 80 percent of all such patients treated with lithium have reported feeling better. As a result, doctors are concluding that lithium may be a remedy for the symptoms of chronic depression. Their reasoning is inductive. There is no guarantee that a patient who takes lithium will feel better, but it is probable that he or she will.

Example 6   Suppose you are given the following sequence of numbers: 1, 4, 7, 10, 13, . . . What is the next number in this sequence? Your answer is 16. How did you arrive at this answer? Are you one hundred percent sure that you are right? Is there any possibility that there may be a different answer?

Now consider the following argument:

1. All musicians have beards.
2. Jane is a musician.
---
3. Therefore, Jane has a beard.

In this example, we are using a different kind of reasoning called *deductive reasoning*. If you agree that statements (1) and (2) are true, then you *must* agree that statement (3) is also true. The three statements together make up an *argument*. Statements (1) and (2) are called the *hypotheses* or *premises* of the argument. Statement (3) is called the *conclusion* of the argument. It is important to note that this argument does not say that the statements (1) and (2) *are* true. It just says that *if* they are true, then so is (3).

Definition 2.2   *If we are given a series of statements with the claim that one* must *follow from the others, then this is called a **deductive argument.***

The statement that follows from the others is called the **conclusion**. The other statements are called the **hypotheses**, or **premises**.

Definition 2.3   *If the conclusion of a deductive argument* does *follow logically from the hypotheses, then we say that the argument is **valid**. If the conclusion does not necessarily follow logically from the hypotheses, then the argument is **invalid**.*

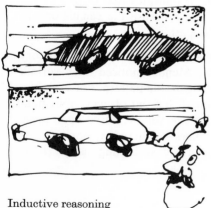  

Inductive reasoning

Is this inductive or deductive reasoning? (From Sir Arthur Conan Doyle's "The Sign of Four," *The Complete Sherlock Holmes*, New York: Doubleday, 1927. Reprinted by permission.)

It was half-past five before Holmes returned. He was bright, eager, and in excellent spirits, a mood which in his case alternated with fits of the blackest depression.

"There is no great mystery in this matter," he said, taking the cup of tea which I had poured out for him; "the facts appear to admit of only one explanation."

"What! you have solved it already?"

"Well, that would be too much to say. I have discovered a suggestive fact, that is all. It is, however, *very* suggestive. The details are still to be added. I have just found, on consulting the back files of the *Times*, that Major Sholto, of Upper Norwood, late of the Thirty-fourth Bombay Infantry, died upon the twenty-eighth of April, 1882."

"I may be very obtuse, Holmes, but I fail to see what this suggests."

"No? You surprise me. Look at it in this way, then. Captain Morstan disappears. The only person in London whom he could have visited is Major Sholto. Major Sholto denies having heard that he was in London. Four years later Sholto dies. *Within a week of his death* Captain Morstan's daughter receives a valuable present which is repeated from year to year and now culminates in a letter which describes her as a wronged woman. What wrong can it refer to except this deprivation of her father? And why should the presents begin immediately after Sholto's death unless it is that Sholto's heir knows something of the mystery and desires to make compensation? . . ."

Comment   In inductive reasoning, the conclusion is never more than probably true. In deductive reasoning, if the hypotheses are accepted as true, then the conclusion is *inescapable*.

Let us consider the following examples.

Example 7   *Hypotheses:* 1. All Brooklynites live in New York State.
2. All people who live in New York State pay high taxes.

Conclusion: All Brooklynites pay high taxes.

Example 8   *Hypotheses:* 1. All cats are dogs.
2. All dogs meow.

Conclusion: All cats meow.

Example 9   *Hypotheses:* 1. All college students are nuts.
2. All nuts talk to themselves.

Conclusion: All college students talk to themselves.

Example 10   *Hypotheses:* 1. All men are women.
2. All women are mammals.

Conclusion: All men are mammals.

Example 11   *Hypotheses:* 1. All worms are Texans.
2. All Texans are U.S. residents.

Conclusion: All worms are U.S. residents.

In Example 7, both the hypotheses and the conclusion are true. In Example 8, both of the hypotheses are false, but the conclusion is true. In Example 9, both the hypotheses and conclusion are false. In Example 11, the first hypothesis is false, but the second hypothesis is true. The conclusion is false. What can you say about the truth or falsity of the hypotheses and conclusion of Example 10? All of the above arguments are *valid*.

Comment   The above examples show that an argument may be valid even if one or more of the statements in it (that is, hypotheses or conclusion) is false. In everyday English we often use the words "true" and "valid" interchangeably. In logic, they have different meanings. A **statement** is either true or false. An **argument** is either valid or invalid.

Inductive
All dogs are small
All cats bark
All dogs bark

Deductive
All Flowers have Purple leaves
A Tulip is a flower
All Tulips have purple leaves

EXERCISES

1. Give two examples each of inductive and deductive logic (do not use any given in the text).

2. Classify each of the following as either inductive or deductive.

   a) George has been given one key to the auditorium. After trying the key a countless number of times without success, he concludes that he has the wrong key.

   b) Since the sun has risen each morning for the last $4\frac{1}{2}$ billion years, it will rise tomorrow.

   c) All early-childhood education majors must take a course in psychology. Then, since Betty is an early-childhood education major, she will be taking a course in psychology.

   d) All fish are swimmers. All swimmers wear bathing suits. Therefore, all fish wear bathing suits.

3. State the hypotheses and conclusions for each of the arguments of Exercise 2.

4. Explain how a doctor uses inductive reasoning.

5. Honest Pete has a pair of dice. If, in fifty successive throws, double sixes came up forty-seven times, would you say that the dice are loaded? Explain.

6. Assume you are offered two jobs, both with a starting salary of $5,000 per year. One job promises an annual raise of $400. The other promises a semiannual raise of $200. Which job would you take and why?

7. What is the next number in each of the following sequences?

   a) 1, 4, 9, 16, 25   b) 25, 21, 17, 13, 9.

   c) 2, 4, 6, 4, 6, 8, 6, 8, 6 d) 2, 3, 5, 7, 11, 13, 15.

   e) 0, 1, 1, 2, 3, 5, 8, 13, 20   f) $1, \frac{1}{2}, \frac{1}{4}, \frac{1}{8}, \frac{1}{16}.$

   g) 5, 5, 5, 5, 5.   h) $\frac{1}{3}, \frac{2}{3}, 1, \ldots$

   i) 1, 2, 1, 4, 1, 6, 1, 8.   j) 8, 5, 4, 9, 1, 7, 6, 11.

8. What is the next picture in the sequence on the left?

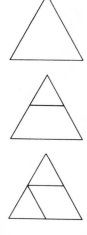

9. a) If the hypotheses of a valid deductive argument are true, *must* the conclusion be true? Explain.

b) If the hypotheses of a valid deductive argument are false, *must* the conclusion be false? Explain.

c) If the conclusion of a valid argument is false, could the hypotheses be true? Explain.

---

3. MATHEMATICAL USES OF INDUCTIVE AND DEDUCTIVE LOGIC

Look at the triangle shown below. Side AC is 4 in. and side BC is 3 in. Now measure side AB.

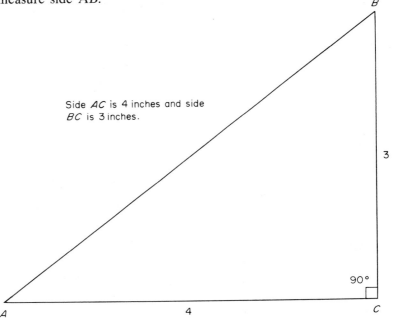

Side AC is 4 inches and side BC is 3 inches.

How many inches did you get? You should get 5 in. If you didn't, try again. Notice that

$(3 \times 3) + (4 \times 4) = 5 \times 5.$

In mathematics 3×3 is abbreviated as 3^2. Similarly, 4×4 is abbreviated as 4^2. The same is true for 5×5, which is written as 5^2. Using this notation, we have

$3^2 + 4^2 = 5^2.$

This is a special case of a well-known theorem in geometry that is known as the Pythagorean theorem. It states the following: *In a right triangle (that is, a triangle with a 90° angle) such as the one in Fig. 1, the sides are related by the formula $a^2 + b^2 = c^2$.*

The Pythagorean theorem is named for the Greek mathematician Pythagoras who lived in the sixth century B.C. Actually, the theorem was known much earlier. According to some historians, the ancient Egyptians knew of at least one special

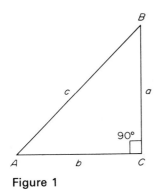

Figure 1

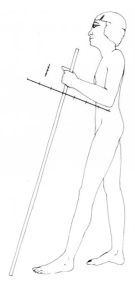

The Egyptian measuring system was based on body
measurements. For example, the cubit (see cubit rod shown) was a unit of length
based on the length of the forearm from the elbow to the tip of the middle finger
(about 18 inches).
Metropolitan Museum of Art, New York, Bequest of W. Gedney Beatty, 1941.

case of the theorem. Tablets from the Babylonian era show that they also knew of
at least one special case of this theorem. In fact, one historian claims that the
Babylonians "did indeed make use of this theorem in its full generality."[1] In
measuring land for tax assessment and in building, some of the basic ideas of
geometry were developed. It seems reasonable to assume that in measuring areas
that were in the form of right triangles, they observed this important relationship
between the sides.

The Pythagorean theorem was discovered inductively by observing special
cases of right triangles. But inductive reasoning can tell us only that it is *probably*
true. In order to *know* that it is *always* true, it must be *proved* deductively.
Remember, in deductive reasoning, the conclusion *must* be true (if the hypotheses
are true and the reasoning is valid). The proof of this theorem, which you may
have studied in high-school geometry, was given by Euclid about 200 years after
Pythagoras. It is not known when it was *first* proved. The proof involves deductive
reasoning.

We see, then, that both inductive and deductive reasoning have a place in
mathematics, the former in *discovering* new truths and the latter in *proving* these
truths.

Example 1   A **prime** number is any whole number larger than 1 that can be evenly divided
only by itself and 1 (assuming we divide only by positive numbers). Some of the
prime numbers are 2, 3, 5, 7, 11, 13, 17, 19, ... Note that

$$6 = 3 + 3$$
$$8 = 5 + 3$$
$$10 = 7 + 3 \quad \text{or} \quad 5 + 5$$
$$12 = 5 + 7$$
$$14 = 7 + 7 \quad \text{or} \quad 11 + 3$$

$$16 = 11 + 5$$
$$18 = 11 + 7 \quad \text{or} \quad 13 + 5$$

1. Asger Aaboe, *Episodes from the Early History of Mathematics.* New York: Random House, 1964,
p. 26.

Observe that all the numbers on the right are prime numbers. All the numbers on the left are *even* numbers larger than four. The mathematician Goldbach, reasoning *inductively*, claimed that *any* even number larger than four can be written as the sum of two odd prime numbers. Although it seems reasonable, as yet no one has been able to prove this *deductively*. Do you agree with his claim?

Example 2   In Exercise 7 at the end of Section 2, inductive reasoning is needed. Explain how.

Example 3   Consider the following arrangement of numbers.

$$
\begin{array}{ccccccccccc}
 & & & & & 1 & & & & & \\
 & & & & 1 & & 1 & & & & \\
 & & & 1 & & 2 & & 1 & & & \\
 & & 1 & & 3 & & 3 & & 1 & & \\
 & 1 & & 4 & & 6 & & 4 & & 1 & \\
1 & & 5 & & 10 & & 10 & & 5 & & 1
\end{array}
$$

Can you find the next row of numbers?

This arrangement of numbers is called *Pascal's Triangle*. Although studied by Pascal (1623–1662), it was known to the Chinese much earlier. Among other things, this triangle has applications in the theory of probability, as we shall see later.

Example 4   Consider the following chart.

Numbers added	Sum	Another way of writing sum
$1 + 2$	3	$\dfrac{2 \times 3}{2}$
$1 + 2 + 3$	6	$\dfrac{3 \times 4}{2}$
$1 + 2 + 3 + 4$	10	$\dfrac{4 \times 5}{2}$
$1 + 2 + 3 + 4 + 5$	15	$\dfrac{5 \times 6}{2}$
$1 + 2 + 3 + 4 + 5 + 6$	21	$\dfrac{6 \times 7}{2}$
$1 + 2 + 3 + 4 + 5 + 6 + 7$	28	
$1 + 2 + 3 + 4 + 5 + 6 + 7 + 8$	36	
$1 + 2 + 3 + \cdots + 20$	210	
$1 + 2 + 3 + \cdots + n$		

Can you complete the third column?

If you get the last entry in the last column correctly, you will have obtained a useful mathematical formula that gives the sum of the first n counting numbers, where n is any counting number. Your answer was obtained by observing the pattern in the last column. You reasoned inductively. This result can be proved deductively.

Example 5 A famous problem of mathematics is concerned with map coloring. When we color maps drawn on flat surfaces (planes), like sheets of paper, two countries having a common border must be colored with different colors. If two countries meet at only one point, they can have the same color. Some examples of maps are shown in Fig. 2. No one has ever been able to draw a map, no matter how complicated, that requires more than four colors. Therefore, by inductive reasoning, it seems probable that *every* map can be colored in *at most* four colors. However, as of now, no one has been able to prove this deductively.[2] This is the famous 4-Color Problem. It *has* been proved that no map needs more than five colors.

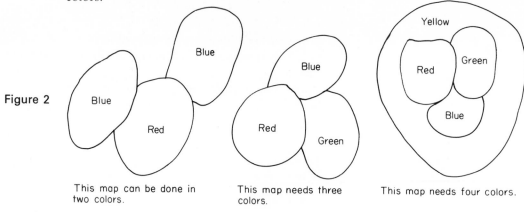

Figure 2

This map can be done in two colors.

This map needs three colors.

This map needs four colors.

2. After this material was written, it was reported that the 4-Color Problem was solved by Kenneth Appel and Wolfgang Haken of the University of Illinois. For further details, see *Science*, 13 August 1976, p. 564, and *Bulletin of the American Mathematical Society*, September 1976.

EXERCISES

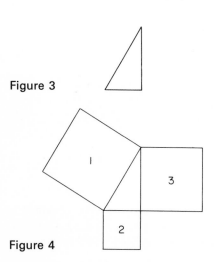

Figure 3

Figure 4

*1. An interesting proof of the Pythagorean theorem was given by the Indian mathematician Bhaskara who lived in the twelfth century. Bhaskara was also an astrologer, and there is a legend that he predicted that his daughter could only marry on a particular day at a given hour. On the wedding day, the eager girl was bending over a water clock and a pearl from her headdress fell into the clock, stopping the flow of water. By the time the accident was noticed, the wedding hour had passed and the girl was doomed to spinsterhood. To console her, Bhaskara named his best known mathematical work after her. It is called the *Lilavati*. But now back to Bhaskara's proof. The Pythagorean theorem is given on p. 7. Look back at it. Now look at the right triangle shown in Fig. 3, and draw squares on each of the three sides as shown in Fig. 4.

The Pythagorean theorem says that (the area of square 1) equals (the area of square 2) + (the area of square 3).

Unlike the proofs of the theorem that are usually given in high-school geometry books, Bhaskara's proof consists only of cutting up square 1 and rearranging it as

*The single asterisk throughout indicates that the section or exercise is somewhat more difficult and requires more time and thought.

shown in Fig. 5. His only explanation was the word "Behold!" Can you explain how this diagram "proves" the Pythagorean theorem? (*Hint:* Copy the diagram on a piece of paper, cut it up, and rearrange the pieces.)

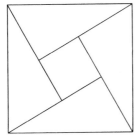
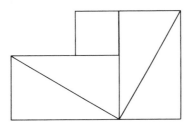

Figure 5

2. Draw maps that satisfy the following conditions.
   a) It has at least four countries and can be colored in at most two colors.
   b) It has exactly five countries and can be colored in no less than four colors.
   c) It has at least six countries and can be colored in at most two colors.

3. Suppose we draw maps, not on a sheet of paper, but on the surface of a donut. Can you draw such a map that
   a) requires only two colors?
   b) requires only three colors, but no less?
   c) requires at least four colors?
   d) requires at least five colors?
   e) requires at least six colors?
   f) requires at least seven colors?

---

4. USING VENN DIAGRAMS TO TEST VALIDITY

Many statements of the English language are of one of the following forms:

1. All *A* are *B*.
2. Some *A* are *B*.
3. Some *A* are not *B*.
4. No *A* are *B*.

For example:

"All songs are sad"   is of type (1).
"Some songs are sad"   is of type (2).
"Some songs are not sad"   is of type (3).
"No songs are sad"   is of type (4).

These statements can be represented by pictures that are known as *Venn diagrams.*

We draw two circles, one to represent songs and one to represent sad things.

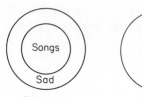

Figure 6

Figure 7

"All songs are sad" is pictured by one of the two possibilities on the left.

In Fig. 6, since "all songs are sad," then anything that is a song must be sad; but there may be things that are sad that are not songs. In Fig. 7, anything that is a song is sad, and also anything that is sad is a song. *Both* of these are acceptable pictures for "all songs are sad." The given statement does not contain enough information to tell us which of these situations is the correct one. The two diagrams represent two different *interpretations* of the statement "All songs are sad." Both of these interpretations are in agreement with the given statement.

"Some songs are sad" is pictured by one of the following possibilities. Notice the "*x*" in each of the diagrams. This will be explained shortly.

Figure 8

Figure 9

Figure 10

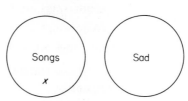

Figure 11

In Fig. 8, there are some things that are both songs and sad. There are also sad things that are not songs, and songs that are not sad things.

Although Fig. 9 implies "*all* songs are sad," this is still a possible picture for "some songs are sad." In mathematics the word "some" does not have quite the same meaning that it does in everyday English. *In mathematics "some" means "at least one and possibly all."* How does this differ from the everyday usage of the word?

In Fig. 10, some songs are sad. Moreover, there are no sad things that are not songs. This is another possible interpretation of "some songs are sad."

In Fig. 11, some songs are sad. As a matter of fact *all* songs are sad and all sad things are songs. This is yet another way of interpreting the statement "some songs are sad."

The statement "some songs are sad" means that there is at least one song that is sad. There may be many songs that are sad. Perhaps all songs are sad. All we can know from "some songs are sad" is that there is at least one song that is sad. Since we do not know which case is correct, we must allow for all the possibilities. The *x* in the diagrams represents the one song that we definitely know is sad.

"Some songs are not sad" is pictured by one of the following possibilities.

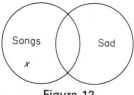

Figure 12

Figure 13

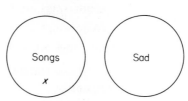

Figure 14

In Fig. 12, there is at least one song, as indicated by the *x*, that is not sad. How does this differ from Fig. 8 which represents "some songs are sad?"

In Fig. 13, there is also at least one song that is not sad. What else does this figure imply?

Figure 14 says that "no songs are sad." The statement "some songs are not sad" does allow for that possibility. The statement "some songs are not sad" does not tell us which of these three pictures is the right one. Therefore, any one of the three is a possible interpretation of "some songs are not sad."

"No songs are sad" can be pictured in only the way shown on the left.

We will now show how to test the validity of arguments by means of these Venn diagrams. Consider the following example.

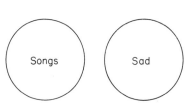

Example 1   *Hypotheses:* 1. All cats are felines.

2. All felines are mammals.

Conclusion: All cats are mammals.

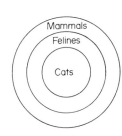

We draw a diagram to represent hypothesis (1). We place the circle of cats within the circle of felines. On the *same* diagram, we picture hypothesis (2). We place the circle of felines within a larger circle of mammals. Observe that in this diagram, the circle of cats is within the circle of mammals. The conclusion "all cats are mammals" is then inescapable. We therefore say that this argument is valid.

Some other ways of drawing the diagrams for these hypotheses are given below. Can you find any others? In each of these cases, the conclusion is inescapable and therefore the argument is *valid*.

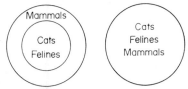

Example 2   *Hypotheses:* 1. All college students are mermaids.

2. All mermaids are human beings.

Conclusion: All college students are human beings.

Figure 15

We represent the hypotheses as shown in Fig. 15.

The conclusion "all college students are human beings" follows from the diagram since the circle of college students is within the circle of human beings. Other diagrams are possible. (Draw as many as you can.) In each case, the conclusion is inescapable. Hence, the argument is *valid*.

Example 3   *Hypotheses:* 1. All gentle people are Republicans.

2. Some dentists are gentle.

Conclusion: Some dentists are Republicans.

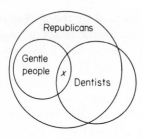

"All gentle people are Republicans" is pictured by placing the circle for "gentle people" within the circle for "Republicans."

Since "some dentists are gentle," the circle for dentists and the circle for gentle people must have some overlap. There must be at least one person who is in both circles. This person is designated by the x in the diagram. Therefore, the conclusion "some dentists are Republicans" follows. There are many other possible diagrams. You should draw at least three of these. In each case, the conclusion follows. Hence, the argument is valid.

Example 4   *Hypotheses:* 1. No children are nuisances.

2. Some children are lovable.

Conclusion: No nuisances are lovable.

Two ways of picturing the hypotheses are shown (Figs. 16 and 17).

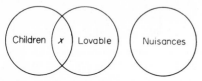

Figure 16

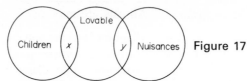   **Figure 17**

In Fig. 16, since "some children are lovable," these two circles must overlap. We are also told that "no children are nuisances." Thus the circle of "children" and the circle of "nuisances" must be separate circles. From Fig. 16, the conclusion follows.

On the other hand, look at Fig. 17 which represents a different interpretation of the hypotheses. In this diagram it is possible for someone to be lovable and also a nuisance. The conclusion "no nuisances are lovable" does not follow.

Other diagrams can be drawn. In some, the conclusion will follow. In others, it will not. In a *valid* deductive argument, the conclusion *must* follow no matter how we interpret the hypotheses (that is, draw different diagrams). Therefore, *in a valid argument the conclusion must follow from all possible pictures. If there is even one diagram that does not agree with the conclusion, then the argument is invalid.*

Getting back to our example, since we have at least one diagram (Fig. 17) that contradicts the conclusion, the argument is *invalid*.

Example 5   *Hypotheses:* 1. Some women are angels.

2. Some angels cry.

Conclusion: Some women cry.

One way of picturing the hypotheses is shown on the left. The diagram does not agree with the conclusion. We then know immediately that the argument is *invalid*.

Note that there may be other diagrams for this argument. However, we do not need to draw them, since we have already found one diagram that does not

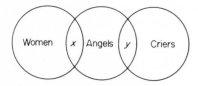

agree with the conclusion. *We draw different diagrams until we find one that contradicts the conclusion. If we cannot find such a diagram, then the argument is valid.*

Example 6   *Hypotheses:* 1. Some men are bullies.

2. No women are bullies.

3. Some crybabies are women.

Conclusion: Some men are crybabies.

One way of picturing these hypotheses is shown in Fig. 18. From the diagram the conclusion "some men are crybabies" does not follow. Hence the argument is *invalid.*

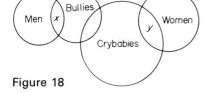

Figure 18

Example 7   *Hypotheses:* 1. Some *A*'s are *B*'s.

2. Some *B*'s are *C*'s.

Conclusion: Some *A*'s are *C*'s.

One way of picturing the hypotheses is shown below. The conclusion does not follow, so the argument is *invalid.*

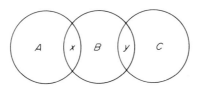

Compare Examples 5 and 7. They are really the same argument, except that in Example 5 we have used words, whereas in Example 7 we have used letters. Mathematicians and logicians often analyze arguments using letters instead of words. Perhaps this seems pointless to you. Much may be gained, however, by using this technique. In using words, your personal feelings may prejudice your decision as to the validity of the argument. No one can get emotionally involved (we hope!) with letters.

EXERCISES   Using Venn diagrams, test the validity of each of the following arguments for each conclusion.

1. *Hypotheses:*   1. All drugs are expensive.
   2. All expensive things are beautiful.

   Conclusion: All drugs are beautiful.

2. *Hypotheses:*   1. All college students are lunatics.
   2. Bill is a lunatic.

   Conclusion: Bill is a college student.

3. *Hypotheses:*   1. All math teachers are lazy.

2. All lazy people sleep until noon.

Conclusion: All math teachers sleep until noon.

4. *Hypotheses:*   1. All husbands are vicious.

2. Some wives are vicious.

Conclusion: Some husbands are wives.

5. *Hypotheses:*   1. No geniuses are college administrators.

2. Some college administrators are tall.

Conclusions: a) Some geniuses are not tall.

b) Some geniuses are tall.

c) Some college administrators are geniuses.

6. *Hypotheses:*   1. Some cars are lemons.

2. All lemons are yellow.

Conclusion: Some cars are yellow.

7. *Hypotheses:*   1. Some psychiatrists are crazy.

2. All crazy people need psychiatrists.

Conclusion: Some psychiatrists need psychiatrists.

8. *Hypotheses:*   1. Bob is a ski bum.

2. No ski bums work.

Conclusion: Bob does not work.

9. *Hypotheses:*   1. Some pilots are afraid of heights.

2. No skydivers are afraid of heights.

Conclusions: a) No pilots are skydivers.

b) Some pilots are skydivers.

c) Some skydivers are not pilots.

10. *Hypotheses:*   1. Some cats are finicky.

2. Some babies are finicky.

3. Morris is a cat.

Conclusions: a) Morris is a baby.

b) Morris is not a baby.

c) Morris is finicky.

11. *Hypotheses:*   1. All Beaver paper towels are absorbent.

2. Some Beaver paper towels are squeezably soft.

3. No absorbent things are blue.

	Conclusions:	a) No Beaver paper towels are blue.
		b) Some Beaver paper towels are squeezably soft.
		c) Some Beaver paper towels are not squeezably soft.
		d) Some squeezably soft things are blue.
		e) Some blue things are not absorbent.

12. *Hypotheses:* 1. Some detectives hate women.

2. Sherlock Holmes is a detective.

3. Some detectives are not married.

Conclusions: a) Sherlock Holmes hates women.

b) Sherlock Holmes does not hate women.

c) Sherlock Holmes is married.

d) Sherlock Holmes is not married.

e) Some people who are married hate women.

13. *Hypotheses:* 1. All barbers have beards.

2. Some men with beards have weak chins.

3. No weak-chinned men are skinny.

Conclusions: a) Some barbers have weak chins.

b) Some barbers are skinny.

c) Some skinny men do not have beards.

14. *Hypotheses:* 1. All politicans are sly.

2. Some politicians smile a lot.

3. All politicians kiss babies.

Conclusions: a) Some people who kiss babies smile a lot.

b) Some people who kiss babies are not sly.

c) Some sly people kiss babies.

d) Some people who kiss babies do not smile a lot.

e) Some sly people smile a lot.

f) Some sly people do not kiss babies.

g) Some people who smile a lot do not kiss babies.

15. *Hypotheses:* 1. All movie stars are irresistible.

2. Some people who wear wigs have false teeth.

3. No people who are irresistible wear wigs.

Conclusions: a) All irresistible people are movie stars.

b) Some people with false teeth are irresistible.

c) No one who wears a wig is irresistible.

d) No irresistible people have false teeth.

e) Some people who wear wigs are movie stars.

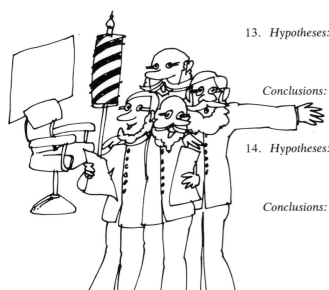

16. *Hypotheses:* 1. Some college presidents hate women.
     2. No charming people hate women.
     3. Some people who talk too much are charming.

Conclusions: a) No college presidents are charming.
     b) Some college presidents are charming.
     c) Some charming people are not college presidents.
     d) Some college presidents do not talk too much.

17. *Hypotheses:* 1. All A's are B's.
     2. All C's are B's.
     3. All A's are D's.

Conclusions: a) All C's are A's.
     b) Some B's are D's.
     c) No C's are B's.
     d) Some C's are B's.
     e) Some C's are not D's.

18. *Hypotheses:* 1. All A's are B's.
     2. All B's are C's.
     3. Some A's are D's.

Conclusions: a) Some B's are D's.
     b) All A's are C's.
     c) Some C's are D's.
     d) No C's are D's.
     e) Some C's are not D's.

19. *Hypotheses:* 1. If John eats too much cake, then he will get sick.
     2. If he gets sick, his mother will give him castor oil.
     3. If his mother gives him castor oil, he will cry.

Conclusion: If John eats too much cake, he will cry.

20. We have mentioned that there are advantages in using letters rather than words in analyzing arguments. We gave one advantage. Can you give any others?

---

*5. ANOTHER WAY OF TESTING VALIDITY—THE STANDARD DIAGRAM

There is another way of using Venn diagrams to test the validity of arguments. This is by using the **standard diagram**. The standard diagram shows all possibilities in one diagram. It has the advantage that when using it, you need only one diagram and no more. However, you may find the diagram a little more difficult to work with than other Venn diagrams.

We will illustrate the standard diagram by using it to test some of the examples of the previous section.

Example 1   *Hypotheses:* 1. All cats are felines.

2. All felines are mammals.

Conclusion: All cats are mammals.

Let *C* stand for cats, *F* stand for felines, and *M* stand for mammals. We draw three intersecting circles to represent *C*, *F*, and *M* as in Fig. 19.

Hypothesis (1) tells us that all *C*'s are inside the *F* circle. And there are *no C*'s *outside* the *F* circle. We indicate this by putting a "∅" (the null set) in *all* parts of the *C* that are *outside* the *F* circle. This is shown in Fig. 20.

Hypothesis (2) tells us that there are no *F*'s outside the *M*'s. We represent this by putting ∅ in *all* parts of the *F*'s that are outside the *M* circle. We add this to the diagram, obtaining the result shown in Fig. 21.

The conclusion says that there are *no C*'s outside the *M* circle. This *does* follow from the final diagram (Fig. 21).

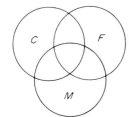

Figure 19

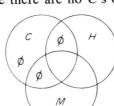

Figure 20

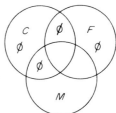

Figure 21

Example 2   *Hypotheses:* 1. All college students are mermaids.

2. All mermaids are human beings.

Conclusion: All college students are human beings.

Let *C* stand for college students, *M* stand for mermaids, and *H* stand for human beings. Again we draw three intersecting circles to represent *C*, *M*, and *H*, as in Fig. 22.

Hypothesis (1) tells us that there are no *C*'s outside the *M* circle. Thus, we put ∅ in every part of *C* that is outside the *M* circle as shown in Fig. 22.

Hypothesis (2) tells us that there is nothing in the *M* circle that is outside the *H* circle. We indicate this by putting ∅ in each part of the *M* circle that is outside the *H* circle. The resulting diagram is shown in Fig. 23.

Looking at this final diagram (Fig. 23), we see that the conclusion does follow since there are no *C*'s outside the *H* circle.

Figure 22

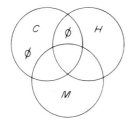

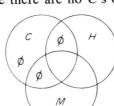

Figure 23

Example 3   *Hypotheses:* 1. All gentle people are Republicans.

2. Some dentists are gentle.

Conclusion: Some dentists are Republicans.

Let *G* stand for gentle people, *R* stand for Republicans, and *D* stand for dentists. Hypothesis (1) is pictured in Fig. 24.

To picture hypothesis (2), we must put an *x* where *D* and *G* overlap. There are two parts where *D* and *G* overlap. Since one of these already has ∅ in it, we know that the *x* can't go there. (The ∅ tells us that there is *nothing* there.) Thus, the *x* must go in the other part. This is shown in Fig. 25.

The conclusion says that there is an *x* where *H* and *R* overlap. The diagram confirms this. Thus, the argument is valid.

Figure 24 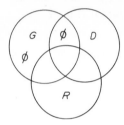   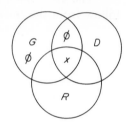 Figure 25

Example 4   *Hypotheses:* 1. No children are nuisances.

2. Some children are lovable.

Figure 26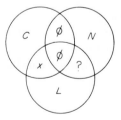

Conclusion: No nuisances are lovable.

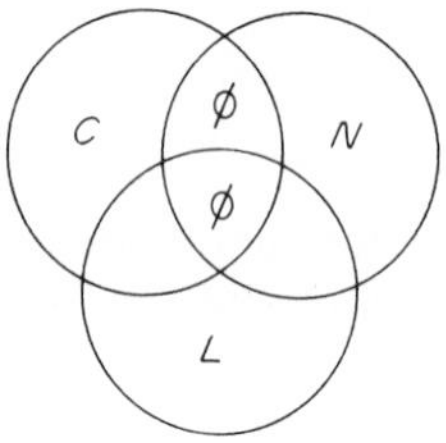

Let *C* stand for children, *N* stand for nuisances, and *L* stand for lovable.

Hypothesis (1) says that there is no one who is both a child and a nuisance. Thus, we put ∅ in each part where *C* and *N* overlap, as in Fig. 26.

Since hypothesis (2) tells us that some children are lovable, we know that we must put an *x* in one of the two parts where *C* and *L* overlap. One of these parts already has a ∅ in it. Thus, we must put the *x* in the other part. This is pictured in Fig. 27.

Figure 27

The conclusion says there is nothing in the overlap of *N* and *L*. In our diagram, the overlap of *N* and *L* has two parts. One of these definitely has a ∅ in it. The other part we know nothing about. In Fig. 27, we indicate this by a "?" The ? tells us that we do not know if there is anything there or not (from the given hypotheses). Thus, the conclusion does not follow, and the argument is *not* valid.

Example 5   *Hypotheses:* 1. Some women are angels.

2. Some angels cry.

Conclusion: Some women cry.

Let *W* stand for women, *A* stand for angels, and *C* stand for criers.

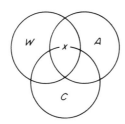

Figure 28

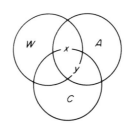

Figure 29

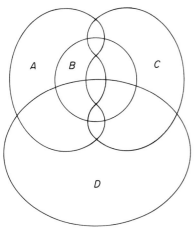

Figure 30

Hypothesis (1) tells us that there is an x in one of the two parts where W and A overlap, but it does not tell us which. Thus, we put it on the border between the two parts as shown in Fig. 28.

Hypothesis (2) tells us that there is a y in one of the two parts where A and C overlap, but again it does not tell us in which of the two parts to put it. Therefore, we again put it on the borderline between A and C, as shown in Fig. 29.

The conclusion says that there is *one x* or *y* that is in *both* W and C at the same time. In the diagram there is no such x or y, since the y is on the border between W and C. It may be in the part where W and C overlap, but, on the other hand, it may be in the part where they do not overlap. We do not know which situation is correct. The same is true for the x. Thus, the conclusion is not necessarily true, and therefore the argument is invalid.

The standard diagram can also be applied when there are more than three letters. If there are four letters, then the diagram looks like the one in Fig. 30. This diagram cannot be made by drawing four circles for A, B, C, and D. In order to get all the possibilities, we have to "wiggle" the A and C, as shown.

EXERCISES

1. Use the standard diagram to test the validity of Exercises 1–19 on pp. 15–18.

Test the validity of each of the following arguments by using the standard diagram.

2. *Hypotheses:* 1. No comedians are sad.
   2. All teachers are sad.

   Conclusion: No teachers are comedians.

3. *Hypotheses:* 1. Some ideas are dangerous.
   2. Some dangerous things are exciting.

   Conclusion: Some ideas are exciting.

4. *Hypotheses:* 1. All math teachers are blonde.
   2. Some blonde people have more fun.

   Conclusion: Some math teachers have more fun.

6. PROPOSITIONS AND TRUTH TABLES

In doing the exercises of Section 4, you probably found that in some exercises (especially the ones with three hypotheses) the number of possible diagrams became very large. Even if you used the standard diagram discussed in Section 5, you may have had difficulty. Perhaps in some cases you failed to obtain the correct answer simply because you did not think of the correct diagram. Fortunately we have available another method of testing validity called **truth tables**. This method is purely mechanical, as we shall see shortly. Nothing is left to chance.

This new technique also has the advantage of working for statements where Venn diagrams do not apply. For example, we are unable to draw a Venn diagram for the statement "Sherry will have either fish or meat for dinner."

We will start our discussion with simple sentences. Examples of these are:

1. All humans breathe oxygen.
2. $1 + 1 = 2$.
3. Hamlet was written by Elvis Presley.
4. Gerald Ford is the King of Iran.
5. How are you?
6. Marijuana is a popular type of hair spray.
7. Shut up!
8. The word "drug" is a four-letter word.
9. $x + 3 = 5$.
10. Right on!

In the above examples, sentences (1), (2), and (8) are definitely true. Sentences (3), (4), and (6) are false. Sentences (5), (7), (9), and (10) are neither true nor false. In sentence (9) if we replace x by 2, then it becomes a true sentence. If we replace x by any other value, then it becomes a false sentence. We will be concerned only with true or false sentences.

Definition 6.1   *Any sentence that is either true or false is called a **proposition** or **statement**. Propositions are denoted by the lower case letters p, q, r, . . .*

Definition 6.2   *If a proposition is a true statement, then we say its **truth value** is true, denoted by T. If a proposition is a false statement, then we say its **truth value** is false, denoted by F.*

The examples given above are simple statements. Each expresses but one idea. If we were to use only simple sentences in everyday speech and thought, then what we could express would be extremely limited. (Try for just one hour to express your ideas using only simple sentences.) For this reason, we will also deal with compound statements.

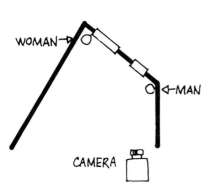

Is it logical to think that the room is rectangular? (From *The Psychology of Consciousness* by Robert E. Ornstein. New York: W. H. Freeman, copyright © 1972. Courtesy of the Exploratorium, San Francisco.)

Definition 6.3   *Simple statements joined together in different ways are called* **compound statements**.

One way of combining statements is by means of the word "and." For example, if we have the simple statements

1. Bruce failed math,

and

2. Bruce passed folk dancing,

then we can form the compound statement "Bruce failed math and passed folk dancing." Such a statement is called a *conjunction*.

Definition 6.4   *A* **conjunction** *consists of two or more statements joined together by means of the word "and." We use the symbol "∧" to stand for the word "and."*

If p stands for "Bruce failed math" and q stands for "Bruce passed folk dancing," then the statement "Bruce failed math and passed folk dancing" can be symbolized by $p \wedge q$.

As another example, if we let p = "all dogs wear glasses" and q = "flowers are intelligent," then the symbol $p \wedge q$ stands for "all dogs wear glasses and flowers are intelligent." Notice that the two simple statements forming the conjunction are not related. There is no rule that the statements must be related in any way.

It is necessary for us to have a method of determining the truth value for a conjunction, depending upon the truth values of its components (that is, the simple statements out of which it is formed). We will make up a table (similar to the multiplication and addition tables of elementary school). This table will show all the different combinations of truth values for p and q, and the resulting value for $p \wedge q$. Such a table is called a **truth table**.

p	q	$p \wedge q$
T	T	T
T	F	F
F	T	F
F	F	F

Conjunction table

Look at the first line of this table. It says that if p is true and q is true, then $p \wedge q$ is true. This seems reasonable, and it agrees with everyday usage. The second line says that if p is true and q is false, then $p \wedge q$ is false. The third line says that if p is false and q is true, then $p \wedge q$ is false. Finally, the last line says that when p is false and q is false, we conclude that $p \wedge q$ is false. These also agree with everyday usage. Thus, the conjunction of two statements is true whenever each of the statements individually is true. In any other case it is not true.

Now, if we are given any conjunction, we do not have to consider the meaning to decide if it is true or false. We can just look at the table and determine *mechanically* its truth value, depending upon the truth of falsity of the individual parts. We shall see in Section 3 how this is done.

If we join together simple statements by means of the word "or," we have what is known as a **disjunction**.

Definition 6.5

*A **disjunction** consists of two or more statements joined together by means of the word "or." We use the symbol "$\vee$" to stand for the word "or."*

If p stands for "Joe will get an A in his math course" and q stands for "Joe will get an A in his English course," then the statement "Joe will get an A in his math course or an A in his English course" can be symbolized by $p \vee q$. In this example, it is quite possible that Joe will get an A in both courses, so that both components of the disjunction may be true. This is referred to as the **inclusive disjunction**.

As another example, let p = "Mickey Mouse will be elected president of the United States in 1992" and q = "Castro will be elected president of the United States in 1992." Then the symbol $p \vee q$ stands for "Mickey Mouse will be elected president of the United States in 1992 or Castro will be elected president of the United States in 1992." In this case, obviously both cannot be true. At most, one of the statements is true. It may turn out that neither is true. If both components of a disjunction cannot be true at the same time, then this is referred to as the **exclusive disjunction**.

In our discussion we will use only the inclusive disjunction. Again, this means the disjunction $p \vee q$ is true if p is true, or if q is true, or if both p and q are true.

It is easy to see how to construct the truth table for disjunction:

p	q	$p \lor q$
T	T	T
T	F	T
F	T	T
F	F	F

Disjunction table (inclusive case)

Look at the first line of this table. It says that if p is true and q is true, then $p \lor q$ is true. This conforms to the inclusive use of the disjunction. The second and third lines say that $p \lor q$ is true if either of the components (p or q) is true. The fourth line says that a disjunction is false if both components are false.

Suppose we are given the statement "it is raining." From this we can form another statement, "it is not raining," by inserting the word "not." This is referred to as the **negation** of the original statement. Alternate ways of writing this are "it is false that it is raining" and "it is not true that it is raining."

Definition 6.6   *The **negation** of statement p is the statement "p is not true" denoted as $\sim p$ (read as "not p"). Often, to avoid awkward English, we form the negation by simply inserting "not" in the appropriate place.*

Example 1   If p is "Sharon is a genius," then $\sim p$ is "Sharon is not a genius."

Example 2   If p is "2 + 2 = 4," then $\sim p$ is "2 + 2 is not equal to 4."

Example 3   If p is "I was a fool to take this course," then $\sim p$ is "I was not a fool to take this course."

Example 4   If p is "I take a bath once a month," then $\sim p$ is "I do not take a bath once a month."

The truth table for the negation is rather obvious. It is

p	$\sim p$
T	F
F	T

Negation table

In everyday language, conjunctions, disjunctions, and negations are often combined in the same sentence. Sentences of this kind can be symbolized as illustrated in the following examples.

Example 5   Symbolize the following:

a) My cat is stuck in the tree, and the firemen are not coming to rescue it.

b) My cat is not stuck in the tree, and the firemen are not coming to rescue it.

Solution   Let p = "my cat is stuck in the tree," and let q = "the firemen are coming to rescue it." Then we can symbolize statement (a) as $p \land (\sim q)$. Using the same notation, we can symbolize statement (b) as $(\sim p) \land (\sim q)$.

Example 6 If p stands for "my car rattles" and q stands for "my mouse squeaks," then express in words each of the following symbolic expressions:

a) $(\sim p) \land q$

b) $(\sim p) \land (\sim q)$

c) $(\sim p) \lor q$

Solution a) My car does not rattle and my mouse squeaks.

b) My car does not rattle and my mouse does not squeak.

c) Either my car does not rattle or my mouse squeaks.

EXERCISES

1. Which of the following are propositions and which are not?

   a) Richard Nixon is the mayor of New York City.

   b) Do you understand the first exercise?

   c) Cool it!

   d) Most people claim that $2 + 2$ is 4.

   e) Gosh, my cat does not know how to do this problem.

   f What time is it?

   g) Let's split!

2. If p stands for "I am a monkey's uncle," and q stands for "I eat bananas," then express in words each of the following symbolic expressions.

   a) $p \land q$ b) $\sim p$ c) $p \lor q$ d) $(\sim p) \land q$

   e) $(\sim p) \land (\sim q)$ f) $p \land (\sim q)$ g) $\sim q$ h) $(\sim p) \lor q$

   i) $(\sim p) \lor (\sim q)$ j) $\sim (p \land q)$

3. If p stands for "men work" and q stands for "women weep," express in symbolic form each of the following.

   a) Men work and women weep.

   b) Men do not work.

   c) Women do not weep and men work.

   d) It is not true that men work and women weep.

   e) Men work or women weep.

   f) Men work or women do not weep.

7. IMPLICATIONS

An important statement that occurs frequently is the **implication** or **conditional**. These are statements such as the following.

1. If Mary likes Pete, then she is insane.

2. If I eat my spinach, then I will be strong.

3. If I were a cat, then I would bark.

4. If I were a bell, then I would ring.

5. If I were you, I'd run.

The implication can be defined in the following way.

Definition 7.1   *An **implication** or **conditional** is any statement of the form "if p, then q." We symbolize this as p → q. This is read as "p implies q" and means that if p is true, then q must be true.*

Definition 7.2   *The p statement of the conditional "p → q" is called the **hypothesis**, and the q statement is called the **conclusion**.*

Example 1   If *p* stands for "Leon passes the Math 5 exam," and *q* stands for "I will eat my hat," then the implication "if Leon passes the Math 5 exam, then I will eat my hat" can be symbolized as *p → q*. The hypothesis is "Leon passes the Math 5 exam." The conclusion is "I will eat my hat."

Example 2   Let *q* stand for "you are out of Schlitz" and *r* stand for "you are out of beer." The implication "if you're out of Schlitz, you're out of beer" can be symbolized as *q → r*. The hypothesis is "you are out of Schlitz," and the conclusion is "you are out of beer." What is *r → q*?

Example 3   Let *p* stand for "it is raining" and *q* stand for "the streets are wet." The implication "if it is raining, then the streets are wet" can be symbolized as *p → q*. What does the symbol *q → p* represent?

Christina's father has made the following promise to her: If she gets A's in all her courses this semester, then he will buy her a new car. There are four different things that can happen.

1. Hypothesis true: She gets A's in all her courses.
   Conclusion true: Her father buys her a new car.

2. Hypothesis true: She gets A's in all her courses.
   Conclusion false: Her father does not buy her a new car.

3. Hypothesis false: She does not get A's in all her courses.
   Conclusion true: Her father buys her a new car.

4. Hypothesis false: She does not get A's in all her courses.
   Conclusion false: Her father does not buy her a new car.

In case (1) her father certainly kept his promise. Thus the implication "if she gets A's in all her courses this semester, then he will buy her a new car" is true. *The hypothesis and conclusion are true and this makes the implication true.*

In case (2), her father definitely did not keep his promise. Thus the implication "if she gets A's in all her courses this semester, then he will buy her a new car" is false. *The hypothesis is true, but since the conclusion is false, this makes the implication false.*

p	q	$p \rightarrow q$
T	T	T
T	F	F
F	T	T
F	F	T

Implication table

In cases (3) and (4), Christina does not live up to her end of the deal. Therefore her father's promise is never put to the test. Whether he buys her a car or not, he cannot be said to have broken his promise. Thus we cannot call his implication false. Since all statements are either true or false, we then classify his implication as true. This leads us to the truth table shown at the left.

The first two lines of the implication table seem to agree with everyday usage. You may not yet be "sold" on the third and fourth lines. This is because in everyday speech we do not make implications that have false hypotheses. To convince yourself that they are reasonable consider the following discussion overheard in the student cafeteria.

Eric: My uncle is a famous movie producer.
José: If your uncle is a famous movie producer, then I am the King of France.

José's comment indicates his strong disbelief in Eric's claim that his uncle is a famous movie producer. José is sure that Eric's statement is false. José knows that his own statement "I am the King of France" is false. He definitely intends his own implication to be true. Thus, José is really saying that "false implies false" results in a true statement. This corresponds to line 4 of the implication table. This is an example of an implication with a false hypothesis being used in everyday conversation. Can you find an example from everyday usage that illustrates line 3 of the implication table?

Variations of the Conditional

Let us look at the implication (or conditional) "if it is raining, then the streets are wet." Now consider the following variations (changes).

Original statement: If it is raining, then the streets are wet.

1. *Converse:*   If the streets are wet, then it is raining.
2. *Inverse:*   If it is not raining, then the streets are not wet.
3. *Contrapositive:* If the streets are not wet, then it is not raining.

If we let p stand for "it is raining" and q stand for "the streets are wet," then the original statement can be symbolized as $p \rightarrow q$.

The **converse** (statement 1) is formed by switching around the hypothesis ("it is raining") and the conclusion ("the streets are wet"). Thus the converse is symbolized as $q \rightarrow p$.

In Example 2 on p. 27, we asked you to find $r \rightarrow q$. This was just the converse of the given implication $q \rightarrow r$. Similarly in Example 3, $q \rightarrow p$ is the converse of the given implication $p \rightarrow q$.

The **inverse** (statement 2) is formed by inserting the word "not" in the hypothesis ("it is raining") and the word "not" in the conclusion ("the streets are wet"). Thus we have "if it is not raining, then the streets are not wet." So the inverse is $(\sim p) \rightarrow (\sim q)$.

The **contrapositive** (statement 3) is formed by switching around the hypothesis and the conclusion *and* then inserting the word "not" in both the hypothesis and the conclusion. Thus we symbolize the contrapositive of $p \rightarrow q$ as $(\sim q) \rightarrow (\sim p)$.

Comment   The contrapositive of an implication is really obtained by first taking the converse of the given statement, and then taking the inverse of the result. Try it for our original statement.

Now obviously the original statement "if it is raining, then the streets are wet" is true. Must the converse also be true? Obviously not. It is possible that the street cleaners have just washed the streets. Thus if the streets are wet, it does not necessarily mean that it is raining. In other words, the converse is not necessarily true. (It may be either true or false.)

What can we say about the inverse? Must it be true? Again your answer should be "no." If it is not raining, it does not automatically follow that the streets are not wet. The street cleaners may have just washed the street. Obviously the inverse may be false.

How about the contrapositive? Everyone would agree that if the streets are not wet, then it can't be raining. Thus the contrapositive must be true.

Summarizing   If the original implication is true, then *only* the contrapositive *must* be true. The converse and inverse may or may not be true. In the next section when we discuss truth tables in more detail, we will *prove* this.

As another example, consider the implication "if John studies hard, then he will pass this course." We have the following.

Original statement: If John studies hard, then he will pass this course.

Converse: If John passes this course, then he will have studied hard.

Inverse: If John does not study hard, then he will not pass this course.

Contrapositive: If John will not pass this course, then he will not have studied hard.

Look at the original statement. The hypothesis is "John studies hard." In forming the converse, we rephrased this as "he will have studied hard." This change in the English was done merely to avoid an awkward sentence. (Had we not made this change, the converse would have read, "if he will pass this course, then John studies hard." Sounds peculiar, doesn't it?) This change does not affect the statement itself. Now look at the inverse and contrapositive. We have made similar changes. What would the inverse and contrapositive be without these adjustments?

Comment   When asked for the converse of "if John studies hard, then he will pass this course," some students answer, "John will pass this course if he studies hard." This is wrong because it says the same thing as the original statement.

It may be helpful to put these results into a table.

		Symbolic form	Truth value of this statement, assuming original statement is true
Original statement:	If p, then q.	$p \rightarrow q$	True
Converse:	If q, then p.	$q \rightarrow p$	May be true or false
Inverse:	If $\sim p$, then $\sim q$.	$(\sim p) \rightarrow (\sim q)$	May be true or false
Contrapositive:	If $\sim q$, then $\sim p$.	$(\sim q) \rightarrow (\sim p)$	*Must* be true

In the rest of this chapter, we will be using the truth tables for conjunction, disjunction, implication, and negation a lot. Therefore, for handy reference, we summarize them here.

p	q	$p \wedge q$
T	T	T
T	F	F
F	T	F
F	F	F

Conjunction table

p	q	$p \vee q$
T	T	T
T	F	T
F	T	T
F	F	F

Disjunction table

p	q	$p \rightarrow q$
T	T	T
T	F	F
F	T	T
F	F	T

Implication table

p	$\sim p$
T	F
F	T

Negation table

EXERCISES

1. If p stands for "they raise the toll on the Verrazano Bridge" and q stands for "I will swim to school," then express in words each of the following symbolic expressions.

   a) $p \rightarrow q$
   b) $q \rightarrow p$
   c) $p \rightarrow (\sim q)$
   d) $(\sim p) \rightarrow q$
   e) $(\sim p) \rightarrow (\sim q)$
   f) $(\sim q) \rightarrow (\sim p)$

2. Let r stand for "I am a mouse" and let s stand for "I am afraid of cats." Express, in symbols, each of the following.

   a) If I am a mouse, then I am afraid of cats.

   b) If I am not afraid of cats, then I am not a mouse.

   c) If I am not a mouse, then I am not afraid of cats.

   d) If I am afraid of cats, then I am a mouse.

3. Write the converse, inverse, and contrapositive of each of the following implications.

   a) If I eat my spinach, then I will be strong.

   b) If I can understand mathematics, then I am a genius.

4. Assume that the following implication is true: "If I get drunk, then I will be fired." Which of the following statements must be true? Explain.

   a) If I am not drunk, then I won't be fired.

   b) If I won't be fired, then I am not drunk.

   c) If I will be fired, then I am drunk.

   d) I will be fired if I am drunk.

8. TAUTOLOGIES AND SELF-CONTRADICTIONS

In everyday language, as well as in mathematics, we usually make statements that combine conjunctions, negations, disjunctions, and implications in different forms. Take, for example, the following statement: If it rains this evening, I will either go to the movies or go bowling." By means of truth tables we can determine when such statements are true or false. These truth tables will involve combinations of $\wedge$, $\vee$, $\sim$, and $\rightarrow$. Let us now look at some truth tables and analyze them.

Example 1   Determine the truth table for $(\sim p) \vee q$.

Solution   Two letters are involved, p and q. So we have a column for each of these.

p	q

Again looking at the given expression, we notice that the negation of p occurs as one of the parts. So we add another column, $\sim p$. We then have:

p	q	$\sim p$

Now we need a column for $(\sim p) \vee q$. This gives us the following:

p	q	$\sim p$	$(\sim p) \vee q$

Comment   We arrive at the last column by working our way across. We start *first* with the simple letters, and *then* add the necessary $\sim$, $\wedge$, $\vee$, and $\rightarrow$ where needed, in the order in which they occur. The parentheses in the above example tell us that we must *first* take the negation of p, and *then* the disjunction of this with q.

Comment   Notice that in this example we do not need a column for $p \vee q$, since this is not part of the original expression. We need a column *only* for $(\sim p) \vee q$.

Getting back to this example, we now complete the table, using the basic truth tables given on p. 30. We work from left to right, filling in the blanks.

p	q	$(\sim p)$	$(\sim p) \vee q$
T	T	F	
T	F	F	
F	T	T	
F	F	T	

Table 1

The metal figure above is made up of unrelated components (e.g., an auto for its head). Is the whole greater than the sum of its parts? Does the unrelatedness affect the logic of the composition?

"Baboon and Young," 1951, bronze, by Pablo Picasso. In the collection of the Museum of Modern Art, New York, Mrs. Simon Guggenheim Fund. Printed by permission.

In the first two columns, we have written all possible combinations of T and F (there are exactly four). The third column is obtained from the negation table. To get the final column, we go to the disjunction table and apply it to the disjunction of $(\sim p)$ and q. Notice that to do this, we cannot use the *headings* of the disjunction table. We just work from the *entries* as follows. We look at the first row in Table 1. The entries for $(\sim p)$ and q are F and T *in that order*. In the disjunction table, we find the entries F and T *in that order*. They appear on the third row. The result for this row is T. So we enter T in the first row of Table 1.

Now we look at the second row of Table 1. The entries for $(\sim p)$ and q are F and F in that order. In the disjunction table, we find the entries F and F in that order. These are on the fourth row and their result is F. So we enter F on the second row of Table 1. We now look at the third row of Table 1. The entries for $(\sim p)$ and q are T and T, in that order. In the disjunction table, these entries are on the first row and their result is T. So we enter T in the third row of Table 1. Finally, on the fourth row of Table 1, the entries for $(\sim p)$ and q are T and F in that order. We look for T and F in that order in the disjunction table, and find them on the second row. The result is T, so we enter T on the fourth row of Table 1.

The final table then looks like this:

p	q	$(\sim p)$	$(\sim p) \vee q$
T	T	F	T
T	F	F	F
F	T	T	T
F	F	T	T

The final column of the table is circled. This means:

1. If p and q are both true, then the whole statement is true.
2. If p is true and q is false, then the whole statement is false.
3. If p is false and q is true, then the whole statement is true.
4. If p is false and q is false, then the whole statement is true.

The truth value of the entire statement may be true or false depending on the truth values of p and q.

Example 2   Determine the truth table for $(p \wedge q) \rightarrow (p \vee q)$.

Solution   The parentheses tell us that we first take $p \wedge q$. Then we take $p \vee q$. Finally we connect $(p \wedge q)$ with $(p \vee q)$ by means of $\rightarrow$. Again two letters, p and q, are involved. So we need a column for each of these. Looking at the expression we see that we also need columns for $p \wedge q$ and $p \vee q$. Finally we need a column for the whole expression [which connects $(p \wedge q)$ with $(p \vee q)$ by means of $\rightarrow$]. The truth table is as follows.

p	q	$p \wedge q$	$p \vee q$	$(p \wedge q) \to (p \vee q)$
T	T	T	T	T
T	F	F	T	T
F	T	F	T	T
F	F	F	F	T

As in the previous example, we must consider all possible combinations of truth values for the letters p and q. This gives us the first two columns. The entries in the third and fourth columns are obtained from the conjunction and disjunction tables, respectively.

The final column is obtained by applying the implication table to the entries in the $(p \wedge q)$ and $(p \vee q)$ columns. This is because the last column states that the third and fourth columns are connected by means of implication. For example, the entries in the second row of the $(p \wedge q)$ and $(p \vee q)$ columns are F and T *in that order*. So we go to the implication table and look for the row containing F and T in that order. It is in the third row. The result is T, so we enter T in the final column of our chart. In a similar manner, we complete the final column.

Again we circle the final column. Notice that in this example all the entries in the final column are T. This means that $(p \wedge q) \to (p \vee q)$ is true regardless of the truth values of p and q. In other words, the statement $(p \wedge q) \to (p \vee q)$ is *always* true. Such a statement is called a **tautology**.

Definition 8.1 *Any statement that is always true is called a **tautology**. This means that in the final column of its truth table there are only T's.*

The first example we gave is not a tautology. Why not?

Example 3 Write the truth table for $(p \to q) \wedge (q \to p)$.

Solution We need columns for p, q, $(p \to q)$, $(q \to p)$, and finally $(p \to q) \wedge (q \to p)$. The truth table is:

p	q	$p \to q$	$q \to p$	$(p \to q) \wedge (q \to p)$
T	T	T	T	T
T	F	F	T	F
F	T	T	F	F
F	F	T	T	T

Example 4 Construct the truth table for $(p \to q) \wedge [p \wedge (\sim q)]$.

Solution Proceeding in the same manner as we did in the first three examples, we make columns for p, q, $\sim q$, $(p \to q)$, $p \wedge (\sim q)$, and finally for $(p \to q) \wedge [p \wedge (\sim q)]$. We then have the table shown on the following page.

p	q	$\sim q$	$p \to q$	$p \wedge (\sim q)$	$(p \to q) \wedge [p \wedge (\sim q)]$
T	T	F	T	F	F
T	F	T	F	T	F
F	T	F	T	F	F
F	F	T	T	F	F

The final column of this table consists only of F's. This statement is *never* true. Such a statement is called a *self-contradiction*.

Definition 8.2   *Any statement that is always false is called a **self-contradiction**. This means that in the final column of its truth table there are only F's.*

To better understand the idea of a self-contradiction, consider the following statement: "I have read *Hamlet* and I have not read *Hamlet*." Clearly this statement cannot be true under any circumstances. We can examine it symbolically by letting p stand for "I have read *Hamlet*." Symbolized, this statement is $p \wedge (\sim p)$. Its truth table is:

p	$\sim p$	$p \wedge (\sim p)$
T	F	F
F	T	F

Since we have only F's in the final column, it is a self-contradiction.

Example 5   Determine the truth table for $[(p \to q) \wedge (q \to r)] \to (p \to r)$.

Solution   In this case, we need columns for p, q, r, $(p \to q)$, $(q \to r)$, $(p \to q) \wedge (q \to r)$, $(p \to r)$, and finally a column for the whole statement $[(p \to q) \wedge (q \to r)] \to (p \to r)$. The truth table for this is as follows.

p	q	r	$(p \to q)$	$(q \to r)$	$(p \to q) \wedge (q \to r)$	$p \to r$	$[(p \to q) \wedge (q \to r)] \to (p \to r)$
T	T	T	T	T	T	T	T
T	T	F	T	F	F	F	T
T	F	T	F	T	F	T	T
T	F	F	F	T	F	F	T
F	T	T	T	T	T	T	T
F	T	F	T	F	F	T	T
F	F	T	T	T	T	T	T
F	F	F	T	T	T	T	T

The table has eight lines since there are 8 possible combinations of T and F for the three letters. If there were four letters, how many lines would we need for the truth table? Notice that this expression is a tautology since we have only T's in the final column.

Example 6 Show by means of truth tables that if an implication is true, then its converse does not necessarily have to be true. (If you have forgotten the definition of "converse," see p. 28.)

Solution Let the implication be $(p \rightarrow q)$. Then its converse is $(q \rightarrow p)$. We compare their truth tables.

p	q	$p \rightarrow q$
T	T	T
T	F	F
F	T	T
F	F	T

p	q	$q \rightarrow p$
T	T	T
T	F	T
F	T	F
F	F	T

Looking at line 3 of each table, we see that if p is false and q is true, then $(p \rightarrow q)$ is true, while $(q \rightarrow p)$ is false. Compare the other lines of the truth tables. They show us that $(p \rightarrow q)$ and $(q \rightarrow p)$ may have different or the same truth values depending upon the truth values of p and q.

EXERCISES

1. Construct truth tables for each of the following.

   a) $\sim(p \rightarrow q)$
   b) $\sim(p \vee q)$
   c) $\sim[p \vee (\sim q)]$
   d) $(p \wedge q) \rightarrow p$
   e) $[p \wedge (\sim q)] \rightarrow p$
   f) $[q \vee (\sim p)] \rightarrow (\sim q)$
   g) $(p \rightarrow q) \rightarrow p$
   h) $[(p \rightarrow q) \wedge p] \rightarrow q$
   i) $\sim[(p \vee q) \rightarrow (q \vee p)]$
   j) $(p \wedge q) \rightarrow (\sim r)$
   k) $(\sim p) \rightarrow (r \vee s)$
   l) $[\sim(r \wedge s)] \rightarrow [(\sim r) \vee (\sim s)]$
   m) $[(p \wedge q) \wedge (\sim r)] \rightarrow r$
   n) $\{[p \vee (\sim q)] \wedge r\} \rightarrow (p \vee r)$
   o) $[\sim(p \wedge q)] \rightarrow [(\sim p) \vee (\sim q)]$
   p) $[(\sim p) \vee (\sim q)] \rightarrow [\sim(p \wedge q)]$
   q) $[\sim(p \vee q)] \rightarrow [(\sim p) \wedge (\sim q)]$
   r) $[(\sim p) \wedge (\sim q)] \rightarrow [\sim(p \vee q)]$

2. Which of the statements of Exercise 1 are tautologies, self-contradictions, or neither?

3. Show by means of truth tables that if the implication $p \rightarrow q$ is true, then its inverse $(\sim p) \rightarrow (\sim q)$ may be false. (You will appreciate this exercise more if you reread the discussion of this on pp. 28–29 before attempting this exercise and the next one.)

4. Show by means of truth tables that if the implication $p \rightarrow q$ is true, then its contrapositive $(\sim q) \rightarrow (\sim p)$ must be true.

*5. a) If an expression involves 4 different letters, how many lines are needed for the truth table?

   b) If an expression involves 5 different letters, how many lines are needed for the truth table?

   c) If an expression involves n different letters, how many lines are needed for the truth table?

*6. The symbol "$p \veebar q$" means *either p or q is true but not both*. This is the exclusive disjunction discussed on p. 24. Write the truth table for $p \veebar q$.

*7. The symbol "$p \downarrow q$" means *p and q must both be false*. This is called the joint denial of p and q. Write the truth table for $p \downarrow q$.

8. Make up two examples each of tautologies and self-contradictions (either verbal or symbolic will do).

9. APPLICATION TO ARGUMENTS

In Section 4 we learned how to test the validity of arguments by means of Venn diagrams. We pointed out then that not only does this method not work in every case, but even when it does work, there may be too many diagrams to consider.

Truth tables provide another method of testing the validity of arguments that overcomes these difficulties. The best way to see how this method works is to actually try it on some examples.

Example 1   Test the validity of the following argument: "If Bigmouth wins the election, then I will leave this state. I am not leaving this state. Therefore Bigmouth will not win the election."

Solution   Let p stand for "Bigmouth wins the election" and let q stand for "I will leave this state." We can rewrite each sentence as follows:

	Verbally	*Symbolically*
Hypotheses:	If Bigmouth wins the election, then I will leave this state.	$p \rightarrow q$
	I am not leaving this state.	$\sim q$
Conclusion:	Bigmouth will not win the election.	$\sim p$

Remember that an argument is valid if the conclusion follows logically from the hypotheses. In other words, it is valid if the hypotheses imply the conclusion. In our example, we must determine whether *the hypotheses $(p \rightarrow q)$ and $(\sim q)$ together imply the conclusion $(\sim p)$*. That is, we are asked to determine if $[(p \rightarrow q) \wedge (\sim q)] \rightarrow (\sim p)$ is always true. (Remember, $\wedge$ stands for "and." We are asking if the conclusion follows from *both* of the hypotheses and not from one or the other alone. Therefore, we connect them by "and.") This gives us the following truth table.

p	q	$\sim p$	$\sim q$	$p \rightarrow q$	$(p \rightarrow q) \wedge (\sim q)$	$[(p \rightarrow q) \wedge (\sim q)] \rightarrow (\sim p)$
T	T	F	F	T	F	T
T	F	F	T	F	F	T
F	T	T	F	T	F	T
F	F	T	T	T	T	T

Since the last column has only T's, the expression is always true. Hence, whether the hypotheses are true or false, the conclusion *always* follows from them. Therefore, the argument is valid.

Example 2   Test the validity of the following argument: "If the teacher talks too long, Arthur gets a headache. Arthur has a headache. Therefore the teacher is talking too long."

Solution   Let r stand for "the teacher talks too long" and s stand for "Arthur gets a headache." We can symbolize each sentence of the argument as shown below.

	Verbally	*Symbolically*
Hypotheses:	If the teacher talks too long, Arthur gets a headache.	$r \to s$
	Arthur has a headache.	s
Conclusion:	The teacher is talking too long.	r

We must now determine whether $(r \to s)$ and s together imply r. We want to know if $[(r \to s) \land s] \to r$ is always true. The truth table is as follows.

r	s	$r \to s$	$(r \to s) \land s$	$[(r \to s) \land s] \to r$
T	T	T	T	T
T	F	F	F	T
F	T	T	T	F
F	F	T	F	T

The final F of line 3 shows us that if r is false and s is true, then the conclusion does *not* follow from the hypotheses. A deductive argument is valid only if the conclusion *always* follows from the hypotheses. Hence this argument is *invalid*.

Comment   In Example 1 we had all T's in the truth table, so the statement was a tautology, and the argument was valid. In Example 2, since we had one F, the statement was not a tautology, and the argument was invalid. *If an argument is valid, its truth table must be a tautology.*

Example 3   Test the validity of the following argument: "If Dave sings, the cat howls. Either the baby cries or the cat howls. The baby is not crying. Therefore Dave is not singing."

Solution   Let p stand for "Dave sings," q stand for "the cat howls," and r stand for "the baby cries." We have:

	Verbally	*Symbolically*
Hypotheses:	If Dave sings, the cat howls.	$p \to q$
	Either the baby cries or the cat howls.	$r \lor q$
	The baby is not crying.	$\sim r$
Conclusion:	Dave is not singing.	$\sim p$

The argument is symbolized as $[(p \rightarrow q) \wedge (r \vee q) \wedge (\sim r)] \rightarrow (\sim p)$. Its truth table is:

p	q	r	$\sim p$	$\sim r$	$p \rightarrow q$	$r \vee q$	$(p \rightarrow q) \wedge (r \vee q)$	$(p \rightarrow q) \wedge (r \vee q) \wedge (\sim r)$	$[(p \rightarrow q) \wedge (r \vee q) \wedge (\sim r)] \rightarrow (\sim p)$
T	T	T	F	F	T	T	T	F	T
T	T	F	F	T	T	T	T	T	F
T	F	T	F	F	F	T	F	F	T
T	F	F	F	T	F	F	F	F	T
F	T	T	T	F	T	T	T	F	T
F	T	F	T	T	T	T	T	T	T
F	F	T	T	F	T	T	T	F	T
F	F	F	T	T	T	F	F	F	T

Since the final column contains one F, this statement is not a tautology. Hence the argument is invalid.

Example 4   Test the validity of the following argument: "If I kiss that frog, it will turn into a handsome prince. If the frog turns into a handsome prince, then I will marry him. Therefore, if I kiss that frog, then I will marry him."

Solution   Let p stand for "I kiss that frog," q stand for "the frog will turn into a handsome prince," and r stand for "I will marry him." Then the argument can be symbolized as:

Hypotheses: $p \rightarrow q$
$$\underline{q \rightarrow r}$$

Conclusion: $p \rightarrow r$

The expression we have to test is $[(p \rightarrow q) \wedge (q \rightarrow r)] \rightarrow (p \rightarrow r)$. The truth table is:

p	q	r	$p \rightarrow q$	$q \rightarrow r$	$p \rightarrow r$	$(p \rightarrow q) \wedge (q \rightarrow r)$	$[(p \rightarrow q) \wedge (q \rightarrow r)] \rightarrow (p \rightarrow r)$
T	T	T	T	T	T	T	T
T	T	F	T	F	F	F	T
T	F	T	F	T	T	F	T
T	F	F	F	T	F	F	T
F	T	T	T	T	T	T	T
F	T	F	T	F	T	F	T
F	F	T	T	T	T	T	T
F	F	F	T	T	T	T	T

Since this statement is a tautology, the argument is valid.

EXERCISES   Write each of the following arguments symbolically and then test its validity.

1. If Bill's car runs out of gas, it will stop running. His car has stopped running. Therefore, Bill's car has run out of gas.

2. If they raise the taxes, Alexis will move out of New York City. Alexis is not moving out of New York City. Therefore, they have not raised the taxes.

3. Mary must make her monthly payments to the finance company or her car will be repossessed. Mary has not been making her monthly payments. Therefore, her car will be repossessed.

4. If Claude gets a raise, he will buy his wife a new watch. If he buys his wife a new watch, she will be happy. His wife is not happy. Therefore, Claude did not get a raise.

5. If Harvey drinks too much, he gets a headache. If Harvey doesn't study math, he doesn't get a headache. Therefore, if Harvey doesn't study math, he doesn't drink too much.

6. If it is hot, John will go swimming. If John goes swimming, then it is hot. Either it is hot or Mary is in town. Mary is not in town. Therefore, John will not go swimming.

7. If the teacher is interesting, then this course is worth taking. This course is not worth taking. Therefore, the teacher is not interesting.

8. To pass this math course, you must be a genius. To pass this math course, you must be crazy. Therefore, if you are crazy, you are a genius.

9. If the wheel needs greasing, then it squeaks. If it squeaks, then Ben gets the chills. The wheel is not squeaking. Therefore, Ben does not have the chills.

10. If Pete burns the hot dog, then George will get indigestion. If George gets indigestion, then either Pete didn't burn the hot dog or George threw it away. George didn't throw it away. Therefore, George will get indigestion.

11. If Leon can do all these problems, then I will eat my hat. If I eat my hat, then my head will be cold. Therefore, if my head is cold, then Leon cannot do all these problems.

12. If nuclear waste is dumped in the Hudson River, then the fish will die. If the water is not polluted, then the fish will not die. The water is not polluted. Therefore, nuclear waste is not being dumped in the Hudson River.

13. If I am a bell then I will ring. Either I am a bell or I am a telephone. Therefore, if I am a telephone then I will ring.

14. If Steve wins the lottery or robs a bank, then he will be rich. If Steve is rich, then he will not drop out of school. Therefore, if Steve does not drop out of school, he will not win the lottery.

15. If Ramon goes to the beach, he will meet Maria. If he meets Maria, he will date her. Ramon doesn't go to the beach. Therefore he won't date Maria.

10. APPLICATION TO SWITCHING CIRCUITS

Open switch

Figure 31

Closed switch

Figure 32

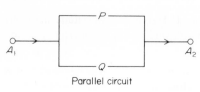

Parallel circuit

Figure 33

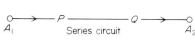

Series circuit

Figure 34

An interesting application of truth tables is to switching circuits.

A **switching circuit** consists of wires and switches through which electricity flows from one point to another. These points are called **terminals**. The purpose of a switch is either to allow or to stop electricity from flowing. The electrical system in your home is an example of a switching circuit. When you turn a light switch on, electrical current will flow, causing the light to burn. When you turn the switch off, the electrical current is interrupted and the light goes off.

In a car when you put the key into the ignition and turn it on, current will flow (the circuit is completed) and the engine starts. When you turn the ignition off, current will no longer flow and the engine stops. (The circuit is broken.) Diagrams of some simple circuits are given (Figs. 31 and 32).

In both Fig. 31 and Fig. 32, current can flow from terminal A_1 to terminal A_2. The arrows indicate the direction of the flow. Both circuits have a switch at P. In Fig. 31 the switch is open. Current *will not* flow from A_1 to A_2. The circuit is broken. In Fig. 32 the switch is closed. Current *will* flow from A_1 to A_2. The circuit is complete.

In Figs. 33 and 34, more complicated circuits are shown. Each has two switches that control the flow of current.

In Fig. 33, current starting at A_1 can go through either switch P or switch Q to get to A_2. Thus, current will flow if either switch is closed. It is not necessary for both switches to be closed. If both switches are closed, then current can flow through either P or Q or both. If both switches are open, then current cannot flow at all from A_1 to A_2. Switches P and Q are said to be connected in **parallel**.

Definition 10.1   *Two switches P and Q are said to be **connected in parallel** if current will flow when either or both switches are closed. Current cannot flow if both switches are open.*

Now look at Fig. 34. The only way for current to flow from terminal A_1 to terminal A_2 is for both switches to be closed. If either switch P or Q is open, then current will not flow. What happens if both switches are open? In this circuit, switches P and Q are said to be connected in **series**.

Definition 10.2   *Two switches P and Q are said to be **connected in series** if current will flow only when both switches are closed. If either or both switches are open, current will not flow.*

To see the connection between switching circuits and truth tables, we use the following notation.

1. If a switch is closed, then we will assign to it the letter T. If a switch is open, we will assign to it the letter F.

2. If current will flow in a circuit, we will assign the letter T to the circuit. Otherwise, we will assign the letter F to the circuit.

In Fig. 31 switch P is open and is assigned F. Current will not flow in the circuit. So we assign F to the circuit.

In Fig. 32 switch *P* is closed. We assign T to it. Current will flow. We assign T to the circuit.

Look at Figs. 33 and 34. Using this notation, we can describe what will happen in each circuit by means of truth tables.

P	*Q*	Circuit
T	T	T
T	F	T
F	T	T
F	F	F

P	*Q*	Circuit
T	T	T
T	F	F
F	T	F
F	F	F

Table for parallel circuit (Fig. 33)   *Table for series circuit (Fig. 34)*

Look at the table for parallel circuits. Line 1 says that if both switches are closed, current will flow. Lines 2 and 3 say that current will flow if either switch is open and the other closed. Line 4 says current will not flow if both switches are open. Does this remind you of another truth table? Refer back to p. 25. You will see that this is exactly the disjunction table. If switches *P* and *Q* are connected in parallel, then another way of describing this is to say *P* ∨ *Q*.

Now look at the table for series circuits. Line 1 says current will flow if both switches are closed. Lines 2, 3, and 4 show that current cannot flow if either or both of the switches are open. Does this table look familiar? It should (see p. 24). This is exactly the conjunction table. Thus, another way of saying that *P* and *Q* are connected in series is to say *P* ∧ *Q*.

More complicated circuits can be constructed by using different combinations of parallel and series circuits. One example is given on the left.

In this circuit switches *P* and *Q* are connected in series. This combination of two switches is itself connected in parallel with *R*. This circuit can be symbolized as $(P \wedge Q) \vee R$. The truth table is as follows:

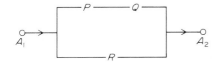

Switching circuits.

P	*Q*	*R*	$P \wedge Q$	$(P \wedge Q) \vee R$
T	T	T	T	T
T	T	F	T	T
T	F	T	F	T
T	F	F	F	F
F	T	T	F	T
F	T	F	F	F
F	F	T	F	T
F	F	F	F	F

Current will flow in the circuit if the switches are set up under the conditions of lines 1, 2, 3, 5, or 7 of the table.

As another example, consider Fig. 35 on the following page.

Notice that one switch has the peculiar label ∼*Q*. This means that this switch is open if switch *Q* is closed, and is closed if switch *Q* is open. *P* and ∼*Q* are

Bill Finch

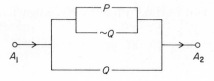

Figure 35

connected in parallel. The combination is connected in parallel with *Q*. This circuit can be symbolized as $[P \vee (\sim Q)] \vee Q$. The truth table is:

P	Q	~Q	$P \vee (\sim Q)$	$[P \vee (\sim Q)] \vee Q$
T	T	F	T	T
T	F	T	T	T
F	T	F	F	T
F	F	T	T	T

Notice that current will *always* flow in this circuit no matter which switches are open or closed. This is because we have only T's in the final column of the truth table. Therefore the switches may be removed from the circuit. They perform no useful function so far as allowing current to flow. They may be needed for other purposes.

Another example of a circuit is shown at the left.

We can symbolize this as $[P \wedge (\sim Q)] \vee (P \wedge Q)$. The truth table is as follows:

P	Q	~Q	$P \wedge (\sim Q)$	$P \wedge Q$	$[P \wedge (\sim Q)] \vee (P \wedge Q)$
T	T	F	F	T	T
T	F	T	T	F	T
F	T	F	F	F	F
F	F	T	F	F	F

There is something interesting about this table. The first and last columns are identical. This means that current will flow if *P* is closed (lines 1 and 2) and will not flow if *P* is open (lines 3 and 4). Thus *P* alone will determine whether current flows. The other switches are unnecessary. Therefore we can **simplify** this circuit by eliminating all switches but switch *P*. The simplified circuit is shown at the left.

Why would we want to simplify the circuit?

As another example consider the circuit shown on the left. In this circuit there are three *Q* switches. This means that when one of them is on, they are all on. When one of them is off, they are all off.

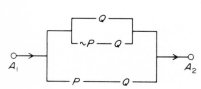

We can symbolize this circuit as $\{Q \vee [(\sim P) \wedge Q]\} \vee (P \wedge Q)$. The truth table is:

P	Q	~P	$P \wedge Q$	$(\sim P) \wedge Q$	$Q \vee [(\sim P) \wedge Q]$	$\{Q \vee [(\sim P) \wedge Q]\} \vee (P \wedge Q)$
T	T	F	T	F	T	T
T	F	F	F	F	F	F
F	T	T	F	T	T	T
F	F	T	F	F	F	F

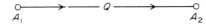

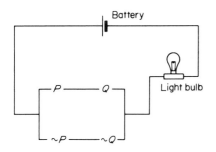

Look at the last column. It should look familiar. It is exactly the same as the column for Q. Thus, the above circuit is exactly the same as the circuit that would be symbolized as Q. This would be a simple circuit as shown at the left.

Since this circuit is much simpler than the original one and it does exactly the same job, we would probably prefer to use it rather than the original. Again we may ask, "Why would a simpler circuit be better?" There are several answers. One obvious reason is that a simpler circuit has fewer switches and is therefore cheaper and easier to construct. Also, if something goes wrong in the circuit, it is easier to locate the trouble and make repairs when there are fewer switches.

Our final example illustrates an interesting application of switching circuits to a familiar game. Everyone has at one time or another played the game "matching pennies." The game works as follows: Two players, Moe and Larry, flip a coin at the same time. If both coins come up heads or both coins come up tails, then Moe wins. Otherwise, Larry wins. A toy manufacturer is interested in making an electrical version of this game. A bulb is to light up if Moe wins. No bulb will light up if Larry wins. Instead of both players flipping coins, they will both push a button that will open or close a switch. We want to design the circuit for this game.

Let P be Larry's switch and Q be Moe's switch. At a given signal, each player pushes a button. This corresponds to flipping a coin. Just as a coin may land heads or tails, pushing a button may open or close a switch. If both switches are in the same position (open or closed), the bulb will light up. Otherwise, it will not light up. The simplest circuit for this is given at the left.

The formula for this circuit is $(P \wedge Q) \vee [(\sim P) \wedge (\sim Q)]$. Its truth table is:

P	Q	$\sim P$	$\sim Q$	$P \wedge Q$	$(\sim P) \wedge (\sim Q)$	$(P \wedge Q) \vee [(\sim P) \wedge (\sim Q)]$
T	T	F	F	T	F	T
T	F	F	T	F	F	F
F	T	T	F	F	F	F
F	F	T	T	F	T	T

This shows that current will flow and the bulb will light up if both switches are open (line 4) or if both switches are closed (line 1). If one switch is closed and one open, then current will not flow and the bulb will not light up (lines 2 and 3).

EXERCISES

1. Symbolize and construct truth tables for each of the following circuits.

a)

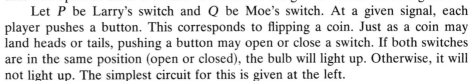

b)

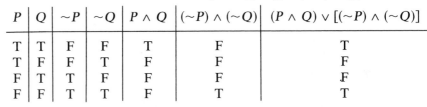

c)

d)

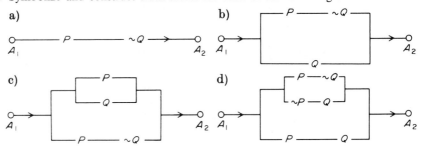

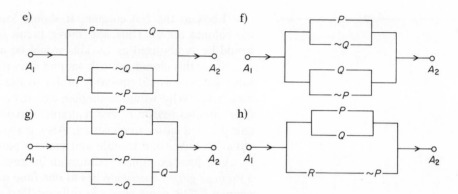

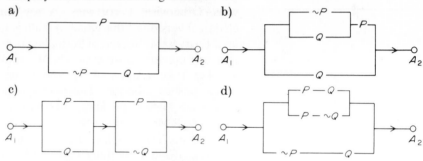

2. Simplify each of the following circuits.

3. Simplify the circuits of Exercises 1(b), 1(c), and 1(d).

4. Draw diagrams for the circuits corresponding to the following formulas.

   a) $[(\sim P) \vee (\sim Q)] \wedge P$

   b) $[P \vee (\sim Q)] \wedge [(\sim P) \vee Q]$

   c) $(P \wedge Q) \wedge (P \vee Q)$

   d) $[(\sim P) \vee Q] \vee [P \wedge (\sim Q)]$

   e) $[(P \wedge Q) \vee (\sim R)] \wedge (\sim P)$

   f) $(P \wedge Q) \vee [R \wedge (\sim Q)]$

   g) $[(P \wedge Q) \wedge (\sim R)] \vee [(\sim P) \wedge Q]$

   h) $[P \wedge (\sim Q)] \wedge [(Q \vee R) \wedge (\sim P)]$

   i) $[(P \wedge R) \vee Q] \vee [((\sim R) \wedge (\sim P)) \vee ((\sim Q) \wedge P))]$

---

11. CONCLUDING REMARKS

In the previous material on logic, a number of interesting points were purposely omitted because they were not in the mainstream of our discussion. However, they are too important to be ignored entirely. Moreover, you may have thought about them yourselves and have unanswered questions. Therefore, we conclude this chapter with these "logical afterthoughts."

Symbolizing Other Kinds of Statements

Consider the statement "there is a man with seven wives and there are men who do not have seven wives." This statement cannot be usefully symbolized by the techniques we have learned so far. If you are wondering why not, then try it.

What about the statements "all teachers, except mathematics teachers, are human," and "not all dogs like children"? If you try to symbolize these statements with what we have learned so far, you will run into trouble. Try it!

You may then wonder if there are statements that cannot be symbolized at all. By more advanced techniques of mathematical logic, it is possible to symbolize the above propositions and others like them. Such statements require the use of other symbols. For the most part, truth tables cannot be used to determine the truth value of these more complicated statements.

Many-Valued Logic

Up to now we have worked with propositions that were either true or false. It probably has occurred to you that one may not always know for sure whether a statement is definitely true or false. There are statements that we may not want to classify as definitely true or false, but rather as maybe. For example, consider the statement "Richard likes spinach." This statement may or may not be true, depending upon the condition in which it is served. For statements like this and others similar to it, we may want to consider three possible truth values: True, False, and Maybe. A logic that allows any statement to have three possible truth values is called a three-valued logic. Truth tables can be constructed accordingly. A simple example of such a table is the negation. If M stands for the truth value of "Maybe," then we have:

p	$\sim p$
T	F
M	M
F	T

Negation table for three-valued logic

A logic system may also be constructed with more than three possible truth values for any statement. Such systems are called many-valued logics. There are some situations to which such systems can be applied, but these are beyond the scope of this book.

Other Applications of Logic

We have discussed only a few of the possible applications of logic. It can be applied to many different fields, some outside mathematics. For example, if you have ever examined an insurance policy, you will find that it has a large number of confusing clauses. Some of the clauses may even appear to be contradictory or repetitious. Insurance companies often hire experts in logic to analyze and simplify policies by the methods of symbolic logic.

Similarly, logic can be used to examine complicated business contracts and legal situations for possible contradictions and repetitions.

Logic can also be used to test the accuracy of surveys and censuses.

EXERCISES
1. Make up a truth table for conjunction in a three-valued logic.

2. Make up a truth table for disjunction in a three-valued logic.

3. Make up a truth table for $p \wedge (\sim q)$ in three-valued logic.

Some of the following problems can be solved by using the techniques we have discussed throughout this chapter. Others can be solved by using informal but correct reasoning. Try to solve as many as you can. Have fun.

*4. Suppose that Mr. Smith is a small-town barber who shaves those men and only those men of the town who do not shave themselves. Which of the following statements are true?

Mr. Smith shaves himself.
Mr. Smith does not shave himself.

*5. Mr. Buick, Mr. Chrysler, and Mr. Ford owned a Buick, a Chrysler, and a Ford (not necessarily in that order.) The Chrysler's owner often beat Ford at cards. Ford was the brother-in-law of the Buick's owner. Chrysler had more children than the Chrysler's owner. Who owned the Buick?[3]

*6. A travel agent has just booked flights for three of her clients, Bill Holland, Pat Canada, and Debbie English. One of them is going to Holland, one is going to Canada, and one is going to England. Bill is not going to Holland, Pat is not going to Canada, and Debbie is not going to England. If Pat is not going to England either, to which country are they each going?

*7. There were three prisoners, one of whom was blind. Their jailor offered to free them all if any one could succeed in the following game. The jailor produced three white hats and two red hats and, in a dark room, placed a hat on each prisoner. Then the prisoners were taken into the light where, except for the blind one, they could see one another. (None could see the hat on his own head.) The game was for any prisoner to state correctly what color hat he himself was wearing. The jailor asked one of those who could see if he knew, and the man answered no. Then the jailor asked the other prisoner who could see, and he answered no. The blind prisoner at this point correctly stated the color of his own hat, winning the game for all three. How did he know?[4]

*8. A father wished to leave his fortune to the most intelligent of his three sons. He said to them: "I shall presently take each one of you away separately and paint either a white or a blue mark on each of your foreheads, and none of you will have any chance to know the color of the mark on his own head. Then I shall bring you together again, and anybody who is able to see two blue marks on the heads of his companions is to laugh. The first of you to figure out his own color is to raise his hand, and on convincing me that his solution is correct, will become my heir." After all three had agreed to the conditions, the father took them apart and painted a white

3. Avron Douglis, *Ideas in Mathematics*. Philadelphia: W. B. Saunders, 1970, pp. 5–6.

4. Irving M. Copi, *Introduction to Logic*. New York: MacMillan, 1961, pp. 16–17. Reprinted by permission.

mark on each forehead. When they met again, there was silence for some time, at the end of which the youngest brother raised his hand, saying: "I'm white." How was he able to deduce the color of the mark on his forehead?[5]

*9. Benno Torelli, friendly host at Jamtrack's most exclusive nightclub, was shot and killed by a racketeer gang because he fell behind in his protection payments. After considerable effort on the part of the police, five men were brought before the attorney, who asked them what they had to say for themselves. Each of the men made three statements, two true and one false. Their statements were:

Lefty: "I did not kill Torelli. I never owned a revolver. Spike did it."

Red: "I did not kill Torelli. I never owned a revolver. The other guys are all passing the buck."

Dopey: "I am innocent. I never saw Butch before. Spike is guilty."

Spike: "I am innocent. Butch is the guilty man. Lefty lied when he said I did it."

Butch: "I did not kill Torelli. Red is the guilty man. Dopey and I are old pals."

Whodunnit?[6]

*10. Of two tribes inhabiting a tropical isle, members of one tribe always tell the truth, and members of the other always lie. A math teacher vacationing on the isle comes to a fork in the road and has to ask a native bystander which branch he should take to reach the nearest village. He doesn't know whether the native is a truth-teller or a liar but nevertheless manages to ask only one true-false question so cleverly phrased that he will know from the reply which road to take. What question could he ask?

*11. In a certain mythical community, politicians always lie, and nonpoliticians always tell the truth. A stranger meets three natives and asks the first of them if he is a politician. The first native answers the question. The second native then reports that the first native denied being a politician. Then the third native asserts that the first native is really a politician. How many of these three natives are politicians?[7]

*12. Mabel tells her friends, "I always lie, I never tell the truth." Is Mabel lying or telling the truth? Explain your answer.

*13. Mrs. Ada Gusher, wife of the oil billionaire Tex Gusher, was found murdered. Rock Head, the dashing private eye, was called in to solve the mystery. He discovered the following clues:

a) The Gushers' maid, Dimples, was not home when the crime was committed.

b) Either Dimples was home or the Gushers' son Rodney was out.

c) If the stereo was on, Rodney was home.

d) If the stereo was not on, Mr. Gusher did it.

Rock Head solved the crime the same day. How? And who did it?

---

5. Max Black, *Critical Thinking*, 2nd ed. New York: Prentice-Hall, © 1952, p. 12. Reprinted by permission.

6. Copi, op. cit., p. 17.

7. Ibid., p. 16.

STUDY GUIDE   In this chapter, we discussed the following ideas concerning logic and its applications.

Basic ideas

Inductive logic (p. 3)
Deductive logic (p. 4)
Hypothesis (premise) (p. 4)
Conclusion (p. 4)
Valid argument (p. 4)
Invalid argument (p. 4)
Mathematical uses of inductive logic (p. 7)
Venn diagrams (p. 11)
Proposition (statement) (p. 22)
Truth value (p. 22)
Simple statement (p. 22)
Compound statement (p. 23)
Conjunction (p. 23)
Disjunction (inclusive and exclusive) (p. 24)
Truth tables (p. 24)
Variations of the conditional (p. 28)
Converse (p. 28)

Inverse (p. 28)
Contrapositive (p. 29)
Tautologies (p. 33)
Self-contradictions (p. 34)
Switching circuit (p. 40)
Parallel circuit (p. 40)
Series circuit (p. 40)
Many-valued logics (p. 45)

Applications were given for:

a) arguments (p. 36),
b) switching circuits (p. 40).

Testing the validity of arguments by

a) Venn diagrams (p. 11),
b) standard diagram (p. 18),
c) truth tables (p. 22).

SUGGESTED FURTHER READINGS

Bell, E. T., *Men of Mathematics.* New York: Simon & Schuster, 1961. Chapter 23 contains a bibliography of G. Boole.

Copi, I. M., *Introduction to Logic.* New York: Macmillan, 1961. This text contains a clear discussion of inductive and deductive logic and the use of inductive logic in science.

Dantzig, T., *Number, The Language of Science.* New York: Macmillan, 1954. Chapter 4 discusses inductive reasoning.

Franklin, P., "The Four-Color Problem." *Scripta Mathematica* 6: 149–156, 197–210 (1939).

Kemeny, J., J. L. Snell, and G. Thompson, *Introduction to Finite Mathematics*, 3rd ed. Englewood Cliffs, N.J.: Prentice-Hall, 1974. Chapter 1 discusses logic.

Mathematics in the Modern World (Readings from *Scientific American*). San Francisco: W. H. Freeman, 1968. Article 1 discusses mathematical discovery, Article 29 discusses symbolic logic, and Article 31 discusses Gödel's proof.

May, K. O., "The Origin of The Four-Color Problem." *Isis* 56: 346–348 (1965).

Nagel, E., "Symbolic Notation, Haddocks' Eyes and the Dog-Walking Ordinance." In James R. Newman, ed., *The World of Mathematics*, Vol. 1, pp. 1878–1900. New York: Simon & Schuster, 1956. This article shows how logic can be applied to everyday language.

Nagel, E., and J. Newman, *Gödel's Proof.* New York: New York University Press, 1958.

Northop, E. P., *Riddles in Mathematics: A Book of Paradoxes.* New York: Van Nostrand, 1944.

Ore, O., *The Four-Color Problem.* New York: Academic Press, 1967.

Stebbing, L. S., *A Modern Elementary Logic.* London: Methuen, 1961. This book contains a detailed discussion of both inductive and deductive logic with applications.

2 Sets

1. INTRODUCTION

In recent years there has been much talk about the "new math." In reality, there is nothing really new about the new math. Rather, it is a different way of looking at traditional math. The new math has its roots in the last half of the nineteenth century. At that time, mathematicians started to look very carefully at the basic ideas and methods of their subject, at the "foundations" of mathematics. They began to ask questions such as: What is a number? What is infinity? How do we know that our methods of reasoning are correct? Moreover, general philosophic questions were considered, questions such as these: What is mathematics? Is mathematics part of science? What is the relationship between mathematics and the "real world"? Do mathematical concepts like "number," "point," and "circle" really exist?

One of the outstanding men involved in these investigations was Georg Cantor. Born in 1845 in St. Petersburg (now Leningrad), he spent most of his life in Germany. He decided at an early age that he wanted to be a mathematician, but his father was determined that he should become an engineer, which he believed to be a better-paying profession. Being an obedient son, Cantor did study engineering as his father wished; however, he was so miserable that when he was seventeen, his father finally allowed him to pursue a career in mathematics.

Cantor became a great mathematician. However, his work was so different and controversial that it was vigorously attacked by the mathematical community. This led to a prolonged and unpleasant dispute with one of his former teachers, Leopold Kronecker (1823–1891), himself an important mathematician. Possibly as a result of the hostility and criticism directed at his work, Cantor was never able to get a teaching position at the University of Berlin, which was his ambition, but taught all his life at a third-rate university. Unable to cope with the attacks and the disappointments, Cantor suffered several mental breakdowns. He died in an insane asylum at the age of seventy-three.

It is not hard to understand why Cantor's work met with such resistance, since many of his results were startling. For example, suppose that you could draw a line from your house to the moon.

Now look at this line:

Georg Cantor

Which line has more points on it? Did you say the line to the moon? Then you are wrong. Cantor showed that both lines contain *the same number of points.* Even Cantor himself found some of his results hard to accept. In a letter to another mathematician, he wrote of one result, "I see it, but I don't believe it."

Basic to Cantor's work is the idea of a set. The word "set" in mathematics has exactly the same meaning that it does in everyday English usage. It is just any collection of things.

In our daily lives we come across sets all the time. Think of all your friends and relatives; they form a set. The pages in this book form a set. If you own a phonograph, your collection of records forms a set. Empty your pockets! The contents form a set.

In what follows, we will see how the ideas of sets are easy to understand and interesting to apply.

2. DEFINITION OF A SET AND ITS ELEMENTS

Since the idea of a set is so important in mathematics, we will state again what we mean by it.

A **set** is any collection of objects.

Example 1   The set of all letters of the English alphabet.

Example 2   The set of all American Indians.

Example 3   The set of all people over 22 ft tall.

Example 4   The set of all months with less than 30 days.

Example 5   The set of all numbers larger than 1.

Definition 2.1   *The objects in the set are called either **elements** or **members** of the set.*

In Example 1 above, the elements are $a, b, c, d, e, f, g, h, i, j, k, l, m, n, o, p, q, r,$ $s, t, u, v, w, x, y,$ and z. In Example 2, some of the elements are Pocahontas, Jim Thorpe, Geronimo, and Sitting Bull. In Example 4, there is only one element, namely, February. How many elements are there in the set of Example 3? of Example 5?

Instead of writing out the words "the set of all letters of the English alphabet," we can write the same thing in shortened form as {all letters of the English alphabet}. Similarly, Example 2 can be written as {all American Indians}. The braces (curly brackets) stand for the set containing whatever is written inside them.

Very often we must refer to a set whose elements we do not know or do not wish to write down. We then denote the set by the capital letter A, or B, or C, etc. We denote the elements of such a set by small letters of the alphabet, a, b, c, etc.

In Example 1 above, we notice that "g" is an element of the set. If we denote this set by the letter A, we write $g \in A$. This means "g is an element of set A," that is, the symbol $\in$ stands for the words "is an element of." Similarly, $a \in A$, $b \in A$, $c \in A, \ldots, z \in A$. Observe that Geronimo is not an element of set A. We symbolize this by writing Geronimo $\notin A$, that is, $\notin$ will stand for "is not an element of." If B denotes the set of Example 2, then Geronimo $\in B$.

Notice that in Example 1 we have merely described the elements of the set without actually naming them. For someone not familiar with the English language, it would not be immediately obvious which elements are in the set. It would then be advisable to actually list the elements of the set, and write it as

$$A = \{a, b, c, d, e, f, g, h, i, j, k, l, m, n, o, p, q, r, s, t, u, v, w, x, y, z\}.$$

This set can also be written in abbreviated form as $\{a, b, c, d, \ldots, z\}$ where the three dots stand for the letters between d and z. This notation assumes that the reader *is* familiar with the letters of the English alphabet. In the set $\{1, 2, 3, 4, \ldots\}$, the three dots stand for 5, 6, 7, 8, and so on.

When we list the elements of a set, we refer to the process as the **roster method**, as opposed to the **descriptive method** in which we *describe* the elements of a set rather than list them. Similarly, the set of Example 4 can be written in the roster method as {February}.

In a West Coast town, the local Clear-Air Committee and the Pure-Water activists decide to merge into one antipollution group. A list is drawn up of the membership of the new group. This list will, of course, contain the names of all those in either of the original groups. Mr. I. M. Concerned belongs to both of the original groups. We write his name only *once* on the combined mailing list.

When working with sets, an element of a set is never listed more than once. If it were, there would be pointless duplication.

Consider the set M = {all great baseball players}. Is Tom Seaver an element of that set or not? Some people may say yes, others may say no. There is no way of determining who is right. Such a set is said to be **not well defined**.

Definition 2.2   *If there is a way of determining for sure whether an object belongs to a set or not, we say the set is **well defined**.*

Example 6   The set of all pretty girls.

Example 7   The set of all tall men.

Example 8   The set of all ripe apples.

Example 9   The set of all 2-door automobiles.

Example 10   The set of all planets on which water can be found.

Examples 6 and 7 are *not* well defined because one may disagree as to whether Mary is pretty or not, or as to whether John, who is 5 ft 8 in., is tall or not. Is Example 8 well defined or not? Example 9 *is* well defined. Example 10 *is* well defined also, because there is a way of determining which planets are to be included in this set (even though, at present, technology has not advanced far enough for us to know.) In mathematics, we use only well-defined sets.

---

3. KINDS OF SETS   Consider the set {all months with less than 30 days}. This set has only one element, namely, February. Such a set is called a **unit** set.

Definition 3.1   *Any set that has exactly one element in it is called a **unit set**.*

Now how many elements are there in the set {all people who weigh a ton} or the set {all people over 30 feet tall}? Your answer, of course, will be "none." Such a set is called a **null** set.

Definition 3.2   *Any set that has no elements in it is called a **null** or **empty set**. We denote such a set by the symbol $\varnothing$ or { }.*

We will show in the next section that there is only one null set.

Let us look again at Examples 1–5 in Section 2. Example 1 has 26 elements in it. Example 2 has a specific number of elements in it, although we may not know the number offhand. (The latest U.S. census figures show that there are 985,000 American Indians.) Example 3 has no elements and is thus a null set. Example 4 has only one element and is thus a unit set. All of these are examples of **finite sets**. How many elements are there in Example 5? Example 5 is called an **infinite set**. Another example of an infinite set is the set of all fractions. Do you think that the set of all grains of sand on the beaches of Florida is a finite or an infinite set?

Stock, Boston

4. EQUAL AND EQUIVALENT SETS

John and Ann were asked to think of a set. John thought of the set of all the vowels in the English alphabet. Ann thought of the set {a, e, i, o, u}. Are they thinking of different sets? Of course not. Both sets obviously have the same elements.

Definition 4.1   *Two sets that have exactly the same elements are called **equal sets**.*

Mabel thought of the set {e, o, a, u, i}. Leon thought of the set {t, o, a, u, i}. Notice that Mabel's set is the same as Ann's and John's set, even though the elements are presented in a different order. *The order in which the elements of a set are written is not important.* What about Leon's set? Is it equal to the others? You probably answered that Leon's set was not equal to the others. However, it has the same number of elements as the others. Let us investigate this further.

Imagine that we have a set of three students: Maurice, Ben, and Arline. We also have a set of three chairs: a red chair, a blue chair, and a green chair. Each of the three students sits down on a chair. Obviously each chair will be occupied and each student will be seated. There will be no vacant chair and no standing student. We have matched each student with a chair and each chair with a student. Such a matching is called a one-to-one correspondence. In more general terms, the following holds.

Definition 4.2   *Set A and set B can be put in **one-to-one** (1–1) **correspondence** if each element of set A can be paired with exactly one element of set B and every element of set B can be paired with exactly one element of set A. No element of either set may be left out.*

Example 1 $A = \{a, e, i, o, u\}$

$B = \{t, o, a, u, i\}$

Sets A and B can be put in 1–1 correspondence in many different ways. One way is

$A = \{a, e, i, o, u\}$

$\updownarrow \updownarrow \updownarrow \updownarrow \updownarrow$

$B = \{t, o, a, u, i\}$.

Another possible way is the following:

$A = \{a, e, i, o, u\}$

$B = \{t, o, a, u, i\}$.

How many other possible ways can you find of putting set A and set B in 1–1 correspondence?

Example 2   Let

$C = \{\triangle, \square, 0, 人\}$ and

$D = \{a, b, c, d, e\}$.

If we attempt to put these sets in 1–1 correspondence, we find that it cannot be done. One element of set D is always left out of any pairing. For example,

$C = \{\triangle, \square, 0, 人\}$

$\updownarrow \quad \updownarrow \quad \updownarrow \quad \updownarrow$

$D = \{a, \quad b, \quad c, \quad d, \quad e\}$.

Observe that element e is not matched with any element of set C. If we try to match e with any element of set C, we will be forced to leave out another element of set D. Thus, sets C and D cannot be put into 1–1 correspondence.

In Example 1 above, set A and set B are not equal, but they can be put in 1–1 correspondence. We say that they are **equivalent**.

Definition 4.3   *Sets A and B are said to be **equivalent** if they can be put into 1–1 correspondence.*

Comment   In everyday English, "equal" and "equivalent" mean the same thing. In our discussion of sets, equal sets and equivalent sets are not the same thing.

How many elements are there in set A and set B of Example 1? What about set C and set D of Example 2? In Example 1, both sets A and B have five elements. On the other hand, in Example 2, set C has four elements, whereas set D has five elements. This leads us to the following definition.

Definition 4.4 *The* ***cardinal number*** *of any set A is defined as the number of elements in set A. This is denoted by the symbol $n(A)$, read as "the number of elements in set A."*

Example 3 The set $T = \{x, y, z\}$ has cardinal number 3, that is, $n(T) = 3$.

Example 4 The set $W = \{$□, △, ☆, 🯅, m, 1, Joe$\}$ has cardinal number 7, that is, $n(W) = 7$.

Example 5 The set $M = \{$all letters of the English alphabet$\}$ has 26 elements; $n(M) = 26$.

Example 6 If $D = \{1, 2, 3, \ldots, 10\}$, what is $n(D)$?

In view of this definition, we could say that equivalent sets are sets which have the same cardinal number.

Comment The following statements all say the same thing.

1. Sets A and B can be put into 1–1 correspondence.
2. Sets A and B are equivalent.
3. Sets A and B have the same cardinal number.

Comment Any two null sets have the same elements, namely, *none*. Thus, any two null sets are equal. This means that essentially there is only one null set. Therefore, we speak of *the* null set.

EXERCISES

1. Describe the following sets of words.
   a) $\{1, 3, 5, 7, 9, \ldots, 19\}$
   b) {November, June, April, September}
   c) {Friday, Saturday, Sunday}
   d) $\{b, e, l, y, u, t, o, n\}$
   e) {tire, horn, fender, battery, carburetor, . . .}
   f) $\{3, 6, 9, 12, 15, \ldots\}$
2. List the elements of the following sets.
   a) The set of letters in the word "supercallyfragilisticexpyalidoshus"
   b) The set of vowels in the word "pfft"
   c) The set of U.S. presidents since Harry Truman
   d) The set of all positive even numbers
   e) The set of all Nobel prize winners in mathematics
3. Are the following sets well defined?
   a) The set of all valuable postage stamps
   b) The set of all 40-game winners among baseball pitchers

c) The set of all boring teachers

d) The set of all beautiful women

4. Classify the following sets as finite, infinite, unit, or null sets.

a) The set of all cells in the human body

b) The set of all heroin addicts

c) The set of all numbers greater than one trillion

d) The set of all cats that speak French

e) The set of all countries that have landed astronauts on the moon

f) $\{1, 2, 3, \ldots, 300000000\}$

Does this picture show a set?

Stock, Boston

5. Classify each of the following as true or false.

a) Funny bone $\in$ {all parts of the human body}

b) The set $\{0\}$ and the null set are equal sets.

c) $\varnothing = 0$

d) Gerald Ford $\in$ {great U.S. presidents}

e) $\{a\} = a$

f) $y \in$ {all vowels of the English alphabet}

6. a) Are equal sets equivalent? Explain.

b) Are equivalent sets equal? Explain.

7. Demonstrate several 1–1 correspondences between the following sets:

$X = \{e, f, 1, 5\}$

$Y = \{k, e, 3, m\}$.

8. Are the following pairs of sets equal, equivalent, neither, or both?

a) {head, tail} and {black, white}

b) {spring, summer, winter, fall} and {four seasons of the year}

c) {all even numbers between 3 and 9} and $\{x, y, z\}$

d) $\{m, n, p\}$ and $\{m, n, p, q\}$

e) {all fish that can read Russian} and {all elephants that can cook}

f) {all male senators} and {all female senators}

9. What is the cardinal number of the following sets?

a) $\{\triangle, 1, \text{👤}, \text{John}, \text{☆}\}$

b) {present presidents of the United States}

c) $\{3, 4, 5, 6, \ldots, 14\}$

d) $\{\ \ \}$

e) $\{1, 2, 3, \ldots\}$

f) {female presidents of the United States}

10. Give examples of each of the following. (Do not use any of the previously given examples.)

   a) A set whose cardinal number is 7

   b) A set whose cardinal number is 0

   c) Two equivalent but not equal sets

   d) A unit set

   e) A finite set

   f) An infinite set

*11. Suppose that Mr. Smith is a small-town barber who shaves *all* the men and *only* those men in his town who do not shave themselves. (Everybody in the town is shaved or shaves daily.)

   Let A = {all men in this town who shave themselves}.
   Let B = {all men in this town who do not shave themselves}.

   a) Is Mr. Smith a member of set A?

   b) Is Mr. Smith a member of set B?

   c) What is $A \cup B$?

   d) Is Mr. Smith a member of $A \cup B$?

*12. a) Is it possible for a set to be an element of itself?

   b) Let A = {all sets that are *not* elements of themselves}. Does A belong to A?

   c) What connection (if any) is there between this exercise and Exercise 11?

---

5. SUBSETS

Consider the set B = {all teachers at this school} and set A = {all math teachers at this school}. We notice that every member of set A is also a member of set B, that is, set A is part of set B. We then say that set A is a subset of set B.

Definition 5.1

*Set A is said to be a **subset** of set B if every element of set A is an element of set B. We denote this by $A \subseteq B$. This symbol is read as "A is a subset of B" or "A is contained in B."*

Example 1   {all basketball players in this school} is a subset of {all athletes in this school}.

Example 2   {a, b, c} is a subset of {a, b, c, d}.

Example 3   {Elton John} $\subseteq$ {all singers}

Example 4   {Moe, Larry, Curly} $\subseteq$ {Curly, Moe, Larry}

Dr. Chang, who teaches math at this school, is a member of the set A = {all math teachers at this school}. He is also a member of set B = {all teachers at this school}. Professor Rivera, who teaches history, is a member of set B, but not of set A. We say that A is a proper subset of B.

How many subsets can you find?
(Printed by courtesy of the U.S. Navy.)

Definition 5.2   *Set X is said to be a **proper subset** of set Y if every element of set X is an element of set Y, and also set Y has at least one other element that is not in X. We denote this by X ⊂ Y. This symbol is read as "X is a proper subset of Y."*

Definition 5.3   *Set X is said to be an **improper subset** of set Y if every element of set X is an element of set Y, but Y does not have any other elements that are not in X. This really means X equals Y, and we denote this as X = Y.*

Comment   If we know definitely that A is a proper subset of B, then we write A ⊂ B. If we know only that A is a subset of B but we do not know if it is proper or improper, then we write A ⊆ B.

Example 5   {Moe, Larry, Curly} is an improper subset of {Larry, Curly, Moe}, and thus we usually write this as {Moe, Larry, Curly} = {Larry, Curly, Moe}.

Example 6   {a, e, u, w} is a proper subset of {a, e, u, w, t}. We write {a, e, u, w} ⊂ {a, e, u, w, t}.

Example 7   {a, e, u, w} is also a proper subset of {all letters of the English alphabet}.

Example 8   {a, e, u, w} is an improper subset of {a, e, u, w}. We write {a, e, u, w} = {a, e, u, w}.

Comment   The null set is assumed to be a subset of *every* set. There are reasons for this that are beyond the scope of this text.

Beware   Do not confuse the symbol "⊂," which stands for "is a proper subset of," with "∈," which means "is an element of." For example, Frank Sinatra ∈ {all singers}.

Diana Ross ∈ {all singers}. However, {Frank Sinatra, Diana Ross} is a proper subset of the set of all singers, i.e., {Frank Sinatra, Diana Ross} ⊂ {all singers}.

Suppose in a journalism class we are discussing newspaper editorials. We might then consider the *New York Times*, the *Washington Post*, the *Atlanta Constitution*, the *San Francisco Examiner*, etc. It would not be appropriate for someone to introduce into the discussion an editorial that he heard on NBC News. The discussion is limited to newspapers only. We call the set of all newspapers the universal set for this discussion.

Definition 5.4   *If all sets in a given discussion are to be subsets of a fixed overall set U, then this set U is called the **universal set**.*

Example 9   If we want to talk about the set of math teachers at this school, the set of English teachers at this school, etc., then a suitable universal set would be {all teachers at this school}.

Example 10   If we are going to discuss positive numbers, negative numbers, fractions, etc., we might choose as our universal set {all numbers}.

Example 11   Suppose we are considering Mickey Mantle, Tom Seaver, Willie Mays, etc. We could use {all baseball players} as our universal set. If we consider them as amateur golfers or tennis players as well as baseball players, then a more appropriate universal set would be {all athletes}.

Comment   The universal set may vary from discussion to discussion.

Comment   Every set is a subset of itself. Would you say that it is a proper or improper subset of itself?

EXERCISES   1. Give two different subsets of each of the following sets.

a) {all Oriental languages}   b) {1, 2, ⚇, △}

c) {all musical instruments}   d) {5}

e) {all green vegetables}   f) {all plays of Shakespeare}

2. What would you say is an appropriate universal set for a discussion involving the following objects?

a) elephants, horses, monkeys, whales, humans

b) Pluto, Mercury, Venus, Mars, Earth

c) Math 15, English 11, History 1

d) blue, yellow, red

e) Chevrolet, Ford, Plymouth

f) 5, 10, 15, 20

g) 2, 3, 5, 7, 11, 13

h) New York, New Jersey, Connecticut, Pennsylvania, Brooklyn, Staten Island

3. State whether set A is a proper or improper subset of set B or neither.

   a) A = {Crest, Ultra Brite, Aim, Colgate} and B = {all brands of toothpaste}

   b) A = {rose, lily, violet} and {violet, lily, rose}

   c) A = {all even numbers divisible by 2} and B = {all even numbers}

   d) A = {all even numbers divisible by 4} and B = {all even numbers divisible by 2}

   e) A = {all even numbers} and B = {all odd numbers}

   f) A = {all people who weigh a ton} and B = {1, 2, 3, 4, . . .}

4. Classify each of the following as either true or false. Explain your answer.

   a) {10, 11, 12, 13} $\subseteq$ {13, 12, 11, 10, . . .}

   b) {10, 11, 12, 13} $\subset$ {13, 12, 11, 10, . . .}

   c) $\varnothing \subset \{a\}$ d) $0 \in \varnothing$

   e) $\varnothing \in \{a\}$ f) $\varnothing \subset \{0\}$

   g) $0 \subset \{0\}$ h) $0 \in \{0\}$

   i) $A \subseteq A$ j) {ships} $\subset$ {submarines}

   k) {all radios over 200 years old} $\subset$ {all cars}

5. List *all* the subsets of each of the following sets. (Be sure you don't leave any out.)

   a) { }   b) {x}

   c) {15, y}   d) $\{20, \frac{1}{2}, \triangle\}$

   e) {eeny, meeny, miny, mo}   f) {Popeye, Swee' Pea, Olive Oyl}

   g) {Charlie Brown, Lucy}

6. a) How many subsets are there for each of the sets of Exercise 5?

   b) Would you say that there is a relationship betwen the number of elements in the set and the number of possible subsets? If your answer is yes, can you find a formula expressing this relationship?

7. How many different committees can be formed from each of the following sets?

   a) {Simon, Garfunkel}   b) {Henry Kissinger}

   c) {Mark, Sharon, Chris}   d) {Jonathan, Leo, Margaret, Lenny}

---

6. SET OPERATIONS

Suppose we wanted to distribute a questionnaire to all students in Math 1 and all students in English 1 at our school. We would have to compile a mailing list of all students in both classes. Some students may be in both classes, but of course we would not list them twice. If we denote the set of students in Math 1 by A and the set of students in English 1 by B, our list would be an example of what is meant by the union of A and B.

Definition 6.1   *The **union** of two sets A and B, denoted as $A \cup B$ (read as "A union B"), means the set of all elements that are either in A or in B or both.*

Example 1   If $A = \{1, x, \triangle\}$ and $B = \{y, \text{\Large\char"263B}\}$, then $A \cup B = \{1, x, \triangle, y, \text{\Large\char"263B}\}$

Example 2   If $A = \{m,$ John, Bob, $y\}$ and $B = \{$Bob, Alice, $x\}$, then $A \cup B =$ $\{m,$ John, Bob, y, Alice, $x\}$.

Observe that even though Bob is in both sets, we list him *only once* in $A \cup B$.

Example 3   If $H = \{$orange, apple, fig$\}$ and $K = \{$fig, orange, plum, apple, banana$\}$, then $H \cup K = \{$orange, apple, fig, plum, banana$\}$.

Example 4   If $X = \{$Hamlet, Macbeth, Julius Caesar$\}$ and $Y = \{$Macbeth, Julius Caesar, Hamlet$\}$, then what is $X \cup Y$?

Example 5   If $M = \{$all months of the year with only 6 days$\}$ and $N = \{$all months of the year with 30 days$\}$, then what is $M \cup N$?

Let us consider again our questionnaire project given at the beginning of this section. We might want to send the questionnaire only to students in *both* classes (if there are any). This set of students is called the intersection of sets A and B (remember $A = \{$Math 1 students$\}$ and $B = \{$English 1 students$\}$).

Definition 6.2   *The **intersection** of two sets A and B, denoted as $A \cap B$ (read as "A intersect B"), means the set of all elements that are in* both *sets A and B at the same time (if there are any.)*

Definition 6.3   *If sets A and B have no members in common, then sets A and B are called **disjoint** sets. This means that if A and B are disjoint, then $A \cap B = \varnothing$.*

Example 6   If $X = \{a, e, \text{\Large ⚲}, \triangle\}$ and $Y = \{\text{\Large ⚲}, \triangle, b, 1, \text{Joe}\}$, then $X \cap Y = \{\text{\Large ⚲}, \triangle\}$.

Example 7   If $M = \{$all convertibles$\}$ and $N = \{$all green cars$\}$, then $M \cap N = \{$all green convertible cars$\}$.

Example 8   If A = {all members of this school's baseball team} and C = {all members of this school's basketball team}, then $A \cap C$ = {all members of this school who are on both the basketball and baseball teams}.

Example 9   If M = {all men who have blond hair} and N = {all women who have blond hair}, then $M \cap N = \varnothing$. In this case, sets M and N are disjoint, since there are no elements common to both sets.

Comment   Note that the intersection is not {blond hair}. Having blond hair is a property of all the elements of each set. Blond hair itself is not an element of either set.

Example 10   If A = {all guitar players} and B = {all musicians}, then $A \cap B$ = {all guitar players}.

Example 11   If X = {all Volkswagens with air-conditioning} and Y = {all Fords with air-conditioning}, what is $X \cap Y$? Note that $X \cap Y$ is not {air-conditioned cars}. Why?

Imagine that we are sponsoring a school dance this Friday at 8:00 P.M. Let A be the set of students who have already bought tickets. Being on the dance committee, we would obviously be interested in the students who have not yet purchased tickets. If our universal set is the set of all students at this college, then we call the set of all students who have not yet purchased tickets the complement of set A.

Definition 6.4   *The **complement** of set A, denoted by A' (read as "A prime" or "A complement"), is the set of all elements in the universal set that are not also in A.*

Example 12   If the universal set = $\{1, 2, 3, 4, 5, 6, 7, 8, 9, 10\}$ and $A = \{1, 4, 6, 8, 9\}$, then $A' = \{2, 3, 5, 7, 10\}$.

Example 13   If the universal set = {all men} and X = {men who are bald}, then X' = {all men who have hair on their heads}.

Example 14   If the universal set = {all American-made cars} and G = {all American-made cars with air-conditioning}, then G' = {all American cars without air-conditioning}.

Example 15   If U = {all humans} and P = {all males}, then P' = {all females}.

Example 16   If the universal set = {all vowels in the English language} and $A = \{a, e, i, o, u\}$, then what is A'?

Let A = {all students in this class} and B = {all students who read *Playboy*}. The set of all students in this class who do not read *Playboy* is called the difference between sets A and B.

Definition 6.5   *The **difference** between sets A and B, denoted as $A - B$ (read as "A minus B"), means the set of all elements that belong to set A but not to set B.*

Example 17   If M = {piano, guitar, drums, clarinet} and N = {piano, saxophone, organ, clarinet}, then $M - N$ = {guitar, drums}. Note that saxophone and organ are not part of the difference since they are not in set M.

Example 18   If H = {3, 6, 9, 12} and J = {3, 6, 9}, then $H - J$ = {12}. What is $J - H$? If your answer is {12}, you are wrong. (Look at the definition again.) The answer is $\varnothing$.

Example 19   If V = {Milton, Bob, Carl} and F = {Bob, Carl, Milton}, then $V - F$ = $\varnothing$. What is $F - V$? The answer is $\varnothing$.

Example 20   If R = {all Mickey Mouse sweatshirts} and Q = {all triangles with 2 sides}, then $R - Q = R$, since Q is obviously the null set. What is $Q - R$? It is $\varnothing$.

Example 21   If S = {9} and T = {3}, then $S - T$ = {9} and $T - S$ = {3}.

EXERCISES

1. For each of the following let the universal set U = {1, 2, 3, 4, 5, 6, 7, 8, 9, 10}, A = {1, 4, 6, 8, 9}, B = {2, 4, 8, 9, 10}, C = {2, 3}. Find:

   a) $A \cup B$ b) $A \cap B$ c) A' d) B' e) $A - B$

   f) $B - A$ g) $A - C$

2. If the universal set U = {all members of the U.S. Congress}, A = {all male members of Congress}, B = {female members of Congress}, C = {Republican members of Congress}, D = {Democratic members of Congress}, and E = {bald members of Congress}, describe in words each of the following.

   a) $A \cup B$ b) $A \cap B$ c) $A \cap C$ d) $B \cap C$ e) C'

   f) $C' \cap A$ g) $C \cup D'$ h) E' i) $E' \cap C'$ j) $A' \cap D$

3. If U = {all letters of the English alphabet} and A = {all vowels}, describe each of the following.

   a) A' b) $A \cup A$ c) $A \cup A'$ d) $A \cup \varnothing$

   e) $A \cap \varnothing$ f) $A - \varnothing$ g) $A \cup U$ h) $A \cap A'$

   i) $A \cap U$ j) $A - U$ k) $U - A$ l) $\varnothing - A$

4. If A is *any* set, what are each of the following?

   a) $A \cup U$ b) $A \cap U$ c) $A \cup \varnothing$ d) $A \cap \varnothing$

5. If A = {$\triangle$, 2, 4}, B = {$\triangle$, 2, 4, 6}, C = {4, 2, $\triangle$}, find each of the following.

   a) $A \cap B$ b) $A \cup B$ c) $A \cap C$ d) $A \cup C$

   e) $A - B$ f) $B - A$ g) $A - C$

6. a) If $A \cap B = A$, how are A and B related?

   b) If $A \cap B = B$, how are A and B related?

7. a) If $A \cup B = A$, how are A and B related?

   b) If $A \cup B = B$, how are A and B related?

8. If $A \cup A = \varnothing$, what can be said about A?

9.

Let X = the set of all points between A and B, including A and B.
Let Y = the set of all points between C and D, including C and D.
Let Z = the set of all points between B and D, including B and D.

a) Find $Y \cup Z$.   b) Find $X \cup Y$.   c) Find $X \cap Y$.

d) Find $X \cap Z$.   e) Find $Y \cap Z$.   f) Find $X \cup Z$.

10. In order to know which cars to stock, a car dealer conducted a survey of 200 prospective car buyers. The features that they were interested in are summarized in the following chart.

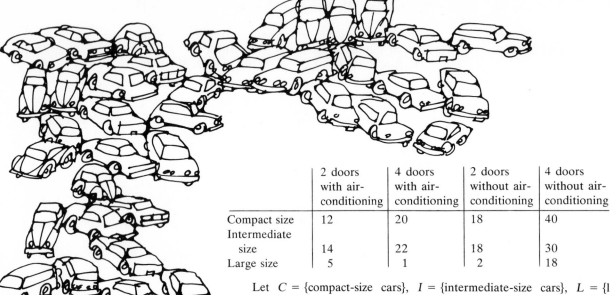

	2 doors with air-conditioning	4 doors with air-conditioning	2 doors without air-conditioning	4 doors without air-conditioning
Compact size	12	20	18	40
Intermediate size	14	22	18	30
Large size	5	1	2	18

Let C = {compact-size cars}, I = {intermediate-size cars}, L = {large-size cars}, T = {two-door cars}, F = {four-door cars}, and A = {all cars that have air-conditioning}. Determine the number of cars in each of the following sets.

a) $C \cap T \cap A$ b) $C \cap T \cap A'$ c) C

d) C' e) T f) $F \cap A$

g) $F \cap L \cap A$ h) $I \cap F \cap A'$ i) $L \cup (I \cap A)$

j) $C - (C \cap F \cap A')$ k) $I \cup (C \cap T \cap A)$

11. In Professor Jones's three classes 150 students were asked to evaluate his performance as a teacher. The results are summarized below.

	Easy tests; explains well	Easy tests; explains poorly	Difficult tests; explains well	Difficult tests; explains poorly
Class A	21	11	16	8
Class B	14	9	24	10
Class C	7	2	12	16

Let A = {students in class A}, B = {students in class B}, C = {students in class C}, E = {students who said tests were easy}, D = {students who said tests were difficult}, W = {students who said he explains well}, and P = {students who said he explains poorly}. Find the number of students who rated Professor Jones in each of the following categories.

a) $A \cap W$ b) $A \cap E \cap W$ c) A d) $C \cap W$ e) D

f) $P \cup C$ g) $B - W$ h) $W - B$ i) $E - D$ j) $A \cup E$

12. The 400 guests at the Rocky Beach Hotel were asked which sport they preferred. Their answers were summarized below.

	Tennis	Golf	Swimming	Shuffleboard
Men	58	133	30	15
Women	45	27	71	21

Let M = {men}, W = {women}, T = {tennis players}, G = {golf players}, S = {swimmers}, and H = {shuffleboard players}. Find the number of people in each of the following sets.

a) $M \cap G$ b) S c) $W \cap H$ d) $(M \cap T) \cup S$

e) $G - M$ f) $M - G$

13. An environmental group made a survey of 1000 people in different age groups to determine which environmental problems were of most concern to them. The following are the results.

	Water pollution	Air pollution	Over-population	Energy conservation
Under 25 yrs	96	64	112	38
25–40 yrs	211	73	27	83
Over 40 yrs	33	183	19	61

Let U = {people under 25 yrs}, T = {people 25–40 yrs}, O = {people over 40 yrs}, W = {people concerned about water pollution}, A = {people concerned about air pollution}, P = {people concerned about overpopulation}, and E = {people concerned about energy conservation}. Find the number of people in each of the following sets.

a) T b) $O \cap E$ c) $U - O$ d) $A \cap T$

e) $T \cup O$ f) $T \cup P$ g) $E - T$ h) $E - P$

i) $(O \cap E) \cup (O \cap A)$ j) $A \cup (E \cap W)$ k) $(A \cup T) \cup U$

7. VENN DIAGRAMS

In your experience with mathematics, you have probably found that the diagram is often a useful device. In working with sets, diagrams can also be very helpful. Such diagrams are called **Venn diagrams**.

The universal set is pictured as a rectangle.

All other sets are drawn as circles within the rectangle. We shade the part that is in set A, as shown in Fig. 1 on the following page.

The universal set U

Set A

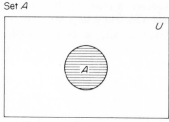

Figure 1

A complement A'

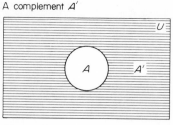

Figure 2

It is easy to see how to picture A'. This time we shade the part that is in set A'. This is shown in Fig. 2.

If A is a *proper subset* of B (see Definition 5.2), we draw the circle for A completely inside the circle for B, as shown in Fig. 3. We do not shade anything here, since we are picturing a **relationship** between two sets. We shade the diagram only when we want to emphasize the set that we are picturing, rather than a relationship between two or more sets.

If A is an *improper subset* of B (remember, this means $A = B$), we draw only one circle and label it as both A and B, as shown in Fig. 4.

Set A is a proper subset of set B. $A \subset B$

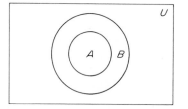

Figure 3

$A = B$ Set A is an improper subset of B.

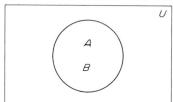

Figure 4

We picture $A \cap B$ as shown in Figs. 5 and 6, depending upon whether $A \cap B = \varnothing$ or $A \cap B \neq \varnothing$:

If A is a subset of B, we picture $A \cap B$ as shown in Fig. 7.

$A \cap B$ is not the null set.

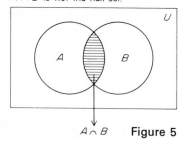

$A \cap B$ **Figure 5**

$A \cap B = \varnothing$ Nothing is shaded.

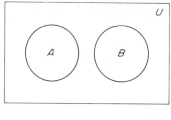

Figure 6

$A \cap B$ when A is a subset of B.

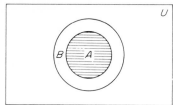

Figure 7

We picture $A \cup B$ in one of the following ways, depending upon the relationship between A and B.

$A \cup B$ when A and B intersect.

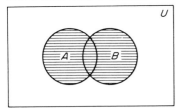

$A \cup B$ when A and B are disjoint.

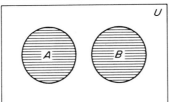

$A \cup B$ when A is a subset of B.

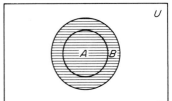

The Venn diagram for $A - B$ or $B - A$ is one of the following (if A and B intersect):

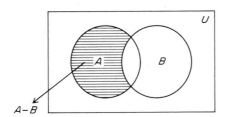

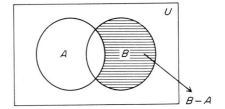

Example 1   Let $U = \{$all people$\}$, $A = \{$all men with blonde hair$\}$, $B = \{$all women with blonde hair$\}$, and $C = \{$all women with blue eyes$\}$. The relationship between these sets is pictured below.

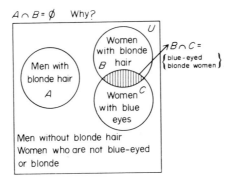

Example 2   If $U = \{$all college students$\}$, $A = \{$all female college students$\}$, $B = \{$all blonde college students$\}$, and $C = \{$all freshmen college students$\}$, then we picture these sets as shown on the following page.

The spheres on the right and left are reflected in the center sphere. The center sphere also contains itself. ("Three Spheres II" by M. C. Escher, from the collection of C. V. S. Roosevelt, Washington, D.C. Printed by permission.)

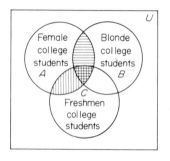

The part that has been shaded horizontally represents $A \cap B$. This is the set of all female blonde college students.

The part that has been shaded vertically represents $A \cap C$. This is the set of all female freshmen college students.

Note that one part has both vertical and horizontal shading. This represents $A \cap B \cap C$. How would you describe this set in words?

Example 3   Using Venn diagrams, show that $(A \cap B)' = A' \cup B'$.

Solution   We make two diagrams, one to represent $(A \cap B)'$ and one to represent $A' \cup B'$. To draw the diagram for $(A \cap B)'$, we first draw $A \cap B$ and then shade everything *outside* $A \cap B$. This is shown below.

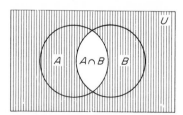

$(A \cap B)'$ is shaded ⫴

A′ ∪ B′

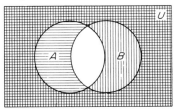

A′ is shaded ||||
B′ is shaded ☰

To draw the diagram for $A' \cup B'$, we first shade in vertically everything outside A. This gives us A'. We then shade in horizontally everything outside B. This gives us B'. The part that has *any* shading is $A' \cup B'$. This is shown on the left.

Observe that in both diagrams the same regions have been shaded. This means that $(A \cap B)'$ and $A' \cup B'$ are equal.

---

8. APPLICATION TO SURVEY PROBLEMS

The student government at State University recently conducted a survey to gather more information on the high-school backgrounds of entering freshmen. There were 100 students interviewed and the following data was collected:

28   took physics,
31   took biology,
42   took geometry,
10   took physics and geometry,
6   took biology and geometry,
9   took physics and biology, and
4   took all 3 subjects.

Based on these figures, can we answer the following questions?

1. How many students took none of the 3 subjects?

2. How many students took physics but not biology or geometry?

3. How many students took biology and physics but not geometry?

Solution

This problem can easily be solved by means of Venn diagrams. Three circles are used to represent the students taking each of the subjects.

In Fig. 8, we first put in the number of students who took all 3 subjects. This was given to be 4. Since we know that 6 altogether took biology and geometry, we must have 6 − 4, or 2, who took biology and geometry *but not* physics. We enter this in Fig. 8. Similarly, we know that 10 took physics and geometry. Therefore, we must have 10 − 4, or 6, who took physics and geometry *but not* biology. We fill this in on the diagram. Also we know that 9 took physics and biology. Therefore, we must have 5 who took physics and biology *but not* geometry. We fill this in on the diagram.

According to the data, 42 took geometry. If we look at the geometry circle, we see that we have accounted for 6 + 4 + 2, or 12. This leaves 42 − 12, or 30 students to be put in the remainder of the geometry circle. We then have the picture shown in Fig. 9 on the following page.

Now 31 took biology. In Fig. 8, we have accounted for 5 + 4 + 2, or 11 students. This leaves 31 − 11, or 20 students to be put in the remaining portion

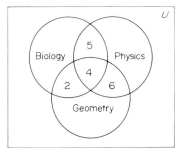

Figure 8

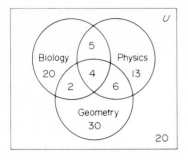

of the biology circle. In a similar manner, there are 28 − 15, or 13 students for the remainder of the physics circle.

We now add together all the numbers appearing in the diagram: 20 + 5 + 4 + 2 + 6 + 13 + 30. This sum is 80. Since the survey involved 100 students, we are left with 100 − 80, or 20, students who did not take any of these subjects. This answers question 1. The answer to question 2 can be read from Fig. 9. It is obviously 13. The diagram also answers question 3. Five took biology and physics, but not geometry.

Figure 9

EXERCISES

1. For each of the following, draw a diagram similar to the diagram below and then shade in the set indicated.

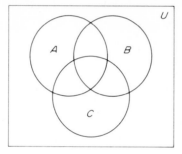

a) $A \cup B \cup C$

b) $A \cap B \cap C$

c) $(A \cup B \cup C)'$

d) $(A \cap B \cap C)'$

e) $A \cup (B \cap C)$

f) $A' \cup (B \cap C)$

g) $A \cap (B \cup C')$

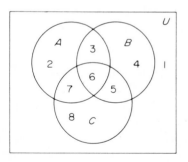

2. In the diagram on the left, we have numbered the different parts from 1 to 8. For example, region 1 corresponds to the set $(A \cup B \cup C)'$. (See your answers to Exercise 1). In a similar manner, describe symbolically the set that corresponds to each of the regions 2, 3, 4, 5, 6, 7, and 8.

3. Draw a Venn diagram to represent each of the following.

a) $A \subset B$

b) $A \subseteq B$ and $B \subseteq A$

c) $A \cap U$

d) $A \cap A$

e) $A \cup \varnothing$

f) A and B are disjoint

g) $A \cup B$ when $A \cap B = \varnothing$

h) $(A \cap B)'$ when A and B are disjoint

i) $A \cap B$ when $A = B$

4. Using the diagram below, find each of the following.

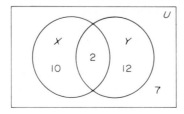

a) $n(X)$

b) $n(X \cap Y)$

c) $n(X \cup Y)$

d) $n(X \cap Y')$

5. Using the diagram below, find each of the following.

   a) $n(A \cup B \cup C)$ b) $n(A \cap B \cap C)$ c) $n(A')$

   d) $n(A \cup B)'$ e) $n(A \cap C')$ f) $n(A \cap B' \cap C')$

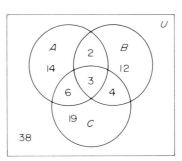

6. In a recent survey of 200 people, the following information was obtained about their investments:

   95 invest in stocks,
   115 invest in bonds,
   55 invest in real estate,
   30 invest in stocks and bonds,
   25 invest in stocks and real estate,
   35 invest in bonds and real estate, and
   10 invest in all three.

   a) How many did not invest in stocks, bonds, or real estate?

   b) How many invested in stocks, but not bonds or real estate?

   c) How many invested in only real estate, but not stocks or bonds?

   d) How many invested in bonds and real estate, but not stocks?

   e) How many invested in stocks and real estate, but not bonds?

7. A travel agency has planned European tours for 400 students.

   110 are going to Paris,
   180 are going to Madrid,
   180 are going to London,
   70 are going to London and Madrid,
   30 are going to London and Paris,
   40 are going to Madrid and Paris, and
   10 are going to London, Paris, and Madrid.

   a) How many students are going to Madrid only?

   b) How many are going to London and Paris but not Madrid?

   c) How many are going to London and Madrid but not Paris?

   d) How many are not going to any of these three cities?

   e) How many are going to Paris only?

Sets and subsets.

Charles Gatewood

8. At a Christmas party, scotch, rye, and bourbon were served to the 500 guests:

230 drank scotch,
115 drank bourbon,
170 drank rye,
 20 drank scotch and rye,
 15 drank rye and bourbon,
 35 drank scotch and bourbon, and
  5 drank all three.

a) How many drank scotch but not bourbon?

b) How many drank rye only?

c) How many drank none of the three?

d) How many drank bourbon and scotch but not rye?

e) How many drank scotch and rye but not bourbon?

9. In the Christmas party of Exercise 8, the 500 guests were asked what kind of music they liked to listen to.

260 liked classical music,
255 liked popular music,
260 liked jazz music,
 75 liked classical and popular music,
115 liked popular and jazz music,
130 liked classical and jazz music, and
 45 liked all three kinds of music.

a) How many people liked classical music only?

b) How many liked jazz and popular but not classical?

c) How many liked jazz and classical but not popular?

d) How many liked jazz only?

e) How many liked none of the three?

10. As the price of meat went up, 500 housewives were surveyed to determine what protein substitute each was using. The following information was obtained:

178 were using poultry,
302 were using soybeans,
232 were using eggs,
 25 were using poultry and eggs,
119 were using soybeans and eggs,
 75 were using poultry and soybeans, and
  7 were using all three.

a) How many were using poultry only?

b) How many were using soybeans only?

c) How many were using soybeans and eggs, but not poultry?

d) How many were using poultry and soybeans, but not eggs?

e) How many were not using any of these substitutes?

11. A car manufacturer's inspector submitted the following report of defects found in a recent shipment of 100 cars. (It was known that all the cars had defects.)

30 had loose engine mounts.
23 had the wrong size tires,
50 had paint defects,
20 had loose engine mounts and paint defects,
 8 had paint defects and the wrong size tires, and
 5 had all three defects.

The inspector was fired. Why?

12. A cosmetics manufacturer claims that of 1000 housewives interviewed, 603 used Lovable soap, 483 used Kissable hand cream, 500 used Close body lotion, 203 used Lovable and Kissable, 71 used all three, 281 used none of these three brands. Should we believe his claim?

13. A stereo manufacturer hired the Wills Agency to determine how people found out about his product. The agency submitted the following report to the manufacturer on the 800 customers whom it interviewed.

385 heard of it through television,
210 heard of it through newspaper ads,
386 heard of it through friends,
 80 through newspaper ads and friends,
 68 through television and friends,
 22 through newspaper ads and television, and
 12 through all three.

The manufacturer refused to pay the agency's bill, claiming that the report was inaccurate. Can you see why?

*14. A small town has a population of 2800 and only one movie house. During the week of November 2–8, the film *Jaws* was shown and 2400 people from the town saw it. The following week the film *The Exorcist* was shown and 1160 people saw it. Find

a) the largest number of people who could have seen both movies.

b) the smallest number of people who could have seen both movies.

---

*9. INFINITE SETS   We have already seen many illustrations of infinite sets. For example,

$A = \{1, 2, 3, 4, \ldots\},$

$B = \{2, 4, 6, 8, \ldots\},$

$C = \{1, 3, 5, 7, \ldots\},$

are infinite sets. However, we have not yet said precisely what we mean by an infinite set. If you ask people what an infinite set is, you will get such answers as "it goes on forever," "it can't be counted," "it has no end," etc. In this section we will try to find a more precise and mathematical description of infinite sets. We will also see that infinite sets behave somewhat differently from finite sets, and as we stated at the beginning of this chapter, many of their properties are very surprising.

We begin by going back to the days of early man where we imagine a caveman who keeps five sheep. Each morning he lets them out of the cave to graze, and in order to keep track of them, he puts a stone in a pile, one for each sheep, as they leave. When they are all gone, he has a pile of stones that looks like the figure on the left.

In the evening when the sheep return, he takes one stone off the pile as each sheep comes back. Afterwards, if he has any stones left, he knows that some of his sheep are missing. On the other hand, if he has no more stones left and another sheep comes in, he knows that a strange sheep has wandered in. Thus, he is **counting** his sheep by using stones.

Let us see what is involved in the caveman's counting system. It is really based on a **one-to-one correspondence** (see p. 54) between the sheep and the stones. This idea is the basis of all counting.

To see how this works, suppose we have a football stadium with 54,327 seats. For a certain game, the stadium is sold out and there are no standees. (Only one person is allowed to sit in each seat.) If we want to know how many spectators are present, we do not have to count them. We know that there are 54,327 people present, because we have a one-to-one correspondence between the seats and the spectators. This one-to-one correspondence enables us to **count** the number of spectators.

Now count the bugs shown on the left.

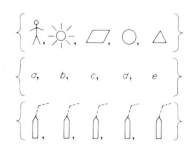

Of course you will get seven by starting at the top left, and counting 1, 2, 3, 4, 5, 6, 7. You too are using a one-to-one correspondence, but you are not doing it with stones or seats. Instead, you are using an "abstraction" of these, namely, numbers.

Now let us go back to the caveman and his five sheep. He has been having trouble with his children playing tricks on him by taking stones off the pile when he is not looking, or by adding stones to the pile. To prevent this, he decides to stop using stones; instead, he makes tally marks, one for each sheep, on the wall (up high where the children can't reach them). Now he counts his five sheep by means of these tally marks:

1   1   1   1   1.

The 1–1 correspondence is now between the sheep and the tally marks, but the idea is the same as before.

Gradually the caveman realizes that he can replace the five tally marks by a single symbol such as V, or 5. (We will have more to say about this in Chapter 3.) These symbols stand for something that the five sheep, the five stones, and the five tally marks all have in common, that is, their "fiveness." Any other set that has five elements in it, such as the ones shown on the left, share this common property of "fiveness." What this means is that all sets with five elements can be put into 1–1 correspondence.

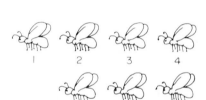

Example 1   Let $A = \{1, 2, \ldots, 100\}$ and $B = \{101, 102, \ldots, 200\}$. Then A and B each have 100 elements and we can put them in 1–1 correspondence as follows:

$$A = \{ \; 1 \; , \quad 2 \; , \quad 3 \; , \; \ldots, \; 100\}$$
$$\updownarrow \quad \updownarrow \quad \updownarrow \qquad \quad \updownarrow$$
$$B = \{101, \; 102, \; 103, \; \ldots, \; 200\}.$$

Example 2   Let $C = \{1, 2, 3, \ldots, 100\}$ and $D = \{2, 4, 6, \ldots, 200\}$. Then C and D each have 100 elements and can be put in 1–1 correspondence as shown:

$$C = \{1, \; 2, \; 3, \; \ldots, \; 100\}$$
$$\updownarrow \; \updownarrow \; \updownarrow \qquad \quad \updownarrow$$
$$D = \{2, \; 4, \; 6, \; \ldots, \; 200\}.$$

It may not be quite so obvious here as it was in Example 1 that each element of set C can be matched with a unique element of set D and vice versa. However, we can make it clear by giving a rule for the matching. The rule is to match each number in set C with twice that number in set D. Thus we have

1 in set C matched with 2 times 1, or 2, in set D

2 in set C matched with 2 times 2, or 4, in set D

3 in set C matched with 2 times 3, or 6, in set D

$$\cdot \qquad\qquad \cdot \qquad\qquad \cdot \qquad\qquad \cdot$$
$$\cdot \qquad\qquad \cdot \qquad\qquad \cdot \qquad\qquad \cdot$$
$$\cdot \qquad\qquad \cdot \qquad\qquad \cdot \qquad\qquad \cdot$$

100 in set C matched with 2 times 100, or 200 in set D.

In other words, we can say that if n is an element of set C, we can match it with 2 times n (which we write as $2n$) in set D as shown:

$$C = \{1, \; 2, \; 3, \ldots, \; n, \ldots, \; 100\}$$
$$\updownarrow \; \updownarrow \; \updownarrow \qquad\qquad \searrow$$
$$D = \{2, \; 4, \; 6, \ldots, \; 2n, \ldots, \; 200\}.$$

To further illustrate this, let us consider another example.

Example 3   Find a 1–1 correspondence between the sets $O = \{1, 3, 5, 7, \ldots\}$ or all odd numbers, and $E = \{2, 4, 6, 8, \ldots\}$ or all even numbers.

Solution   We cannot show the correspondence by actually matching each element of O with an element of E because the sets are infinite and we could never finish. Therefore, to show that such a 1–1 correspondence exists, we must find a rule for matching all the elements of O with unique elements of E and vice versa. It makes sense to match as follows:

$$O = \{1, \; 3, \; 5, \; 7, \ldots\}$$
$$\updownarrow \; \updownarrow \; \updownarrow \; \updownarrow$$
$$E = \{2, \; 4, \; 6, \; 8, \ldots\}.$$

Looking at the first few matchings above, we see that a general rule would be "a number in O is matched to the same number plus 1 in E." To put it in symbols, each number n in O is matched to the number $n + 1$ in E. So the complete 1–1 correspondence is

$$O = \{1, \ 3, \ 5, \ 7, \ldots, \ n, \ldots\}$$
$$\updownarrow \ \updownarrow \ \updownarrow \ \updownarrow \qquad \updownarrow$$
$$E = \{2, \ 4, \ 6, \ 8, \ldots, \ n + 1, \ldots\}.$$

This means that even though O and E are infinite sets, they still have the same number of elements. Remember that sets that can be put in 1–1 correspondence have the same cardinal number (that is, they have the same number of elements). It may seem strange to speak of "the number of elements" in an infinite set, but there is no good reason why we cannot. We call the number of elements in the sets O and E by the name $\aleph_0$ (pronounced aleph-null). That is, $\aleph_0$ is the cardinal number of these sets.

Example 4   Put the sets $A = \{1, 2, 3, 4, \ldots\}$ and $B = \{2, 4, 6, 8, \ldots\}$ in 1–1 correspondence.

Solution   You may say immediately that this cannot be done. For if A and B can be put in 1–1 correspondence, they each have the same number of elements. But it seems clear that A has more elements than B. After all, A contains everything in B and, also, all the odd numbers 1, 3, 5, etc. We see that B is a proper subset of A.

But let us ignore common sense for a while and try to put these two sets in 1–1 correspondence anyway. To do so, we must find a "matching rule." Let us try this: Match every number, n, in A with twice n (written as $2n$) in B. This gives us

$$A = \{1, \ 2, \ 3, \ldots, \ n, \ldots\}$$
$$\updownarrow \ \updownarrow \ \updownarrow \qquad \updownarrow$$
$$B = \{2, \ 4, \ 6, \ldots, \ 2n, \ldots\}.$$

This rule matches each element in A with a unique element in B and vice versa. No element of either set has been left out. So we do have a 1–1 correspondence. This is startling because, as we have said, it means that *A and B have the same number of elements, even though B is a proper subset of A*. How can part of A have the same number of elements as all of A? Perhaps you think that we have played a mathematical trick on you! Of course, we haven't. What is happening here is that we are dealing with *infinite sets*. It is not possible for a *finite* set to have the same number of elements as one of its proper subsets. However, for infinite sets, it is quite possible. In fact, this strange property is exactly the difference between finite and infinite sets! We can use it, therefore, to give a precise definition of infinite sets.

Definition 9.1   *A set is **infinite** if it can be put in 1–1 correspondence with a proper subset of itself.*

Example 5   Prove that the set $A = \{10, 20, 30, \ldots\}$ is infinite.

Solution   Let $B = \{20, 30, 40, 50, \ldots\}$. B is a proper subset of A. Note that A is

$$\left\{ \begin{array}{llll} 10 \text{ times } 1, & 10 \text{ times } 2, & 10 \text{ times } 3, & \ldots \\ \quad \text{or } 10 & \quad \text{or } 20 & \quad \text{or } 30 & \end{array} \right\}.$$

We can put A and B in 1–1 correspondence as follows:

$A = \{10, 20, 30, \ldots, 10n, \ldots\}$

$B = \{20, 30, 40, \ldots, 10n + 10, \ldots\}.$

So A can be put in 1–1 correspondence with a proper subset of itself. Therefore, A is infinite.

Comment Remember that in a 1–1 correspondence, each element of the first set must be matched with a *unique* element of the second set, and vice versa. If any elements of either set are left out of the matching, it is *not* a 1–1 correspondence.

Example 6 Let AB be part of a line as shown below. Several points on this line are shown, P, Q, R, S. Show that the set of all points on this line is infinite.

Solution We take a part of this line from Q to R and redraw it above the line AB as shown.

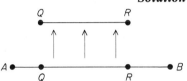

Now erase the letters P, Q, R, S from the *original* line AB. Join A to Q and B to R as shown, and extend the joining lines so that they meet in point E, forming a triangle as shown in Fig. 10.

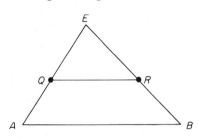

Figure 10

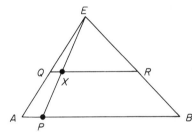

Figure 11

Let P be any point on AB. We match it with a point on QR by joining E and P as in Fig. 11. The line from E to P meets the line QR in X. So X is the point on QR that matches P.

Every point on line AB can be matched in a similar way to a *unique* point on QR and vice versa. Some of these matchings are shown in Fig. 12.

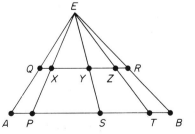

P matches with X
S matches with Y
T matches with Z
A matches with Q
B matches with R

Figure 12

Therefore, the set of points on line AB is in 1–1 correspondence with the set of points on the line CD (which is a proper subset of the line AB). So the set of points on line AB is infinite.

We have now seen many examples of infinite sets. The sets in Examples 3, 4, and 5, namely,

$\{1, 3, 5, 7, \ldots\}$

$\{2, 4, 6, 8, \ldots\}$

$\{1, 2, 3, 4, \ldots\}$

$\{10, 20, 30, 40, \ldots\}$

all had cardinal number $\aleph_0$. What about the set in Example 6? Does it also have cardinal number $\aleph_0$? Since it is also infinite, you might think that the answer must be yes. However, it is not. This fact is known as the *Second Diagonal Proof,* and it is quite remarkable. You can find the proof in the references at the end of this chapter. The cardinal number of the set in Example 6 (that is, the set of points on line AB) is called $\aleph_1$ (pronounced aleph-one).

You will remember that in Example 6, we showed that the set of points on line CD was in 1–1 correspondence with the set of points on line AB. So the set of points on line CD also has cardinal number $\aleph_1$. It can be shown that *any* line, even one from your house to the moon, also has exactly $\aleph_1$ points on it!

$\aleph_0$ and $\aleph_1$ are called *transfinite cardinal numbers.* We might now ask if there are any other transfinite cardinal numbers. The answer is yes. In fact, there are infinitely many, which we call $\aleph_2$, $\aleph_3$, $\aleph_4$, etc. It has been shown that $\aleph_0$ is the smallest of these "infinite" numbers.

Transfinite cardinal numbers can be added and multiplied just as finite numbers are, but they cannot be subtracted or divided. Moreover, the arithmetic of transfinite numbers has strange results, for example,

$\aleph_0 + \aleph_0 = \aleph_0$ $\aleph_0 \text{ times } 2 = \aleph_0$

$\aleph_1 + \aleph_0 = \aleph_1$ $\aleph_1 \text{ times } \aleph_1 = \aleph_1$

$\aleph_1 + 7 = \aleph_1$ $\aleph_0 \text{ times } \aleph_1 = \aleph_1.$

EXERCISES

*1. Put each of the following pairs of sets in 1–1 correspondence. Give a "matching rule" in each case.

  a) $\{1, 3, 5, 7, \ldots\}$ and $\{4, 12, 20, 28, \ldots\}$

  b) $\{3, 6, 9, 12, \ldots\}$ and $\{4, 8, 12, 16, \ldots\}$

  c) $\{1, 2, 3, 4, \ldots\}$ and $\{1, \frac{1}{2}, \frac{1}{3}, \frac{1}{4}, \ldots\}$

  d) $\{1, 2, 3, 4, \ldots\}$ and $\{1, 4, 9, 16, \ldots\}$

  e) $\{2, 4, 8, 16, \ldots\}$ and $\{2, 4, 6, 8, \ldots\}$

*2. In the diagram on the left, we have a circle inside a square. Put the set of points on the circle in 1–1 correspondence with the set of points on the square.

*3. In the diagram below, put the set of points on the inside figure in 1–1 correspondence with the set of points on the outside figure.

*4. In the triangle on the left, put the set of points on side AB in 1–1 correspondence with the set of points on side CB.

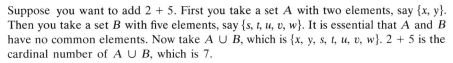

*5. In the "new math" approach to arithmetic, children are taught to add by the following method.

Suppose you want to add $2 + 5$. First you take a set A with two elements, say $\{x, y\}$. Then you take a set B with five elements, say $\{s, t, u, v, w\}$. It is essential that A and B have no common elements. Now take $A \cup B$, which is $\{x, y, s, t, u, v, w\}$. $2 + 5$ is the cardinal number of $A \cup B$, which is 7.

a) In the above, why is it important that $A \cap B = \varnothing$?

b) Use this technique to compute $1 + 3$.

c) Use this technique to compute $4 + 4$.

d) Use this technique to compute $2 + 0$.

e) Use this technique to compute $\aleph_0 + 3$.

f) Use this technique to compute $\aleph_0 + 0$.

g) Use this technique to compute $\aleph_0 + \aleph_0$.

h) Use this technique to compute $\aleph_1 + \aleph_1$.

*6. A very important politician arrives at a hotel and asks for a room. The hotel is full but the manager, who does not want to offend the politician, is determined to find a room for him. He does it by moving the guest in room 1 to room 2, the guest in room 2 to room 3, and so on. In this way, the manager is able to provide a room for the politician and for each of the original guests. There is no doubling up. How many rooms does the hotel have?

STUDY GUIDE In this chapter, the following ideas about sets were introduced.

Basic ideas

Set (p. 50)
Elements or members of
  a set (p. 51)
Set notations (p. 51)
Roster method (p. 51)
Descriptive method (p. 51)
Well-defined sets (p. 52)
Aleph-null (p. 76)
Aleph-one (p. 78)
Transfinite cardinal numbers (p. 78)

1–1 correspondence (p. 54)
Cardinal number (p. 54)
Venn diagrams (p. 65)

Kinds of sets

Unit (p. 52)
Null or empty (p. 52)
Finite (p. 53)
Infinite (p. 53)
Equal (p. 53)

Equivalent (p. 54)
Subsets (p. 57)
Proper subsets (p. 58)
Improper subsets (p. 58)
Universal (p. 59)
Disjoint (p. 61)

Set operations

Union (p. 60)
Intersection (p. 61)
Complement (p. 62)
Difference (p. 62)

Applications of the ideas of sets were given for:

a) survey problems (p. 69),
b) cardinal numbers and
    infinite sets (p. 73).

SUGGESTED FURTHER READINGS

Bell, E. T., *Men of Mathematics*. New York: Simon & Schuster, 1961. Chapter 29 contains a biography of G. Cantor.

Dantzig, T., *Number, The Language of Science*. New York: Macmillan, 1954. Contains a discussion of infinity.

Dinkines, F., *Elementary Theory of Sets*. New York: Appleton-Century Crofts, 1964. Contains a basic but thorough discussion of sets with applications.

Fraenkel, A., *Abstract Set Theory*, 3rd ed. Atlantic Highlands, N.J.: Humanities Press, 1966. A good discussion of set theory.

Gamow, G., *One, Two, Three, Infinity*. New York: Viking Press, 1963. Chapter 1 contains a delightful discussion of infinity.

Hahn, H., "Infinity." In James R. Newman, ed., *The World of Mathematics*, Vol. III, pp. 1593–1611. New York: Simon & Schuster, 1956.

Kamke, E., *Theory of Sets*. New York: Dover, 1950. A slightly more advanced treatment of sets.

Kline, M., *Mathematical Thought from Ancient to Modern Times*. New York: Oxford University Press, 1972. Pages 994–1003 discuss infinite sets.

Mathematics in the Modern World (Readings from *Scientific American*). San Francisco: W. H. Freeman, 1968. Article 28 discusses paradoxes and Article 30 discusses a set theory different from Cantor's.

Russell, B., *Introduction to Mathematical Philosophy*. London: George Allen & Unwin, 1919. Chapters 2 and 8 define finite and infinite numbers.

Stoll, R., *Sets, Logic and the Axiomatic Method*. San Francisco: W. H. Freeman, 1974. A good discussion of set theory.

Zippin, L., *Uses of Infinity*. New York: Random House, 1962. This book contains a discussion of infinity and how it comes up in different mathematical situations.

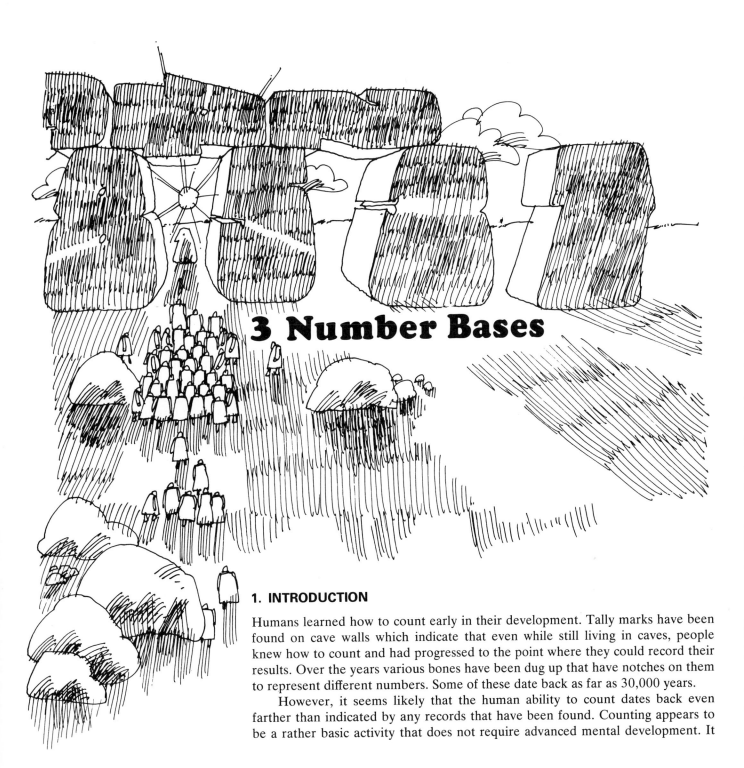

3 Number Bases

1. INTRODUCTION

Humans learned how to count early in their development. Tally marks have been found on cave walls which indicate that even while still living in caves, people knew how to count and had progressed to the point where they could record their results. Over the years various bones have been dug up that have notches on them to represent different numbers. Some of these date back as far as 30,000 years.

However, it seems likely that the human ability to count dates back even farther than indicated by any records that have been found. Counting appears to be a rather basic activity that does not require advanced mental development. It has been demonstrated that even some animals and birds (for example, crows) have the ability to distinguish between groups of up to four objects.[1]

Originally, early man distinguished only between the numbers one and two. Everything else was just many. Thus if Ug had three wives and his brother Og had ten wives, each would say that he had "many" wives. Even today there are some primitive tribes that still count this way.

As soon as people learned how to distinguish between larger numbers, they found it more convenient to count by groups than by ones. Today we count in groups of 10. Historically, counting has been done using groups of 2's, 3's, 5's, 10's, and other numbers. Groupings by 2's and 3's were widely used earlier in time. However, these were almost always replaced by groupings of 5's and 10's. Having five fingers on each hand, it was natural for humans to count in groups of 5's and 10's. From this came our present decimal system (from the Latin *decem*, meaning ten).

As we pointed out earlier (in Section 9 of Chapter 2), piles of stones were often used in counting. For example, six stones represented the number "six." Because piles of stones can easily be disturbed, numbers were sometimes recorded by carving notches in rocks, bones, and sticks. In the nineteenth century, a wolf bone was found in Czechoslovakia with 55 notches in it, arranged in groups of 5.

In some cultures groups of twenty were used. These systems, based on twenty, are still to be found in various parts of the world. The French word for eighty is *quatre-vingts*, which means "four 20's." This suggests that at some time in the past, counting in some parts of France was done in groups of twenty.

It is easy to see how systems developed based on 2, 5, 10, and 20 (remember, cavemen didn't wear shoes). However, it is surprising to find evidence that some cultures may have counted in groups of four or eight. The Latin word for "nine" is *novem*, which may be connected with the Latin word *novus* meaning "new." This suggests that nine was the start of a new group.

In the next section, we will examine one early number system. Then we will examine the decimal system, which we all count with today. We will also discuss other number systems and their applications.

---

2. THE BABYLONIAN NUMBER SYSTEM

The ancient Babylonians counted in groups of 60. Although it might seem strange that such a large number was used, we shall see shortly that it has many advantages. In the late nineteenth century tablets were found, most of which date back to around 1700 B.C. From these tablets, we have been able to learn much about Babylonian mathematics.

In the Babylonian system, two basic symbols were used, a vertical wedge Υ and a corner wedge $\dashv$. The vertical wedge represented the number 1 and the

---

1. See Levi Conant, "The Number Concept. Its Origin and Development," 1923. In James R. Newman, ed., *The World of Mathematics*, pp. 432–441. New York: Simon and Schuster, 1956. Cf. H. Kalmus, "Animals as Mathematicians," *Nature* 202: 1156–1160 (1964).

An early system of writing, in Mesopotamia, was made up of small, simple drawings called pictograms. Each pictogram stood for an object or idea, or sometimes for several words. For numerals, astronomers used a base of 60, repeating number signs up to nine times.

corner wedge represented the number 10. For example, the symbol YYY meant the number 3, the symbol ◄YY meant the number 12, and the symbol ◄YYYY meant the number 14. The Babylonians wrote all the numbers 1, 2, 3, . . . , 59 in this manner. When they got to 60, they moved over one place to the left, as we do in our decimal system when we get to 10. Thus the number 72 would be written as Y ◄YY . The first Y symbol represents not "1," but one "60," because of its position. The last two Y's (on the right) represent two "1's" because of their position.

The number 146 would be written as

The first two YY symbols represent two 60's or 120. The two ◄ ◄ symbols mean two 10's and the YYY YYY symbols mean 6. Thus we have 120 + 20 + 6 or 146.

When the Babylonians got to 60×60 or 3600, they moved over two places to the left as we do in our decimal system when we get to 10×10 or 100. Thus the number 4331 would be written as

$$Y \qquad ◄YY \qquad ◄Y$$
one 3600   twelve 60's   eleven 1's

The first Y means one 60×60 or one 3600. The symbols ◄YY mean twelve 60's or 720. Finally the symbol ◄Y on the right of the number means eleven 1's. This gives 3600 + 720 + 11 = 4331.

You will notice that in this system, the value of a symbol is determined by its position. The same is true in our decimal system, as we will see later in this chapter.

One important difference between the Babylonian system and our decimal system is that they never really had a symbol for zero as we use it today. Thus the symbol Y might stand for 1, 60, or 60×60, etc. In later texts, they did use a sign that looked like this ∧ to indicate the empty spaces if they occurred *inside* numbers. Thus, this symbol acted as a "place-holder," much as our number 0 does. However, they never used this symbol at the end of a number. So, as we have said, the symbol could stand for 1, 60, or 60×60, etc.

To reemphasize the point, the importance of the Babylonian system was that the value of a number was determined by its position. This idea of positional notation is fundamental in our own decimal system as we shall see in the next section.

The Babylonians used 60 as a grouping number for several reasons. One reason was that 60 can be evenly divided by many numbers. For example, 60 can be divided by 2, 3, 4, 5, ... This made division and work with fractions much easier. The choice of 60 may also have been due to the Babylonian interest in astronomy. As a matter of fact, the Babylonian year was divided into 12 months of 30 days each, with an additional 5 feast days (12 times 30 equals 360, which can be evenly divided by 60). It has also been suggested by some historians that 60 was used as a natural combination of two earlier systems, one using 10 and the other using 6.

The Babylonian system was taken over by the Greek astronomers. In fact, it was used for many mathematical and practically all astronomical calculations as late as the seventeenth century. Many traces are to be found even today. For example, hours are divided into 60 minutes and minutes are divided into 60 seconds. In geometry, angles are divided into degrees. Each degree is divided into 60 minutes.

EXERCISES   The early Egyptians used the following symbols to represent numbers (the order or position of symbols was not important).

Number	1	2	3	4	5	6	7	8	9
Symbol	I	II	III	IIII	III II	III III	IIII III	IIII IIII	III III III

Number	10	11	12	13	14	15	16
Symbol	∩	I∩	II∩	III∩	II II ∩	III II ∩	III III ∩

Number	17	18	19	20	100	1000	10,000
Symbol	IIII III ∩	IIII IIII ∩	III III III ∩	∩∩	𓏤	𓆸	𓂭

For example, the symbol 𓆸𓆸𓆸∩∩II would represent the number 322. The symbol 𓆸 IIII III ∩ would represent the number 119. The number 21,318 could have been written by the Egyptians as

$$\text{𝖌𝖌 𓆸 } \overset{IIII}{\underset{IIII}{}} \cap \\ \text{𝖌𝖌𝖌}$$

1. What do the following numbers in the Egyptian system represent in our system?

   a) 𓏤 𝍤 𓎆

   b) 𓎆𓎆𓎆 𝍤

   c) 𓏤𓏤𓏤 𝍤 𓎆

   d) 𓆼 𝍤 𓎆

   e) 𓏤𓏤𓏤𓏤 𝍤 𓎆

   f) 𝍤 𓎆 𓏤

   g) 𓏤 𓎆𓎆𝍤 𓏤𓏤

   h) 𝍤 𓆼 𓎆𓎆

2. Write the following numbers in the Egyptian system.

   a) 112   b) 278   c) 86   d) 519

   e) 389   f) 637   g) 1234   h) 40,444

Addition and subtraction of numbers was not difficult in the Egyptian system. For example, 1231 and 3412 would be added as follows:

$$\text{𓆼 𓏤𓏤 𓎆𓎆𓎆 𝍤}$$

plus 𓆼𓆼𓆼 𓏤𓏤𓏤𓏤 𝍤 𓎆

$$\overline{\text{𓆼𓆼𓆼𓆼 𓏤𓏤𓏤𓏤𓏤𓏤 𓎆𓎆𓎆 𝍤}}$$

Sometimes regrouping (or "carrying") was needed when we obtained more than nine strokes ꞁꞁꞁꞁꞁꞁꞁꞁꞁ. We just replace ten of these strokes (or heelbones) by the symbol 𓎆. Thus, we have the following:

$$\text{𓏤 𓎆𓎆 𝍤}$$

plus 𓏤𓏤 𓎆𓎆𓎆𓎆𓎆𓎆𓎆 𝍤

$$\overline{\text{𓏤𓏤𓏤 𓎆𓎆𓎆𓎆𓎆𓎆𓎆𓎆𓎆 𝍤 𝍤}}$$

More simply, the sum is

𓏤𓏤𓏤𓏤 𝍤 𓎆

3. Add the following numbers in the Egyptian system and check your answer by converting the numbers to our decimal system.

a)

𓏲 ∩∩ 𝟗𝟗 |||

plus 𓏲 𝟗𝟗𝟗𝟗𝟗𝟗𝟗 ||||
                        ||||

b)

𓍶 𓏲 𓏲 𓏲 𝟗𝟗𝟗𝟗 |||

plus 𓏲 𝟗𝟗𝟗𝟗𝟗𝟗𝟗 ∩∩∩|

4. Subtract the following numbers in the Egyptian system and check your answer by converting the numbers to our decimal system.

a)

𝟗𝟗∩∩∩∩∩∩∩∩∩|||

minus   𝟗∩∩ ||||
                 |||

b)

𓏲 ∩∩ ||

minus 𝟗𝟗𝟗∩∩∩||||

5. Explain why the Egyptian system had no need for a symbol for zero.

---

3. BASE 10 NUMBER SYSTEM

Bruce Anderson

In this section we will discuss the decimal system that is commonly used throughout the world today. We will start by looking at the Roman numeral system. In this system the following symbols are used:

I which stands for 1,   C which stands for 100,

V which stands for 5,   D which stands for 500,

X which stands for 10,   M which stands for 1000,

L which stands for 50,

and so forth.

The number 43 would be written as XLIII.

The number 63 would be written as LXIII.

The position of the numerals is important. In 43, the X goes before the L to indicate that it is 10 less than 50. In 63, X goes after the L. This means that 10 is to be added to 50. As we will see shortly, in our number system the position of the digits 0, 1, 2, 3, 4, 5, 6, 7, 8, 9 is even more important than in the Roman system. For example, the number 31 is different from the number 13 even though both numbers contain the same digits, 1 and 3. It is the use of position or *place* which makes our system so convenient to work with. Note that the use of position in the Roman system is different from our use of position.

To better appreciate *our* system (which is known as the **Hindu-Arabic system**), try to add the numbers 43 and 63 *using Roman numerals.*

XLIII
+ LXIII
_____

What is your answer? It should be CVI. If you did get this answer, was it by adding the Roman numerals? Or did you add 43 and 63 in our system and convert the answer to Roman numerals? If you did, then it is understandable, since doing arithmetic in the Roman system is quite complicated. As we look at our system in detail, we will see exactly why it makes arithmetic so much easier.

To get started, consider the number 4683. This means

4(thousands) + 6(hundreds) + 8(tens) + 3(ones).

This can be restated as

4(1000)   + 6(100)   + 8(10)   + 3(1).

Notice something interesting about this:

$1000 = 10 \times 10 \times 10$.

This is usually abbreviated as 10^3 which means 10 multiplied by itself 3 times. Similarly, $100 = 10 \times 10$. This is abbreviated as 10^2 which means 10 multiplied by itself. Moreover, 10 can be written as 10^1.

We can now write the number 4683 as $4(10^3) + 6(10^2) + 8(10^1) + 3$. This can be neatly summarized in the following chart.

10^3	10^2	10^1	1's
4	6	8	3

In each of these columns, one of ten possible **digits** can appear. These digits are 0, 1, 2, 3, 4, 5, 6, 7, 8, 9. Every number can be expressed as some combination of these digits. Why are these ten sufficient? Why don't we need more?

Comment

In this system of writing numbers, the *place* of the number determines its value. In the example above, the *place* of the 4 tells us it stands not for 4 ones, 4 tens, or 4 hundreds, but for 4 thousands. Similarly, the place of the 8 tells us it stands for 8 tens, and so on.

As another example, consider the number 20,694. This stands for

2(ten-thousands) + 0(thousands) + 6(hundreds) + 9(tens) + 4(ones),

or $2(10^4) + 0(10^3) + 6(10^2) + 9(10^1) + 4$. This can be written as

10^4	10^3	10^2	10^1	1's
2	0	6	9	4

The zero that appears in the 10^3 column is very important. If we leave it out, the number would read 2694 and that is not the number we want. The zero is called a **place holder**.

Now let us see how we add these two numbers. For convenience we put them in one chart.

	10^4	10^3	10^2	10^1	1's
		4	6	8	3
+	2	0	6	9	4
Sum	2	5	3	7	7

In the 1's column, we add 3 and 4 and get 7 1's.

In the 10^1 column, we add 8 and 9 and get 17. This means that we have 17 tens or 10(tens) + 7(tens). We know that 10(tens) is 10×10, or 10^2. Thus we have $1(10^2)$ and $7(10^1)$. So the 1 must be **carried over** to the 10^2 column where it belongs. The 7 remains in the 10^1 column.

In the 10^2 column, we add 6 and 6 and the 1 that was carried. This gives $13(10^2)$, which equals $10(10^2) + 3(10^2)$. Now $10(10^2)$ is $10 \times 10 \times 10$, which equals 10^3. Thus $13(10^2)$ equals $1(10^3) + 3(10^2)$. The 3 is left in the 10^2 column and the 1 is *carried over* to the 10^3 column.

This procedure is repeated for each column until we are finished. We can summarize this procedure as follows:

1. In each column, the only digits allowed are 0, 1, 2, 3, 4, 5, 6, 7, 8, 9.

2. In any column, when the sum is more than 9, we carry everything over 9 to the next column.

This convenient method of addition works because of the way in which numbers are written in the Hindu-Arabic system. It will not work for the Roman system. In the Hindu-Arabic system, the value of each digit depends on its *place*. For example, 53 means 5 tens plus 3, whereas 35 means 3 tens plus 5.

In doing multiplication we use a technique similar to this. For example,

$$
\begin{array}{r}
23 \\
\times\ 45 \\
\hline
115 \\
92 \\
\hline
1035
\end{array}
$$

In doing this multiplication, we first multiply the 23 by 5, giving 115. Then we multiply the 23 by 4. This gives 92. We write the 92 under the 115 as shown. We then add and get 1035 as our final answer. Why did we move the 92 one place to the left?

Division and subtraction are done similarly.

The system we have just described is also called the **decimal system**. The decimal system was not widely accepted until the late Middle Ages. It took a long time for this system to be put into everyday use. In fact, in the late thirteenth century, the city of Florence in Italy passed laws against the use of the Hindu-Arabic numerals. This was done to protect its citizens from the dishonest persons who did such things as interchanging the numbers 0, 6, and 9.

Because the decimal system uses groupings of 10's, we refer to it as a *base 10 number system.*

EXERCISES

1. Write each of the following numbers in the Roman numeral system.

   a) 74   b) 86   c) 43   d) 632

2. Write each of the following Roman numerals in the Hindu-Arabic system.

   a) CXI   b) LXXI   c) CCCIV   d) LXII

3. Try to add the following in the Roman numeral system.

   a)   XIV   b)   XXXVI
      + CLI   + LXXXI

4. Look up the Hindu-Arabic numeration system. (Try encyclopedias and books on the history of mathematics.) Find out all you can about its origin and development.

5. Another way of doing multiplication is by the so-called **Gelosia method**. It works as follows: Suppose we want to multiply the two numbers, 257 and 49. We set it up as shown on the left.

   First we multiply 7 by 4. This gives 28 which is placed as shown below.

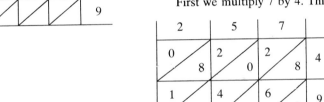

Then we multiply 9 by 7. This gives 63. We put this in as shown (second line, right column). Now we multiply 5 by 4. This gives 20 which we write in (first line, middle column). We complete the diagram in this manner. Note that 2 times 4 is entered as 08 (first line, left column). Now we add along the diagonals starting from the lower right corner, carrying where necessary. This gives:

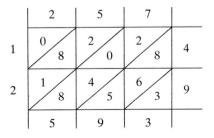

The first diagonal contains only 3. So 3 is its sum. The next diagonal contains 5, 6, and 8, which add to 19. We put the 9 down and carry the 1. The next diagonal contains 8, 4, 0, 2, and the 1 we carried. These add to 15, so we put down 5 and carry the 1. We continue, in a similar manner, until finished. Our answer is then read off as 12,593.

a) Check the answer to the above problem by regular multiplication.

b) Do the following multiplication problems by the Gelosia method.

i) 93	ii) 618	iii) 622	iv) 4172
× 25	× 23	× 834	× 987

v) 2396	vi) 7524
× 8541	× 466

c) Check each of the answers to part (b) by regular multiplication.

*d) Explain why Gelosia multiplication works.

6. The English mathematician John Napier (1550–1617) used the Gelosia system of multiplication to construct what we would today call a computing machine. His gadget is referred to as Napier's rods or Napier's bones. We can construct a variation of Napier's bones using only 10 popsicle sticks and numbering them as shown below.

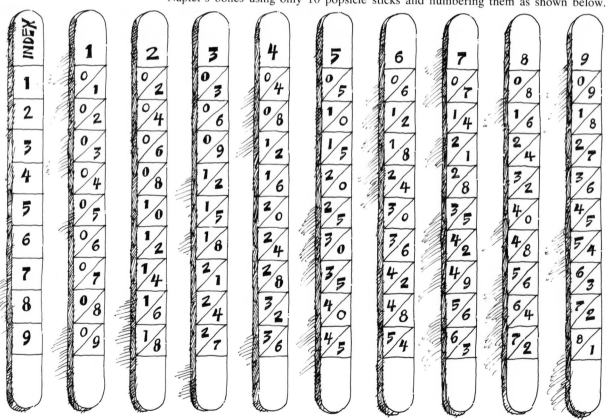

Napier's bones.

Courtesy IBM

The first stick is called the index and lists all the digits from 1 through 9. On top of each of the other sticks we write one of the digits 1 through 9. Each stick gives the product of an index number with the number on the stick. The ones and tens of each product are separated by a diagonal line as in Gelosia multiplication. The different products are separated by horizontal lines.

To see how we can use these sticks to multiply two numbers, let us multiply 435×8. We place the index alongside the sticks headed by the digits 4, 3, and 5 as shown on the left.

Next we locate the row of numbers that are on the same line as 8 on the index.

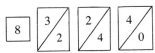

Finally we add along the diagonals as we do in Gelosia multiplication.

Our answer is 3480. Thus $435 \times 8 = 3480$.

Using a similar procedure, we find that the product of 435 and 9 is 3915. Thus, $435 \times 9 = 3915$.

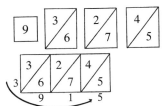

We can combine the two previous results to obtain the product 435×89. We have

$$
\begin{array}{r}
435 \\
\times 89 \\
\end{array}
\qquad
\begin{array}{l}
9 \times 435 = 3{,}915 \\
80 \times 435 = 34{,}800 \, . \\
\hline
89 \times 435 = 38{,}715
\end{array}
$$

Thus $435 \times 89 = 38{,}715$.

Using popsicle sticks make up a set of Napier rods as described above and use them to find each of the following products.

a) 54×9 b) 23×6 c) 234×56 d) 583×75

e) 678×95 f) 456×78 g) 2345×67

4. OTHER BASES

In Section 1 we mentioned that some primitive people counted in groupings of fives. Suppose we had continued this method today and still counted by groupings of fives. What would numbers look like in such a system? The setup is the same as that for the base 10 system. The only difference is that now we work with groupings of fives. We will have only five digits. These are 0, 1, 2, 3, 4.

Let us consider the number written as 324 in the **base 5 system**. We can represent it in a chart as

5^2	5^1	1's
3	2	4

This means that we have $3(5^2) + 2(5^1) + 4(1\text{'s})$.

To indicate that we are working in base 5, we write this as $324_{(5)}$. This is read as "three-two-four" and not as "three hundred and twenty-four." The words "thousand," "hundred," "twenty," "tens," etc., are words that are used only in the base 10 system.

In base 10, when we write 324, we mean

10^2	10^1	1's
3	2	4

This would mean $3(10^2) + 2(10^1) + 4(1\text{'s})$. This, of course, we read as "three hundred and twenty-four."

Comment Although we use the same symbol 324 in both bases, they have different meanings; 324 in base 5 represents a different quantity than 324 in base 10. The symbol 324 is a *numeral* which represents different *numbers* in different bases. A **number** is a quantity, whereas a **numeral** is a symbol used to represent it.

To understand better the difference between number and numeral, consider the *number* seven. We can represent it by any of the following numerals:

VII, 7, 7, IIIIIII (Egyptian)

As another example of a number written in a different base, consider 3203 in base 5, written as $3203_{(5)}$. This means

5^3	5^2	5^1	1's
3	2	0	3

$3(5^3) + 2(5^2) + 0(5^1) + 3(1\text{'s})$.

Let us now add $324_{(5)}$ and $3203_{(5)}$ in base 5.

	5^3	5^2	5^1	1's
	3	2	0	3
+		3	2	4
Numbers carried	1		1	
Answer	4	0	3	2

The procedure is exactly the same as in base 10, except that we are now working with groupings of fives.

First we add 3 and 4. This gives 7, which is the same as 5 plus 2 or $1(5^1) + 2(1\text{'s})$. The $1(5^1)$ must be "carried" to the next column, which is the 5^1 column. The 2 remains in the 1's column.

Next we add 2 and 0 and the 1 we carried in the 5^1 column. This gives 3. No carrying is needed. We then add 2 and 3 in the 5^2 column. This gives $5(5^2)$ or $1(5^3)$. We put down 0 in the 5^2 column and carry a 1 into the 5^3 column.

Finally, we have in the 5^3 column a 3 and the 1 carried which gives 4. Our final answer is $4032_{(5)}$.

We can summarize the procedure as follows.

1. In each column, the only digits allowed are 0, 1, 2, 3, and 4.

2. In any column, when the sum is more than 4, we carry everything over 4 to the next column.

Compare this to the technique used in base 10, summarized on p. 87.

To further illustrate base 5 arithmetic, we give several other examples.

Example 1 Add $1204_{(5)}$
$+ \ 332_{(5)}$
_____
$2041_{(5)}$

Solution In the 1's column, $2 + 4$ gives 6 which is $1(5^1) + 1$. We carry a 1 and leave 1. In the 5^1 column, $0 + 3 +$ the 1 carried gives 4. Since there is no carrying needed, we just leave 4. In the 5^2 column, $3 + 2$ gives 5 which is $5(5^2)$ or $1(5^3)$. We carry a 1 and leave 0. In the 5^3 column, we have a 1 and the 1 carried, which gives 2. There is nothing to carry to the next column. Our answer is $2041_{(5)}$.

Example 2   Multiply $23_{(5)}$
$$\times\ 34_{(5)}$$
$$202$$
$$124$$
$$\overline{1442_{(5)}}$$

Solution   First we multiply 3 by 4 which gives 12. We have $2(5^1)$ to carry and 2(1's) left over. Next we multiply 4 by 2 which is 8. Adding the carried 2, we have 10 altogether. This gives $2(5^1)$ to carry, and 0 left over. So on the first line we have 202. Now we multiply the 23 by the 3. This gives 124. Notice that we move the 124 one space to the left. Compare this to the procedure we use for base 10. Why do we move one space to the left?

Finally, we add and get $1442_{(5)}$.

Example 3   Subtract $123_{(5)}$
$$-\ 104_{(5)}$$
$$\overline{14_{(5)}}$$

Solution   We cannot subtract 4 from 3 so we have to borrow. Since we are working in base 5, we borrow a 5. (Compare this to what we do in base ten.) Now we have 4 from 3 + the borrowed 5, or 8. This leaves 4. The 2 in the second column is now a 1 (why?). 1 minus 0 is 1. Finally, we have 1 minus 1 in the last column, which gives 0. Our final answer is $14_{(5)}$. (As in base 10, a zero at the beginning of a number is not written down.)

Many primitive peoples have only two numbers. Such a system is called a **base 2** or **binary system**. Let the two digits be 0 and 1. How would we do arithmetic in the binary system? Remember that we will use groupings of "twos."

Example 4   Let us consider the number 110110 in the base 2 system. We can represent it in a chart as

2^5	2^4	2^3	2^2	2^1	1's
1	1	0	1	1	0

This means that we have $1(2^5) + 1(2^4) + 0(2^3) + 1(2^2) + 1(2^1) + 0(1\text{'s})$.

Example 5   Add $101_{(2)}$ and $110_{(2)}$.

Solution
$$101_{(2)}$$
$$+\ 110_{(2)}$$
$$\overline{1011_{(2)}}$$

First we add 1 and 0 which gives 1. Then we add 0 and 1 which gives 1. Finally we add 1 and 1 which gives 2. We carry this "2" (as a 1) to the next column and leave behind 0. Our final answer is $1011_{(2)}$. (Since the only digits are 0 and 1, we cannot have 2 in any column.)

Example 6   Multiply $110_{(2)}$
$$\times 11_{(2)}$$
$$\underline{110}$$
$$110$$
$$\overline{10010_{(2)}}$$

Example 7   Subtract $1101_{(2)}$
$$-111_{(2)}$$
$$\overline{110_{(2)}}$$

It is possible to use any number larger than 1 as a base. The following examples illustrate some of these other bases.

Example 8   Add $456_{(7)}$
$$+\,324_{(7)}$$
$$\overline{1113_{(7)}}$$

Example 9   Multiply $357_{(8)}$
$$\times\,65_{(8)}$$
$$\underline{2253}$$
$$2632$$
$$\overline{30573_{(8)}}$$

Example 10   Subtract $3101_{(4)}$
$$-233_{(4)}$$
$$\overline{2202_{(4)}}$$

Example 11   Multiply $468_{(9)}$
$$\times\,57_{(9)}$$
$$\underline{3632}$$
$$2574$$
$$\overline{30472_{(9)}}$$

In a base 12 system, we need 12 digits. (Why?)

Example 12   Let the digits of a base 12 system be 0, 1, 2, 3, 4, 5, 6, 7, 8, 9, t, and e. We have created two new symbols: t which equals $9 + 1$, and e which equals $9 + 2$. Let us add $12e_{(12)}$ and $t1_{(12)}$.

Solution $12e_{(12)}$
$$+t1_{(12)}$$
$$\overline{210_{(12)}}$$

Example 13   Multiply $12e_{(12)}$
$$\times\ t1_{(12)}$$
$$\begin{array}{r} 12e \\ 1052 \\ \hline 1064e_{(12)} \end{array}$$

Example 14   In some base b, $57_{(b)}$ equals 52 in base 10. Find b.

Solution   We know that

$57_{(b)}$ means $5(b^1) + 7(1's)$.

Therefore, if $57_{(b)} = 52_{(10)}$, we have

$57_{(b)} = 5 \cdot b + 7 = 52$.

By trial and error you find that $b = 9$. (If you are familiar with algebra, you can solve it directly.)

Comment   Why should we study numbers written in bases other than 10? After all, the decimal (base 10) system is used throughout the world today.

There are several reasons. As we shall see in Section 7, all modern computers work in base 2, 8, or 16. So if you ever work with computers, you will need this background. In Section 8, we will see how some popular games, many of them thousands of years old, can be analyzed mathematically using base 2. Furthermore, as we have seen, the difficulties we have in doing arithmetic in other bases are similar to those a beginner has in base 10. So if you ever teach a child arithmetic, either as a parent or a teacher, you will understand his or her struggles and be better able to help. We are so familiar with base 10 that we really have to study other bases to get a better understanding of base 10.

EXERCISES   1. Perform the indicated operations in the given base.

a) $\begin{array}{r} 10101_{(2)} \\ + 10010_{(2)} \\ \hline \end{array}$

b) $\begin{array}{r} 20112_{(3)} \\ + \ 2012_{(3)} \\ \hline \end{array}$

c) $\begin{array}{r} 8754_{(9)} \\ - 6815_{(9)} \\ \hline \end{array}$

d) $\begin{array}{r} 432_{(5)} \\ \times\ \ 43_{(5)} \\ \hline \end{array}$

e) $\begin{array}{r} t59e_{(12)} \\ + \ 6te_{(12)} \\ \hline \end{array}$

f) $\begin{array}{r} 2331_{(4)} \\ + \ 323_{(4)} \\ \hline \end{array}$

g) $\begin{array}{r} 512_{(6)} \\ \times\ \ 24_{(6)} \\ \hline \end{array}$

h) $\begin{array}{r} 2654_{(7)} \\ - 1563_{(7)} \\ \hline \end{array}$

i) $\begin{array}{r} 7456_{(8)} \\ - 5637_{(8)} \\ \hline \end{array}$

j) $\begin{array}{r} e94_{(12)} \\ \times 435_{(12)} \\ \hline \end{array}$

k) $\begin{array}{r} 11100_{(2)} \\ \times 10001_{(2)} \\ \hline \end{array}$

l) $\begin{array}{r} 12122_{(3)} \\ - \ 2212_{(3)} \\ \hline \end{array}$

m) $\begin{array}{r} 413_{(5)} \\ - 244_{(5)} \\ \hline \end{array}$

n) $\begin{array}{r} 872_{(9)} \\ \times\ \ 65_{(9)} \\ \hline \end{array}$

o) $\begin{array}{r} 476_{(8)} \\ 325_{(8)} \\ + 654_{(8)} \\ \hline \end{array}$

2. In each of the following find the base b in which the numbers are written.

  a) $65_{(b)} = 53$ b) $54_{(b)} = 49$ c) $12_{(b)} = 5$ d) $15_{(b)} = 25$

3. Why must the base of a number system be greater than one?

4. Make up complete addition and multiplication tables for base 2 and base 4.

5. What's wrong with the following calculation?

$$3671_{(8)}$$
$$+\ \ 145_{(8)}$$
$$\overline{\ \ 3836_{(8)}\ }$$

6. In base 10, by looking at the last digit we can tell whether a number is even or odd (how?).

  a) In base 2, is it possible to tell whether a number is even or odd by looking at the last digit? Explain your answer.

  b) What about in base 3?

7. In base 10, the number 46 is even. Are there any bases in which this number is odd? If your answer is yes, what are these bases? If your answer is no, explain why not.

8. Can the number 33 ever be even in any base?

9. Suppose we were working in base 4, but instead of using the digits 0, 1, 2, 3, we use the letters L, O, V, E, where L represents 0, O represents 1, V represents 2, and E represents 3. Make up an addition table for this system using the letters L, O, V, E.

10. An unusual way of doing multiplication, which was used by the ancient Egyptians and until recently by the Russian peasants, is known as the *Russian peasant method*. In order to use it, you only need to know how to multiply and divide by 2. We illustrate the technique by multiplying 67×18. This is shown on the left.

  What we did is the following. In the left column of the picture on the left we divide 67 by 2, *disregarding the remainder*. This gives 33. Then we divide 33 by 2, disregarding the remainder. This gives 16. We repeat the same procedure until we get 1 on the left. On the right, we double 18 and get 36. Then we double 36 and get 72, etc.

  Finally we cross off numbers on the right that are opposite *even* numbers on the left. We add what is left in the right column. The sum, 1206, is our answer.

  a) Check the answer by ordinary multiplication.

  b) Using Russian peasant multiplication, multiply the following.

i)	63	ii)	423	iii)	325	iv)	223
	$\times\ 55$		$\times\ 78$		$\times\ 68$		$\times\ 47$

v)	49	vi)	782	vii)	69	viii)	103
	$\times\ 78$		$\times\ 53$		$\times\ 223$		$\times\ 68$

ix)	45	x)	219
	$\times\ 216$		$\times\ 17$

67	18
33	36
16	~~72~~
8	~~144~~
4	~~288~~
2	~~576~~
1	1152
	1206

11. a) The following is an addition problem in base 3. However, instead of using the digits 0, 1, and 2, we have used the letters A, D, and M. Each letter stands for the same digit every time it is used. Find out which of the digits 0, 1, and 2 is represented by each letter.

    ```
      MAMA
    + DADA
    ------
     MAMAA
    ```

   b) The following is an addition problem in base 4. However, instead of using the digits 0, 1, 2, and 3, we have used the letters A, D, M, and P. Each letter used stands for the same digit every time it is used. Find out which of the digits 0, 1, 2, and 3 is represented by each letter.[2]

    ```
      MAMA
    + DADA
    ------
      PAPA
    ```

12. Pick a number from 1 to 9. You can multiply it by 9, using *finger multiplication*. Suppose we are multiplying 4 by 9. Hold up your hands. We count off, in succession, 9 fingers and then 4 fingers (after you get to the tenth finger start again) going from right to left. When you finish counting, bend down the next finger.

    We can now read off our answer. The first digit is the number of fingers to the right of the bent finger. This is 3. The second digit is the number of fingers to the left of the bent finger. This is 6. Our answer then is 36.

   a) Multiply 7 by 9 using this method.

   b) Multiply 1 by 9 using this method.

   c) Multiply 9 by 9 using this method.

   d) Explain why this method works.

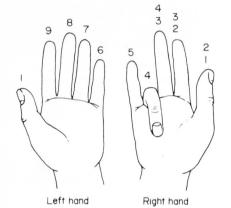

Left hand   Right hand

5. CONVERTING FROM BASE 10 TO OTHER BASES

Imagine one person working in base 10 and another person working in base 5. In order to communicate with each other, they would need a method for changing a base 10 number to a base 5 number and vice versa. In this section we will discuss how we convert a base 10 number to any other base.

We will illustrate the technique by an example. Let us convert 258 written in base 10, to base 5. We write down 258 and divide it by 5. We get 51 with a remainder of 3. This is written as shown.

		Remainder
5	258	3
5	51	1
5	10	0
5	2	2
	0	

2. Problem suggested by Mark Lake.

Then we divide the 51 by 5. This gives 10 with a remainder of 1. We write the 1 under the 3 (as shown) in the remainder column. Now we divide 10 by 5. This gives 2 with a remainder of 0. The remainder of 0 is written under the 1 in the remainder column. Finally we divide the 2 by 5. Since 5 does not "go" into 2, we write down 0 with 2 in the remainder column. We now read off the answer *from bottom to top in the remainder column.* Our answer is 2013. Therefore,

$$258_{(10)} = 2013_{(5)}.$$

This technique will work for conversion from base 10 to *any* other base.

Example 1   Convert $152_{(10)}$ to base 5.

Solution

		Remainder
5	152	2
5	30	0
5	6	1
5	1	1
	0	

Our answer, $152_{(10)} = 1102_{(5)}$.

Example 2   Convert $78_{(10)}$ to base 4.

Solution

		Remainder
4	78	2
4	19	3
4	4	0
4	1	1
	0	

Our answer, $78_{(10)} = 1032_{(4)}$.

Example 3   Convert $167_{(10)}$ to base 2.

Solution

		Remainder
2	167	1
2	83	1
2	41	1
2	20	0
2	10	0
2	5	1
2	2	0
2	1	1
	0	

Our answer, $167_{(10)} = 10100111_{(2)}$.

Example 4   Convert $138_{(10)}$ to base 3.

Solution   Remainder

3	138	0
3	46	1
3	15	0
3	5	2
3	1	1
	0	

Our answer, $138_{(10)} = 12010_{(3)}$.

Example 5   Convert $428_{(10)}$ to base 9.

Solution   Remainder

9	428	5
9	47	2
9	5	5
	0	

Our answer, $428_{(10)} = 525_{(9)}$.

Example 6   Convert $134_{(10)}$ to base 12.

Solution   Remainder

12	134	2
12	11	e
	0	

Our answer, $134_{(10)} = e2_{(12)}$.

Beware   For this method to work you must read *up* the remainder column. If you read down the remainder column, your answer will be wrong.

EXERCISES   1. Convert each of the following base 10 numbers to the indicated base.

a) $335_{(10)}$ to base 5   b) $841_{(10)}$ to base 4   c) $253_{(10)}$ to base 2

d) $749_{(10)}$ to base 12   e) $238_{(10)}$ to base 3   f) $111_{(10)}$ to base 9

g) $312_{(10)}$ to base 8   h) $391_{(10)}$ to base 6   i) $629_{(10)}$ to base 7

j) $370_{(10)}$ to base 12   k) $147_{(10)}$ to base 2   l) $111_{(10)}$ to base 3

m) $325_{(10)}$ to base 8   n) $329_{(10)}$ to base 16

2. Why does the technique discussed in this section work?

3. Write the number $345_{(10)}$ first in base 2 and then in base 8. Now do the same thing for the number $225_{(10)}$. By examining your answer can you find any relationship between base 2 and base 8?

4. Mysterio, the magician, holds up four cards marked as shown on the left. He asks someone in the audience to think of a number from 1 through 15 and to tell him on which of the cards it appears. The first person tells him that the number he is thinking of is on cards 1, 3, and 4. Mysterio then correctly tells him that the number is 13.

    The second person then tells him that he is thinking of a number that appears only on cards 1 and 2. Mysterio tells him that his number is 3.

a) How does Mysterio do it?

b) Why does the trick work? (*Hint:* Write all the numbers 1 through 15 in the binary system.)

5. What are some advantages of a small number base as opposed to a large number base? What are some of the disadvantages?

6. Since base 2 involves only the digits 0 and 1, we can count in base 2 on our fingers. This can be done as follows. Keep all fingers down to represent 0 and a finger up to represent a 1. The position of the "up" finger denotes the position of the "1" in the base 2 representation. Thus we can represent the first six counting numbers as shown below.

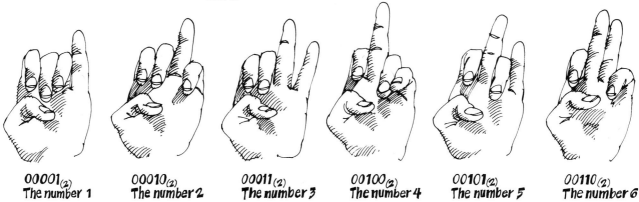

$00001_{(2)}$
The number 1

$00010_{(2)}$
The number 2

$00011_{(2)}$
The number 3

$00100_{(2)}$
The number 4

$00101_{(2)}$
The number 5

$00110_{(2)}$
The number 6

    Using a similar procedure, show how we can represent the numbers 17 through 31 on one hand.

6. CONVERTING FROM OTHER BASES TO BASE 10

In the last section we learned how to convert a number from base 10 to any other base. In this section we will show how to convert a number written in any other base back into base 10.

Remember what 123 in base 5 really means. To refresh your memory we will write it in the following form.

5^2	5^1	1's
1	2	3

This means that we have $1(5^2) + 2(5^1) + 3(1's)$.

Since $5^2 = 5 \times 5 = 25$, we then have

$$1(25) + 2(5^1) + 3(1's)$$
$$25 + 10 + 3$$
$$38.$$

Therefore $123_{(5)} = 38_{(10)}$.

Some further examples will be helpful.

Example 1   Convert $314_{(6)}$ to base 10.

Solution   We rewrite 314 in base 6 in the following form.

6^2	6^1	1's
3	1	4

This means that we have $3(6^2) + 1(6^1) + 4(1's)$.

Since $6^2 = 6 \times 6 = 36$, we have

$$3(36) + 1(6) + 4(1's)$$
$$108 + 6 + 4$$
$$118.$$

Our answer, $314_{(6)} = 118_{(10)}$.

Example 2   Convert $111011_{(2)}$ to base 10.

Solution   We rewrite $111011_{(2)}$ as

2^5	2^4	2^3	2^2	2^1	1's
1	1	1	0	1	1

Since $2^5 = 2 \times 2 \times 2 \times 2 \times 2 = 32$,

$2^4 = 2 \times 2 \times 2 \times 2 \quad = 16,$

$2^3 = 2 \times 2 \times 2 \quad\quad = 8,$ and

$2^2 = 2 \times 2 \quad\quad\quad\quad = 4,$

we then have

$1(32) + 1(16) + 1(8) + 0(4) + 1(2) + 1(1\text{'s})$

$\qquad 32 + 16 + 8 + 0 + 2 + 1$

$\qquad\qquad 59.$

Our answer, $111011_{(2)} = 59_{(10)}$.

Example 3   Convert $673_{(8)}$ to base 10.

Solution   Rewrite $673_{(8)}$ as

8^2	8^1	1's
6	7	3

Again $8^2 = 8 \times 8 = 64$, so we have

$6(64) + 7(8^1) + 3(1\text{'s})$

$\qquad 384 + 56 + 3$

$\qquad\quad 443.$

Our answer, $673_{(8)} = 443_{(10)}$.

Example 4   Convert $158_{(12)}$ to base 10.

Solution   Rewrite $158_{(12)}$ as

12^2	12^1	1's
1	5	8

Since $12^2 = 12 \times 12 = 144$, we get

$1(144) + 5(12^1) + 8(1\text{'s})$

$\qquad 144 + 60 + 8$

$\qquad\quad 212.$

Our answer, $158_{(12)} = 212_{(10)}$.

EXERCISES

1. Convert each of the following numbers from the indicated base to a base 10 number.

   a) $415_{(6)}$ b) $110110_{(2)}$ c) $324_{(5)}$ d) $432_{(5)}$

   e) $111011_{(2)}$ f) $764_{(8)}$ g) $2121_{(3)}$ h) $267_{(9)}$

   i) $578_{(9)}$ j) $1011_{(3)}$ k) $315_{(20)}$ l) $2134_{(5)}$

   m) $eee_{(12)}$ n) $1035_{(6)}$ o) $613_{(7)}$ p) $216_{(8)}$

   q) $202_{(4)}$ r) $543_{(6)}$

2. Sharon tells her mother that she is going to marry Mark who is $1012_{(3)}$ yr old. Since Sharon is only $11_{(25)}$ yr old, her mother is shocked. After she has been revived, she realizes that everything is OK. What is the age of each?

7. APPLICATIONS TO COMPUTER

It has been said that we live in the computer age. Computers are involved in almost every aspect of our lives. Let us look at a day in the life of John Doe.

He wakes up in the morning and turns on the light. Nothing happens. Due to a computer error in processing his electric bill, his electricity has been turned off. On the way to work, he uses his credit card to buy gas for which the computer will bill him at the end of the month. Unfortunately, stopping for gas has made him late. Speeding to work, he is stopped by a policeman who asks to see his computerized license. He is then issued a computerized summons. At work, he "punches in" on a computerized card. At the end of the day, he receives his pay check. You guessed it; the computer made a mistake and underpaid him five dollars.

John decides to register with a computer dating service and at lunchtime he mails them a check. (All checks are processed by computer.) After work, John registers for an evening course at Staten Island Community College. Of course, the entire registration process is computerized. When John gets home, he finds a letter from the Internal Revenue Service stating that the computer has found an error in his income tax return.

While we are all affected by computers, many of us are unaware of what a computer really is, how it works, and what it can and cannot do.

A computer can do *nothing* that a human could not do if given enough time. The usefulness of computers is in their tremendous speed and accuracy. Computer

In 1804, Joseph Marie Jacquard made a loom for weaving intricate patterns. It used punched cards, pressed against the bank of needles, for each pass of the shuttle. Certain needles were pushed back by the card; others stayed in place because of holes in the card. This use of punched cards foreshadowed our present-day computers.

Bruce Read

errors occur only when there is a mechanical failure or when the human operator makes a mistake.

How does a computer do things so quickly? It works on electrical impulse and electricity travels at the speed of light (which is about as fast as you can get).

Most computers work in base 2 because of its simplicity. (Remember, in base 2 there are only two digits, 0 and 1.) Furthermore, the fact that there are only two digits makes it easy to represent numbers on the computer in many different ways.

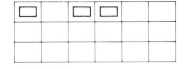

1. If we are using light bulbs on the console (the front of the computer), then a bulb off means 0 and a bulb on means 1. If the console shows the bulbs at left, this means 101100 in base 2, or 44 in base 10.

2. If we use a punched card, then we can represent 1 with a hole and 0 with no hole. Thus, 101100 in base 2 would appear as at the left.

3. If we use magnetic tape or discs, then we can represent 1 by a magnetized spot and 0 by no magnetic spot.

4. If we use switches, then switch on means 1 and switch off means 0.

5. There are still other ways of representing base 2 numbers using magnetic cores and the direction of the current in a circuit.

In the modern-day computer, information is fed in using base 10 (this is usually done by means of punch cards, tape, or typewriter). Words are also converted into base 10 numerals. The information is then converted into base 2 and stored for future use in the computer's "memory." Some computers store information in base 8 or 16 to save space. A typical computer with 100,000 "spaces" uses about 80,000 of these spaces for memory, that is, storage of information. Only about 20,000 spaces are actually used for computations.

In order to use the computer, we must feed information into the memory and then give precise instructions as to the computations to be performed with this information. We must also tell the machine how to write down the result. This is the job of the computer programmer.

All information and instructions must be given to the computer in a specific form. If even a comma is left out, the computer will not be able to understand what to do. The computer *cannot think* on its own. It merely follows human-given directions.

Suppose on an exam you were given the problem: "Dvide 6 by 2." You of course would know that the word "divide" is misspelled. This would not stop you from doing the problem. Since a computer cannot reason, it would not understand what "dvide" meant. It would stop and await further instructions.

Computers cannot do our thinking for us. However, their ability to do computations accurately at lightning speed does relieve us of much of the time-consuming and boring calculation associated with mathematical and scientific projects. Many mathematical problems have been solved with the help of computers doing the "dirty work." Some problems involving long truth tables can be analyzed in a fraction of a second using the computer.

In a later chapter we will discuss computers in more detail.

EXERCISES

Open hole   Closed hole

1. Suppose you have just dropped on the floor 1000 cards numbered 0 to 999. You must pick them up and put them back into numerical order. It will be a long, tedious job, although a computer could do it in a fraction of a second. Surprisingly, by imitating the computer's technique you too could do the job, not as fast as the machine, but in a fairly short time. We will illustrate the method, not on 1000 cards, but on 32 cards numbered 0 to 31. First write the number of each card in base 2. Next cut holes in each card as shown.

A "closed" hole does not go through to the top of the card as does an "open" hole. A closed hole represents 1, an open hole represents 0.

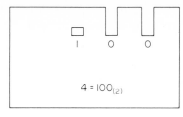

$4 = 100_{(2)}$

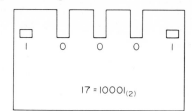

$17 = 10001_{(2)}$

Each card will have 6 holes. Now take a long, thin object like a knitting needle, pass it through the first hole on the right, and lift it up. The cards with closed holes in that place will be lifted with it. Place these behind the other cards. Repeat this on the second hole, then the third, fourth, fifth, and sixth. When you are finished, the cards will be sorted in numerical order from 0 to 31. The sorting takes only 6 "passes" through the pile.

a) Make a set of cards and perform this experiment.

b) Why does each card have 6 holes?

c) Explain why this procedure works.

d) How many passes would be needed if there were only 16 cards? (*Hint:* How many holes would each card have?)

e) How many passes are needed for 100 cards?

2. Read one of the articles 44–50 in *Mathematics in the Modern World* (Readings from *Scientific American*). San Francisco: Freeman, 1968.

---

*8. APPLICATION TO THE GAME OF *NIM*

Nim is a game that has been played in different forms for many centuries. Some time ago, a mechanical version for children was marketed in which the child played against a plastic mechanical computer. The word "Nim" comes from the German word "nehmen" meaning "to take." We will learn how to play the simplest version of the game.

The Game of Nim

We start with several piles of matches. Each pile can have any number of matches, and each pile can have a different number of matches. For example, we may have five piles with 4, 7, 3, 2, and 2 matches. The rules of the game are as follows.

1. There are two players who take alternate turns.

2. Each player, on his turn, takes as many matches as he wants from *one and only one* pile. He must take at least one match and may take them all if he wants. On his next turn he may select the same pile or a different pile.

3. The person who takes the last match wins.

To illustrate how *Nim* works, consider the following sample game between Joe and Frances. There are three piles with 4, 2, and 7 matches.

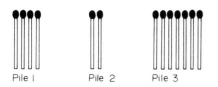

Pile 1   Pile 2   Pile 3

Move	Situation after move		
Start of game	Pile 1	Pile 2	Pile 3
Frances takes 2 matches from pile 3.	Pile 1	Pile 2	Pile 3
Joe takes all the matches in pile 1.	Pile 1	Pile 2	Pile 3
Frances takes 1 match from pile 2.	Pile 1	Pile 2	Pile 3
Joe takes 1 match from pile 3.	Pile 1	Pile 2	Pile 3

Move	Situation after move
Frances takes 1 match from pile 2.	

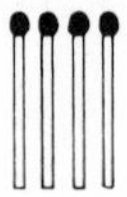

Pile 1   Pile 2   Pile 3

Joe takes the 4 matches from pile
3 and wins.

Pile 1   Pile 2   Pile 3

It may seem to you that winning the game is a matter of skill. However, this is not entirely so. Someone who knows the secret of the game can win, no matter how clever his or her opponent is, provided he or she is allowed to choose who goes first. Gamblers who know this secret have won a lot of money from innocent "suckers."

How to Win at *Nim*   The explanation is lengthy, but not difficult. It is given in three parts. Read each of parts I, II, and III carefully before going on to the next part.

Part I   Write the number of matches in each pile in the binary system. In our sample game this would be 100, 10, and 111. We write these numbers in a column and then add them *as if they were in base 10:*

$$
\begin{array}{r}
100 \\
10 \\
111 \\
\hline
221
\end{array}
$$

If the result has all even digits, then we call it a **bad combination**. Otherwise, we call it a **good combination**.

Example 1   If there are 3, 5, 8, and 4 matches, then these numbers converted to base 2 and added are

$$
\begin{array}{r}
11 \\
101 \\
1000 \\
100 \\
\hline
1212
\end{array}
$$

This is a good combination.

Example 2 If there are 10, 13, 2, 9, and 6 matches, then these numbers converted to base 2 and added are

```
 1010
 1101
   10
 1001
  110
 ────
 3232
```

Again this is a good combination.

Example 3 If there are 11, 13, and 6 matches, then these numbers converted to base 2 and added are

```
 1011
 1101
  110
 ────
 2222
```

This is a bad combination.

Part II Notice that just before the final winning play is made, there is only one pile. This means that there is only one number (written in base 2) in the sum of part I. Thus the total will consist just of this one number. Since numbers in base 2 contain only the digits 0 and 1, this total will also contain only the digits 0 and 1. Thus we must have a *good combination* (since it contains an odd number, 1). In our sample game, before Joe makes the last move, there are four matches in pile 3. Remember that 4 in base 10 is 100 in base 2 so we have the following:

```
Pile 1      0
Pile 2      0
Pile 3    100
        ─────
Sum       100
```

This is a good combination.

Summarizing *The final winning move can only be made from a good combination.*

Part III Therefore, our strategy is to keep our opponent from ever getting a good combination and to make sure at the end we ourselves get one. This can be accomplished in the following way.

  1. *A play from a bad combination will always leave a good combination.* The reason for this is that in any given move you can change only one number in the sum (since you can take only from one pile). Thus in each column of the sum you

can change only one digit (and you must change at least one digit). The only digit change can be from 0 to 1 or from 1 to 0 (because we are in base 2, these are the only digits we have). Since the sum contained all even digits to begin with, this change will make some of them odd. This will result in a good combination.

To see how this works, consider the bad combination of 11, 13, and 6 matches given in Example 3 above. This combination was

$$
\begin{array}{r}
1011 \\
1101 \\
110 \\
\hline
2222
\end{array}
$$

Suppose we take 7 matches from pile 1. This changes the first number to 100, leaving

$$
\begin{array}{r}
100 \\
1101 \\
110 \\
\hline
1311
\end{array}
$$

This is a good combination. Try making the following plays on the original combination and verify that you always get a good combination.

a) Take 2 from pile 3.   b) Take 8 from pile 2.   c) Take all of pile 1.

2.   *A play from a good combination can always be made into a bad combination.* We do this as follows: Go to the first odd column from the left of the sum in the good combination. Pick any number that has a 1 in this column. Circle this 1. Now go to the next odd column. Circle the digit on the same line in this column. Repeat this for all odd columns and only odd columns. Now change all the circled digits. Since we are working in base 2, we can only change 0's to 1's and 1's to 0's. Change *only* circled digits. After we change this number we play to *leave* this amount of matches in this pile.

Comment   We will not explain here why this strategy works. It is given as an exercise.

Example 4   Suppose we are playing the good combination of Example 1, p. 108.

$$
\begin{array}{r}
11 \\
101 \\
1000 \\
100 \\
\hline
1212
\end{array}
$$

The first and third columns from the left (in the sum) are odd. There is only one number with a 1 in the first column. We must use it. We circle it. In column 3, we circle the digit on the same line. We change these and get 0010 which is 2 in base 10. Thus we take all but 2 in pile 3. Notice that after we play, we leave

```
  11
 101
  10
 100
 ───
 222
```

This is a bad combination.

Summary

1. Determine whether the original combination is good or bad.

2. If the original is good, choose to go first. Otherwise, let your opponent go first.

3. Whenever you move, use the strategy discussed to leave your opponent with a bad combination.

4. No matter what your opponent does, he or she must leave you a good combination.

EXERCISES

1. Which of the following are good combinations and which are bad?

   a) 7, 6, 3   b) 8, 4, 5, 3   c) 9, 8, 6, 7   d) 12, 8, 3

2. Using the strategy given, change the following good combinations to bad combinations.

   a) 7, 4, 19   b) 5, 9, 7, 9   c) 13, 10, 15

3. Play the game of *Nim* with an opponent using the following combinations.

   a) 5, 8, 7   b) 3, 4, 6, 8, 9   c) 5, 4, 3, 7, 6, 1

4. Why does the strategy for changing a good combination to a bad combination given on p. 110 work?

STUDY GUIDE

In this chapter we introduced the following ideas about number bases.

Basic ideas

The Babylonian number system (p. 82)
Positional notation (p. 83)
Egyptian system (p. 84)
Base-10 number system (p. 86)
Roman numeral system (p. 86)
Hindu-Arabic system (p. 86)
Base-10 arithmetic (p. 87)
Zero as a place holder (p. 87)

Decimal system (p. 88)

Other bases (p. 92)

Binary system (p. 94)

Arithmetic in other bases (p. 93)

Converting from base 10 to
  other bases (p. 98)

Converting from other bases
  to base 10 (p. 102)

Good combination (p. 108)

Bad combination (p. 108)

Different ways of doing multiplication

Gelosia multiplication (p. 89)

Napier's bones (p. 90)

Russian peasant multiplication (p. 97)

Finger multiplication (p. 98)

The ideas of number bases were applied to:

a) computers (p. 104),

b) the game of *Nim* (p. 106).

SUGGESTED FURTHER READINGS

Aaboe, A. S., *Episodes From the Early History of Mathematics.* New York: Random House, 1964. Chapter 1 discusses the Babylonian number system.

Bergamini, D., et al., *Life, Mathematics* (*Life* Science Library). New York: Time-Life Books, 1970.

Computers and Computation (Readings from *Scientific American*). San Francisco: W. H. Freeman, 1971.

Conant, L. L., "Counting." In James R. Newman, ed., *The World of Mathematics*, Vol. I, Part III, Ch. 2. New York: Simon and Schuster, 1956. This article discusses the ability to count in humans and animals.

Gardner, M., *New Mathematical Diversions from the Scientific American.* New York: Simon and Schuster, 1966. See Chapters 1, 12, and 19.

Gillings, R. J., *Mathematics in the Time of the Pharaohs.* Cambridge, Mass.: MIT Press, 1962. A highly readable discussion of Egyptian mathematics.

Mathematics in the Modern World (Readings from *Scientific American*). San Francisco: W. H. Freeman, 1968. See articles 13, 44, 45, and 50.

Ore, O., *Graphs and Their Uses.* New York: Random House, 1963. Pages 73–75 discuss *Nim*.

Ore, O., *Number Theory and its History.* New York: McGraw-Hill, 1948. See pp. 34–37 for a discussion on how to convert from one base to another.

Smith, David Eugene, and J. Ginsburg, "From Numbers to Numerals and From Numerals to Computation." In James R. Newman, ed., *The World of Mathematics*, Vol. 1., Part IV, Ch. 3. New York: Simon and Schuster, 1956.

Van der Waerden, B. L., *Science Awakening.* Groningen, Holland: Noordhoff Ltd, 1954. Chapters 1 and 2 discuss how ancient civilizations wrote numbers and did computations with these numbers.

Yoshino, Y., *The Japanese Abacus Explained.* New York: Dover, 1963.

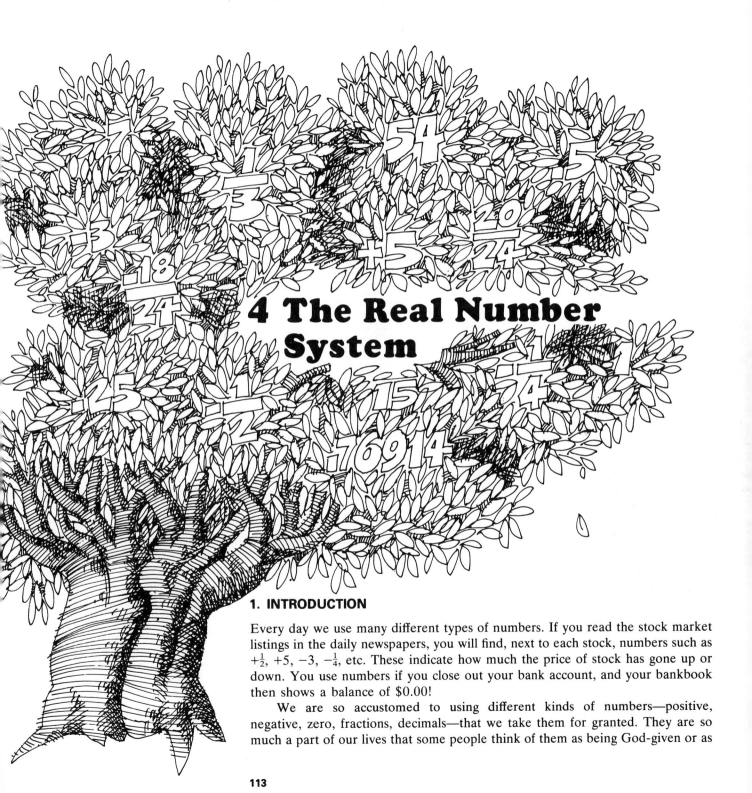

4 The Real Number System

1. INTRODUCTION

Every day we use many different types of numbers. If you read the stock market listings in the daily newspapers, you will find, next to each stock, numbers such as $+\frac{1}{2}$, $+5$, -3, $-\frac{1}{4}$, etc. These indicate how much the price of stock has gone up or down. You use numbers if you close out your bank account, and your bankbook then shows a balance of $0.00!

We are so accustomed to using different kinds of numbers—positive, negative, zero, fractions, decimals—that we take them for granted. They are so much a part of our lives that some people think of them as being God-given or as

113

Here is a listing of New York Stock Exchange transactions on the day of the big crash in 1929. Negative numbers indicate the extent of the disaster.

NEW YORK TIMES, WEDNESDAY, OCTOBER

NS ON THE NEW YORK STOCK

TUESDAY, OCTOBER 29, 1929.

	Day's Sales.	Monday.	Saturday.	A Year Ago.	Two Years Ago.
	16,410,030	9,212,800	2,087,660	3,483,770	1,676,570

Same Period

Year to Date.	1928.	1927.	1926.	1925.
950,797,190	708,649,607	464,944,575	376,924,360	363,064,122

being part of nature. Actually this is not true. Numbers are man-made Different kinds of numbers were created by humans at various stages of their mathematical development to meet their changing needs. First came the counting numbers 1, 2, 3, ... Negative numbers, however, were invented much later and were not generally accepted until the sixteenth century. The concept of a fraction arose during the Bronze Age, but it was not fully developed for some time. As late as the year 1650 B.C.,[1] the Egyptians, who were rather good mathematicians, did not use fractions as we use them today. They used mostly unit fractions, that is, fractions whose numerators are 1 such as $\frac{1}{2}$, $\frac{1}{3}$, $\frac{1}{4}$, etc., and the fraction $\frac{2}{3}$. Other fractions were written as combinations of unit fractions. Although the Egyptians had an idea of the number zero, they had no symbol for it. Symbols for zero were invented at various times by different civilizations.

In this chapter, we will discuss how and why our present-day number system developed.

---

1. Our knowledge of Egyptian mathematics is derived from the Rhind Papyrus which dates from approximately 1650 B.C.

2. THE NATURAL NUMBERS

Humans must have invented natural numbers shortly after they learned how to count. At first they had only the numbers 1 and 2. A few more talented societies also had the number 3. Everything else was "many." Later as society became more complex, it became necessary to count accurately larger quantities. At this point numbers larger than 2 and 3 were introduced. Thus, the natural numbers were born. Because they are used in counting, the natural numbers are also called the counting numbers.

Herds of cattle in this Egyptian tomb painting are being led past for census taking. Scribes are writing down the count.

Definition 2.1   *The **natural numbers**, or **counting numbers**, are the numbers* 1, 2, 3, 4, 5, 6, 7, 8, 9, 10, 11, ...

Addition   Let us go back 25,000 years in history. Zaftig, the caveman, has 3 sheep. His brother has just given him 2 sheep. Zaftig now counts his sheep and discovers that he has 5 altogether. Zaftig, who is the "brain" of the tribe, sees that if you have 3 things and you then get 2 more, your total will *always* be 5 things. This is an example of what we call addition. Roughly speaking, by **addition** we mean that we combine *two numbers* to obtain a third number called the **sum**. This sum must be *unique*. (This means that there is *only one* answer. There can never be more than one answer.) For example,

4 + 7 = 11 (11 is the only answer), and

5 + 8 = 13 (13 is the only answer).

Do the following simple addition problem.

$$
\begin{array}{r}
7 \\
8 \\
\underline{9} \\
24
\end{array}
$$

Your answer is 24, of course. How did you get it? You could add from bottom to top as follows: 9 + 8 equals 17, and then 17 + 7 equals 24. We could also add from top to bottom as follows: 7 + 8 equals 15, and then 15 + 9 equals 24.

Notice that no matter which way we add, *we can only add two numbers at a time.* To add three numbers, we first add any two, and then add the third to the sum. Addition is an example of a **binary operation**.

Definition 2.2   *A **binary operation** on numbers is a process by which **two** numbers are combined to obtain a **unique** third number.*

Suppose you were to ask a small child who is just learning arithmetic how much 2 + 3 is. The child would first put up 2 fingers and then 3 fingers and would count the total, getting 5. If you now ask the child to add 3 + 2, he or she would put up 3 fingers and then 2 more, again getting 5. The child soon discovers that

3 + 2 = 2 + 3.

This is known as **the commutative law of addition**. What this tells you is that the *order* in which you add two numbers doesn't matter. Your answer is always the same. We state this formally as the following law.

Commutative Law of Addition

If a and b stand for any natural numbers, then a + b = b + a.

This law may seem rather obvious to you. You may even feel that it doesn't say much. However, the commutative law is not always true for operations different from addition. For example, subtraction is not commutative:

5 − 3   is not equal to   3 − 5.

When you add a column of numbers, you may add from the top down (or vice versa). The usual way of checking is to add again in reverse order. You know that if your addition is correct, the answer should be the same in both cases. You are using the commutative law of addition, which says that the order in which numbers are added is not important.

In everyday life situations, the order in which we do things may or may not make a difference. For example, combing your hair and brushing your teeth in the morning is commutative. On the other hand, putting on your shoes and putting on your socks is not commutative.

Ask a chemist whether mixing water and sulfuric acid is commutative. That is, does it make a difference if we pour water into sulfuric acid or sulfuric acid into water?

In a certain town there are two schools, a two-story high school and a six-story elementary school. Recently the following headline appeared in the local newspaper: "High School Building Burns Down." Which one was it? The meaning is unclear since we do not know whether the word "school" goes with "high" or with "building." If "school" is grouped with "high," then obviously the high school (two-story building) burned down. In other words, the grouping will make a big difference in the meaning. In English, to avoid this confusion we can put a hyphen between the words "high" and "school" when we mean the two-story building.

Now consider Mr. Jones who has decided to buy a Volkswagen. Since the local bank is advertising "small car loans," he applies to the bank for a $2000 loan. The bank officer tells him that their maximum car loan is $300. Mr. Jones protests that they are advertising loans to buy small cars and that you cannot buy any small car for $300. The bank official responds that Mr. Jones is misinterpreting the advertisement which offers small loans to buy cars. Who is right? In this case the disagreement is also about grouping. Mr. Jones is grouping "car" with "small." The bank official is grouping "car" with "loan."

In these two examples grouping made a difference in the meaning. This is not always the case. In the phrase "my friend Klutz the mechanic," it does not make any difference whether we group the word "Klutz" with "friend" or with "the mechanic."

In mathematics, grouping may or may not be important. Suppose we are asked to add

$2 + 3 + 4,$

without changing the order of the numbers. Since we can only add two numbers at a time, this can be done in two ways.

We have either

$(2 + 3) + 4$ (We always do what is in the parentheses first. In this case we first add the $2 + 3$.)

$\quad 5 + 4$

$\qquad 9$

or

$2 + (3 + 4)$ (This time, we first add the $3 + 4$.)

$\quad 2 + 7$

$\qquad 9.$

Note that in both cases our answer is the same, that is,

$(2 + 3) + 4 = 2 + (3 + 4).$

In other words, it makes no difference whether we group the 3 with the 2 or with the 4. This is called **the associative law of addition**. We state it here.

Associative Law of Addition   *If a, b, and c are any natural numbers, then $(a + b) + c = a + (b + c)$.*

Comment   Since the position of the parentheses does not matter in doing addition, we usually leave them out and write, for example,

$$2 + 3 + 4$$

instead of $(2 + 3) + 4$ or $2 + (3 + 4)$.

Beware   Do not confuse the associative law for addition with the commutative law for addition. The associative law keeps the numbers in the same order but merely groups them differently. The commutative law involves changing the order of the numbers.

Although the associative law holds for addition, it does not work for all the binary operations. In particular, the associative law does not hold for subtraction. For example, $(15 - 4) - 3$ is not equal to $15 - (4 - 3)$. This is so because

$$(15 - 4) - 3 = 11 - 3 = 8.$$

Remember, we do what is in the parentheses first. On the other hand,

$$15 - (4 - 3) = 15 - 1 = 14.$$

Thus, $(15 - 4) - 3$ is not equal to $15 - (4 - 3)$.

Suppose you were restricted for some reason to using *only* the odd numbers 1, 3, 5, 7, 9, 11, etc. Then it would be impossible to add. If you add any two odd numbers, you will always get an *even* number. Since we are limited to *only* the odd numbers, this answer is not acceptable.

Consider now *all* the natural numbers. If we add any two natural numbers, our result will also be a natural number. This is called the **closure property for addition of natural numbers**. We say that the natural numbers are closed under addition. The odd numbers are not closed under addition.

Law of Closure for Addition   *If a and b are any natural numbers, then the sum, a + b, is also a natural number.*

Example 1   If we are restricted to the set of numbers $\{1, 2, 3\}$, then this set is *not* closed under the operation of addition. The numbers 2 and 3 are in the set, but their sum, which is 5, is not in the set.

Example 2   Let $S = \{$all even numbers$\}$. Then S is closed under the operation of addition, since when we add two even numbers, the sum is *also* an even number.

Example 3   Let $S = \{1\}$. S is not closed under addition since $1 + 1$ equals 2, and 2 is not in S.

Comment   Are the natural numbers closed under the operation of subtraction? That is, if we subtract one natural number from another natural number, is our answer also a natural number?

Multiplication   Let us return to our friend Zaftig, the caveman. He now has a herd of cattle. Each day he takes them out to pasture, walking them 5 abreast.

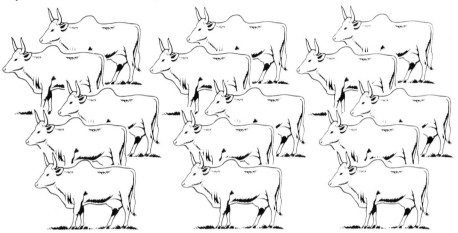

As they return from pasture, Zaftig counts them to make sure that they are all there. Instead of counting them one by one, he discovers that it is easier to count them in groups of 5's. He has three groups of 5 each, that is, he has

$$\underbrace{5 + 5 + 5}_{\text{3 groups of 5}}$$

This means that he has 15 cattle. A convenient way of expressing this is to invent a new operation called **multiplication**, which is symbolized in one of the following ways: $3 \cdot 5$, $3(5)$, 3×5, or $(3)(5)$. Any one of these means

$5 + 5 + 5$.

Similarly $7 \cdot 4$, 7×4, $7(4)$, or $(7)(4)$ means

$4 + 4 + 4 + 4 + 4 + 4 + 4$.

In general, $a \cdot b$ [also written as ab, $a(b)$, $a \times b$, or $(a)(b)$], means

$$\underbrace{b + b + b + \cdots + b}_{a \text{ of them}}$$

Note that since we can only multiply two numbers at a time, **multiplication is also a binary operation**.

According to our definition of multiplication, $2 \cdot 3$ means 2 groups of 3 things each. This can be pictured as it is at the left. On the other hand, $3 \cdot 2$ means 3 groups of 2 things each. This we picture as

In both cases, we have 6 things altogether. Therefore

$2 \cdot 3 = 3 \cdot 2$.

This is an example of the commutative law of multiplication.

Commutative Law of Multiplication   *If a and b are any natural numbers, then $a \cdot b = b \cdot a$.*

Multiplication of natural numbers is also associative. If you multiply $2 \times 4 \times 5$, this can be done in two ways (without changing the *order* of the numbers). One way is

$2 \times (4 \times 5)$ 4 is grouped with 5

  2×20

   40.

Another way is

$(2 \times 4) \times 5$ 4 is grouped with 2

  8×5

   40.

In both cases our answer, of course, is the same. This shows that the grouping does not matter and that

$2 \times (4 \times 5) = (2 \times 4) \times 5$.

In general terms, we have the following law.

Associative Law of Multiplication   *If a, b, and c are any natural numbers, then $a(bc) = (ab)c$.*

Comment   As with addition (see p. 115), since the parentheses do not matter, it is customary to leave them out and write abc instead of $a(bc)$ or $(ab)c$.

If we multiply any two natural numbers, our result will also be a natural number, that is, the natural numbers are closed under multiplication. This we state as the following law.

Law of Closure for Multiplication   *If a and b are any natural numbers, then ab is also a natural number.*

Up to now we have mentioned subtraction without considering precisely what is meant by it. If you spend $3 in a store and give the salesclerk a $10 bill, he will usually give you your change as follows: He says "three" and then, handing you the money, counts out the change, "four, five, six, seven, eight, nine, ten dollars." The clerk has to subtract 3 from 10, and he does it by starting with 3 and then figuring how much change must be added to 3 to make 10. Thus the subtraction

$10 - 3 =$ change

becomes

$3 +$ change $= 10.$

In other words, the change is the amount that must be added to 3 to make 10. This procedure shows us what **subtraction** really is.

Definition 2.3   *If a and b are natural numbers, then a − b (read as "a minus b") means the number that must be added to b to obtain a. If we call this number x, we have a − b = x, if b + x = a.*

Example 4 $4 - 1$ means some number x that must be added to 1 to make 4.
Thus $1 + x = 4.$
Therefore $x = 3.$
Then we have $4 - 1 = 3.$

Example 5 $17 - 6$ means the number x that must be added to 6 to make 17.
Thus $6 + x = 17.$
And $x = 11.$
So, $17 - 6 = 11.$

Comment   Subtraction is also a **binary operation**, since it involves *two* numbers, the number we are subtracting and the number from which we are subtracting.

We have already pointed out that subtraction of natural numbers is *not* closed, commutative, or associative, but the following examples should emphasize these facts.

Example 6 $5 - 3 = 2.$ However, $3 - 5$ is *not* equal to 2. $3 - 5$ equals a *negative* number -2, which we will discuss further in Section 4. Thus

$5 - 3$ is *not* equal to $3 - 5,$

which shows that the operation of subtraction is *not* commutative.

Example 7   In the above example, we saw that $3 - 5 = -2$, which is *not* a natural number. Thus subtraction of natural numbers is *not closed*. (This means that when you subtract one natural number from another, the result is not always a natural number.)

Example 8 $9 - (6 - 4)$ equals
            $9 - 2$ which equals (remember, do what is in parentheses first)
            7.

On the other hand,

$(9 - 6) - 4$ equals
      $3 - 4$ which equals
      $-1.$

Therefore, $9 - (6 - 4)$ is *not* equal to $(9 - 6) - 4$, so that subtraction of natural numbers is *not* an associative operation.

The final binary operation that we will consider is **division**. If you were asked to divide 6 by 2, you would immediately give the answer 3. How do you get this answer? Undoubtedly you would ask yourself, "What number multiplied by 2 will give 6?" With this in mind, we define division in general as follows.

Definition 2.4 *If a and b are natural numbers, then a ÷ b, also denoted as a/b (read as "a divided by b"), means the number x, such that b · x = a.*

Example 9 $39 ÷ 13$ means a number x such that $13 · x = 39$.
Thus $x = 3$.
We have, then, $39 ÷ 13 = 3$.

Example 10 $24 ÷ 6$ means a number x, such that $6 · x = 24$.
Thus $x = 4$.
Therefore $24 ÷ 6 = 4$.

Example 11 $100/20$ means a number x, such that $20 · x = 100$.
Thus $x = 5$.
Therefore $100/20 = 5$.

It can easily be seen that division of natural numbers is not closed, commutative, or associative. This is illustrated by the following examples.

Example 12 $3 ÷ 6$ equals $\frac{1}{2}$. This is not a natural number. Thus we see that when we divide one natural number by another, the result is *not* always a natural number. This shows that division of natural numbers is not closed.

Example 13 We know that $6 ÷ 3 = 2$. However, $3 ÷ 6$ is not equal to 2. Thus $6 ÷ 3$ is *not* equal to $3 ÷ 6$, which shows that division is *not* commutative.

Example 14 $24 ÷ (6 ÷ 2)$ equals
    $24 ÷ 3$ which equals
     8.

However,
$(24 ÷ 6) ÷ 2$ equals
    $4 ÷ 2$ which equals
     2.

Thus we see that $24 ÷ (6 ÷ 2)$ is *not* equal to $(24 ÷ 6) ÷ 2$, which shows that division is *not* associative.

Comment We have seen that subtraction and division are not associative operations. Therefore, when we subtract or divide more than two numbers, we *must* insert parentheses to show which numbers are to be subtracted or divided first. For example, we must not write $24 ÷ 6 ÷ 2$, since (as we saw in Example 14) this can have two different answers, depending on where the parentheses are inserted.

EXERCISES

1. Which property for natural numbers justifies each of the following?

   a) $12 + 3 = 3 + 12$

   b) $(12 \cdot 3)4 = 12(3 \cdot 4)$

   c) $4 + (5 \cdot 9) = 4 + (9 \cdot 5)$

   d) $2 + 5$ is a natural number

   e) $4 + (9 + 7) = (4 + 9) + 7$

   f) $4 + (9 + 7) = 4 + (7 + 9)$

   g) $3 \times 8 = 8 \times 3$

   h) $3(5 \cdot 4) = 3(4 \cdot 5)$

   i) $4 + 5 = 9$

   j) $4 \cdot 5 = 20$

2. Which of the following activities are commutative and which are not?

   a) Jumping out of a plane and putting on a parachute

   b) Making a right turn, and then a left turn in your car

   c) Dusting the furniture and vacuuming the rugs

   d) Loading a gun and firing it

   e) Getting undressed and taking a shower

   f) Studying for a math exam and taking the exam

   g) Changing the oil and rotating the tires on your car

   h) Filling the swimming pool and diving in

3. Are the following sets of numbers closed under the indicated operations?

   a) $\{1, 2, 3\}$; addition

   b) $\{1\}$; multiplication

   c) {even numbers}; multiplication

   d) {odd numbers}; multiplication

   e) $\{3, 6, 9, 12, 15, 18, 21, \ldots\}$; addition

   f) {natural numbers less than 10}; multiplication

   g) $\{1, 2\}$; addition

$\mid$	x	y	z
x	x	y	z
y	y	y	z
z	z	z	z

*4. Consider an operation that we will call $\mid$ given by the table at the left.

   a) Is $\mid$ a closed operation?

   b) Is $\mid$ a commutative operation?

   c) Is $\mid$ an associative operation?

$/$	x	y	z
x	x	z	z
y	y	x	z
z	z	z	x

*5. Consider an operation that we will call $/$, given by the table at the left.

   a) Is $/$ a closed operation?   b) Is $/$ a commutative operation?

   c) Is $/$ an associative operation?

*6. An operation called $\#$ is given by the rule

$$a \# b = a + b + 2.$$

In other words $a \# b$ is obtained by first adding $a + b$ and then adding 2 to the result. For example, if a and b are 5 and 7, respectively, then $a \# b = 5 + 7 + 2 = 14$. If a and b are 8 and 6, respectively, then $a \# b = 8 + 6\,2 = 16$.

   a) Is the operation $\#$ commutative?   b) Is the operation $\#$ associative?

3. THE NUMBER ZERO

If ever a number was misunderstood, then zero is it. Ask any five people (nonmathematicians, of course) what zero means to them. When we did this, some of the responses we got were the following.

1. Zero is nothing.
2. Zero is not just plain nothing; it is a something nothing.
3. Zero is not a number at all.
4. Zero is the absence of anything.
5. Zero is the null set.
6. Zero is either the beginning or end of something.

Despite beliefs to the contrary, zero is a number just like any other number. However, it has certain properties that other numbers do not have. In this section and the following ones, we will discuss these properties.

One source of confusion is the fact that zero plays different roles in different situations. In the number 405, zero acts as a "place holder." If you say sadly, "My bank balance is zero," then zero is used as a quantity indicating that there is nothing left.

Zero probably appeared first as a place holder. The Babylonians, who used a base 60 number system, at first had no symbol to represent empty places. Later in their history, they used the symbol ✦ as a place holder. This is probably the first appearance of any symbol for zero. The Mayans also had invented zero symbols. Some of these are shown on the left.

Later, the Greeks invented their own symbol for zero, and developed the *concept* of zero to represent nothingness.

Earlier civilizations do not appear to have used zero as a *number*. Later the Hindus developed the concept of and a symbol for zero. Present evidence indicates that zero was used in India from the ninth century on, and possibly earlier. The Hindus appear to have used zero not merely as a place holder or as a concept for nothing, but also as a *number*. India was invaded by the Arabs around 700 A.D. The Arabs later introduced the Hindu numerals to Europe. In the Western world, zero was not completely accepted and used as a number until much later, probably around the sixteenth century.

Let us investigate the role of zero as a number.

What can we say about 0? Perhaps the most important property that zero has is that when zero is added to any number, it does not do anything to that number.

For example,

$$5 + 0 = 5,$$
$$0 + 4 = 4,$$
$$117 + 0 = 117, \text{ and}$$
$$0 + 0 = 0.$$

In general, if a is any number, then $a + 0 = a$ and $0 + a = a$.

A "do-nothing" number of this type is called an **identity**. Thus, **zero is the identity for addition**. Would you say that zero is the "do-nothing" number for multiplication? Obviously not. Consider $5 \cdot 0$. This means

$$\underbrace{0 + 0 + 0 + 0 + 0}$$

5 of them.

Thus we have $5 \cdot 0 = 0$. Generally, if a is *any* number, then $a \cdot 0 = 0$ and $0 \cdot a = 0$.

How about division? What can we say about division that involves zero? Very often, division by zero and division into zero are confused. To help us understand the difference, suppose your Uncle Sam dies and leaves a will dividing his money equally among 20 people. Upon investigation, it is found that his entire fortune is in Confederate money and is therefore worthless. So there is really nothing to divide. How much money do these 20 people get? Obviously nothing. Thus

$$\frac{0}{20} = 0.$$

To look at it another way, let us say that 0 divided by 20 is "something." So,

$$\frac{0}{20} = \text{something}.$$

Now what could this something be? Remember what division means. (See p. 122).

$$\frac{0}{20} = \text{something}$$

means

$$0 = 20 \cdot \text{something}.$$

If the something is 1, then $0 = 20 \cdot 1$, which is obviously wrong. If the something is 2, then $0 = 20 \cdot 2$, which is again wrong. If the something is 3, then $0 = 20 \cdot 3$, which is clearly wrong. Obviously, if the something is anything other than 0, the statement

$$0 = 20 \cdot \text{something}$$

cannot be true. We conclude that the something must be zero.

In other words, we can say that if x is any nonzero number, then x divided into 0 is 0, that is,

$$\frac{0}{x} = 0.$$

On the other hand, what would we mean by $\frac{20}{0}$?

In a mathematics class recently, students were asked this question. Various answers were given. Some said that the answer was zero. Others believed that the answer was 20. Still others claimed that the answer was 1. Some even thought that the answer was "infinity."

To see who is right, let us check each of the suggested answers. If we claim that $\frac{6}{3} = 2$, then we can easily check this, since $6 = 3 \cdot 2$ so that $\frac{6}{3} = 2$ is true. Review the definition of division on p. 122.

For those who claimed that $\frac{20}{0} = 0$, is it true that

$$20 = 0 \cdot 0?$$

Obviously not; since we know that 0 times anything is 0 and not 20.

For those who claimed that $\frac{20}{0} = 20$, again this doesn't work, since it must follow that $20 = 0 \cdot 20$. But $0 \cdot 20$ is not 20.

What about those who thought that $\frac{20}{0} = 1$? Were they right? Again no, since 20 is not equal to $0 \cdot 1$.

For those who believed the answer to be infinity, we point out here only that the answer to a division problem (if an answer exists) must be a specific number. "Infinity" (usually denoted by ∞) is *not* a number.[2]

In fact, *it is not possible to divide by 0 at all.* Why? Suppose it were possible. There are two cases:

Case 1. A nonzero number divided by 0, and

Case 2. Zero divided by 0.

For case (1), just to be specific, let the nonzero number be 20. Then as we have seen on the previous page, this is not possible. It would be the same had we used any other nonzero number besides 20. Try it to convince yourself.

For case (2), suppose 0 divided by 0 equals something. Thus,

$$\frac{0}{0} = \text{something}.$$

This means $0 = 0 \cdot$ "something" (from our definition of division). What could this something be? Any number will do. For example, $0 = 0 \cdot 5$, or $0 = 0 \cdot 17$, or $0 = 0 \cdot 141$. Thus the something is not specific. The answer to a division problem *must* be a specific number.

Summarizing   We can summarize division involving 0 as follows.

1. Division into 0 always gives 0 (assuming we are not dividing *by* 0).
2. Division by zero is not possible.

It is interesting to find that the question of dividing by zero was actually considered by Aristotle more than 2000 years ago. He concluded that division by zero was impossible. This is surprising, because it is so much like the modern

---

2. For those who want to know more about ∞, we suggest you consult any elementary calculus text or your teacher. The discussion of this symbol is beyond the scope of this book.

approach to the problem. Division by zero was also considered by the Indian mathematician Bhaskara. He believed that when you divided by zero the result was an "unchangeable infinity" that was almost religious in nature. In fact, he compared this infinity with the unchanging nature of God.

In the following sections, we will have more to say about zero and its properties.

EXERCISES

1. Is there an identity for multiplication? If yes, what is it? If no, why not?

2. Is zero the identity for subtraction?

3. On p. 125 we showed that if a is any number, then $a \cdot 0 = 0$. We also stated that $0 \cdot a = 0$, without giving any justification. Explain why $0 \cdot a = 0$.

The following problems are suggested only for those who have had some algebra. If you have not had any algebra before, then skip these problems and go on to the next section.

†4. What is wrong with the following?

Given: $a = b$

Multiply both sides by b: $ab = b^2$.

Subtract a^2 from both sides: $ab - a^2 = b^2 - a^2$.

Factor $ab - a^2$ as $a(b - a)$.

Factor $b^2 - a^2$ as $(b + a)(b - a)$, so that

$a(b - a) = (b + a)(b - a)$.

Divide both sides by $b - a$: $\dfrac{a(b - a)}{b - a} = \dfrac{(b + a)(b - a)}{b - a}$.

We then have $a = b + a$.

Since we know $a = b$, we then have $a = a + a$ or that $a = 2a$.

Divide both sides by a, and we have

$$\frac{a}{a} = \frac{2a}{a}$$

or that $1 = 2$.

†5. What is wrong with the following?

Let a be any number but 1.

Also let $a = b$.

Add 1 to both sides: $a + 1 = b + 1$.

Multiply both sides by $a - 1$ so that

$(a + 1)(a - 1) = (b + 1)(a - 1)$.

Simplifying, this gives $a^2 - 1 = ab - b + a - 1$.

---

† The dagger (†) throughout indicates those exercises involving algebra.

Adding $+1$ to both sides and simplifying, we get

$a^2 = ab - b + a.$

This can be further simplified by subtracting ab from both sides. We get

$a^2 - ab = a - b.$

Factoring, we have

$a(a - b) = a - b.$

Divide both sides by $a - b$. We get $a = 1$.

This contradicts the assumption that a was any number but 1.

*6. How is the number 0 used in our decimal system?

4. THE INTEGERS

The natural numbers that we introduced in Section 2 (p. 115) are also known as the **positive integers**. They are sometimes written as

$+1, +2, +3, +4, +5, \ldots$

instead of

$1, 2, 3, 4, 5, \ldots.$

Are the positive integers and 0 sufficient for all our everyday needs? Definitely not! To see why, recall that in Section 2, we showed that subtraction of natural numbers is not a closed operation. This means that when we subtract one natural number from another natural number, our answer is not always a natural number. For example, $5 - 2$ is the natural number 3. On the other hand, $2 - 5$ is *not* a natural number. Thus, if we had only the natural numbers to work with, we would not always be able to subtract one number from another.

This situation is very inconvenient, and we can remedy it by introducing new numbers, which are called *negative integers*. We have the following.

Definition 4.1   *The **negative integers** are the numbers $-1, -2, -3, -4, -5, \ldots.$*

This leads us to the following definition of integers.

Definition 4.2   *The set of **integers** consists of the positive integers, zero, and the negative integers, that is, the set of integers is $\{\ldots, -5, -4, -3, -2, -1, 0, +1, +2, +3, \ldots\}$.*

Comment   Do not confuse the "$-$" sign in front of each negative integer with the "$-$" sign that represents the operation of subtraction.

We already know how to add, multiply, subtract, and divide positive integers, since these are just the natural numbers. The following chart shows us how to perform these operations with any kind of integers. After each case, we give several examples. Study them carefully. For a more detailed discussion of the reasons behind these rules, refer to any standard elementary algebra text.

To find

If	a is positive or 0, b is positive or 0	a is positive or 0, b is negative or 0	a is negative or 0, b is positive or 0	a is negative or 0 b is negative or 0
$a + b$	Add the numbers and put a plus sign in front of the answer.	Momentarily disregard the signs of the numbers. Subtract smaller number from larger number and put sign of larger in front of answer.	Momentarily disregard the signs of the numbers. Subtract smaller number from larger number and put sign of larger in front of answer.	Add the numbers and put a minus sign in front of the answer.

$$\begin{array}{cc} +7 & +3 & 0 \\ +3 & +7 & +7 \\ \hline +10 & +10 & +7 \end{array}$$

$$\begin{array}{cccc} +7 & +7 & +7 & 0 \\ -3 & -8 & -2 & -7 \\ \hline +4 & -1 & +5 & -7 \end{array}$$

$$\begin{array}{cccc} -7 & -7 & -7 & -7 \\ +3 & +8 & +2 & 0 \\ \hline -4 & +1 & -5 & -7 \end{array}$$

$$\begin{array}{ccc} -7 & -3 & -5 \\ -3 & -7 & 0 \\ \hline -10 & -10 & -5 \end{array}$$

| $a - b$ | Change the sign of "b" and add according to the above rules. | | | |

$$\begin{array}{cc} +7 & +7 & 0 \\ +3 & +8 & +8 \\ \hline +4 & -1 & -8 \end{array}$$

$$\begin{array}{cccc} +7 & +7 & +7 & 0 \\ -3 & -8 & -2 & -7 \\ \hline +10 & +15 & +9 & +7 \end{array}$$

$$\begin{array}{cccc} -7 & -7 & -7 & -9 \\ +3 & +8 & 0 & +7 \\ \hline -10 & -15 & -7 & -16 \end{array}$$

$$\begin{array}{cc} -7 & -7 \\ -3 & -8 \\ \hline -4 & +1 \end{array}$$

| $a \cdot b$ | Multiply the numbers and put a plus sign in front of the product. | Multiply the numbers and put a minus sign in front of the product. | Multiply the numbers and put a minus sign in front of the product. | Multiply the numbers and put a plus sign in front of the product. |

$$\begin{array}{ccc} +7 & +3 & +7 \\ +3 & +7 & 0 \\ \hline +21 & +21 & 0 \end{array}$$

$$\begin{array}{ccc} +7 & +3 & 0 \\ -3 & -7 & -7 \\ \hline -21 & -21 & 0 \end{array}$$

$$\begin{array}{ccc} -7 & -3 & -3 \\ +3 & +7 & 0 \\ \hline -21 & -21 & 0 \end{array}$$

$$\begin{array}{cc} -7 & -3 \\ -3 & -7 \\ \hline +21 & +21 \end{array}$$

| $a \div b$ | Divide the numbers and put a plus sign in front of the answer. | Divide the numbers and put a minus sign in front of the answer. | Divide the numbers and put a minus sign in front of the answer. | Divide the numbers and put a minus sign in front of the answer. |

$(+8) \div (+4) = +2$
$(+20) \div (+5) = +4$
$0 \div (+8) = 0$
$(-8) \div 0$ not defined

$(+8) \div (-4) = -2$
$(+20) \div (-5) = -4$
$0 \div (-8) = 0$

$(-8) \div (+4) = -2$
$(-20) \div (+5) = -4$
$(-8) \div 0$ not defined

$(-8) \div (-4) = +2$
$(-20) \div (-5) = +4$

Comment To subtract -3 from $+7$, we have written it vertically as

$$\begin{array}{r} +7 \\ - \ -3 \\ \hline +10. \end{array}$$

This same problem can be written horizontally as $(+7) - (-3) = +10$. Note the "$-$" sign between the parentheses represents the operation of subtraction. The "$-$" sign in (-3) represents the fact that (-3) is a negative integer.

How much is $(+7)$ and (-7)? If we try to apply the rule for $a + b$ (where a is positive and b is negative), we see that when we consider the numbers without their signs, there is no larger and so the rule does not apply. What do we do now? To answer this question, suppose Charlie, who is on a diet, gains seven pounds one week and loses seven pounds the following week. His net result after two weeks is no change in weight. Thus we have

$$(+7) + (-7) = 0.$$

We call (-7) the **additive inverse** of $(+7)$. In general, if b is any number, then

$$(+b) + (-b) = 0.$$

Definition 4.3   *The **additive inverse** of a number b is the number which when added to b gives 0. This additive inverse is $-b$.*

Example 1   The additive inverse of $(+15)$ is (-15), since $(+15) + (-15) = 0$.

Example 2   The additive inverse of (-10) is $(+10)$, since $(-10) + (+10) = 0$.

Example 3   We said that (-7) is the additive inverse of $+7$. Is $(+7)$ the additive inverse of (-7)?

EXERCISES

1. Calculate each of the following:

   a) $(+16) + (+7)$ b) $(+7) + (-3)$ c) $(-6) + (+2)$

   d) $(-8) + (-6)$ e) $(-4) - (-7)$ f) $(-4) - (+7)$

   g) $(+6) - (+7)$ h) $(+6)(+7)$ i) $(-6) - (-8)$

   j) $(0) - (-3)$ k) $(-6)(+7)$ l) $(-8) \cdot (-6)$

   m) $(+3) \cdot (-8)$ n) $(+7) - (0)$ o) $(+7) \cdot (0)$

   p) $(-5) \cdot (0)$ q) $(+9) + (-9)$

*2. The rule for addition of a positive and a negative integer (given in the chart) can be stated formally as follows:

   $(+a) + (-b) = a - b$ if a is larger than b,

   $(+a) + (-b) = -(b - a)$ if b is larger than a.

   Illustrate this rule by selecting different values for a and b and calculating $(+a) + (-b)$. Check your answer by using the rule on p. 129.

3. The commutative, associative, and closure laws hold for addition and multiplication of integers. For example, the commutative law for addition states:

   If x and y are any integers, then $x + y = y + x$.

   State, in symbols, each of the other laws.

4. What is the identity for addition of integers?

5. The **distributive law of multiplication over addition** states that if a, b, and c are any integers, then

$a(b + c) = ab + ac$.

For example,

$5(4 + 3) = 5 \cdot 4 + 5 \cdot 3$

$5(7) \quad = \quad 20 + 15$

$35 \quad = \quad 35$.

Is it also true that $(a + b)c = ac + bc$? Give examples to support your answer.

6. Let us calculate the following:

$3 + 5 \cdot 6$

$3 + 30$

33.

Now compute

$3(5 + 6)$

$3 \cdot 11$

33.

Thus we see that $3 + 5 \cdot 6$ is equal to $3(5 + 6)$. From this can we conclude that if a, b, and c are natural numbers, then $a + bc = a(b + c)$?

*7. The rule for multiplying a positive integer and a negative integer can be justified by the following argument. We will show that $(+5) \cdot (-10)$ must equal (-50).

$5[10 + (-10)]$

$= 5(10) + 5 \cdot (-10)$ (Distributive law)

$= 50 + 5(-10)$

On the other hand, we know that

$5[10 + (-10)]$

$= 5(0)$ because $10 + (-10)$ is 0.

$= 0$

We see that $50 + 5(-10) = 0$.

From this it follows that $5(-10)$ is the additive inverse of 50. We already know that the additive inverse of 50 is (-50). Thus,

$5(-10) = (-50)$.

Construct a similar argument showing that $(-a) \cdot (-b) = +(ab)$.

8. Which law justifies each of the following?

   a) $(+5) + (+9) = (+9) + (+5)$

   b) $(-4)[8 + 7] = (-4) \cdot (8) + (-4) \cdot (7)$

   c) $(-7)[8 + 4] = (-7)[4 + 8]$

   d) $(+4) + [(-8) + (-7)] = [(+4) + (-8)] + (-7)$

   e) $(-4)(+3) = (+3)(-4)$

   f) $(+5) \cdot (-3)$ is an integer.

   g) $(+4) + (-4) = 0$

   h) $[(+4) \cdot (-3)](+7) = (+4)[(-3)(+7)]$

   i) $(-4) \cdot 0 = 0$

9. Are the integers closed for division? Explain.

10. The **absolute value** of an integer is defined as the number itself, disregarding its sign. For example, the absolute value of -9 is 9. The absolute value of $+9$ is 9. The symbol for absolute value is $|\quad|$. Thus $|-9|$ is read as "the absolute value of -9." Find each of the following.

   a) $|+5|$ b) $|-5|$ c) $|2 - 6|$ d) $|0|$

   e) $|6 - 2|$ f) $|(5 - 4) - 3|$ g) $|14 - 14|$

†*11. Show that $(-1) \cdot a = -a$.

12. The symbol a^n stands for

   $$a^n = \underbrace{a \cdot a \cdot a \cdot \quad a}$$

   a multiplied by itself n times
   (n is a positive integer)

   If a is a negative integer, will a^n be positive or negative when

   a) n is even? Explain.

   b) n is odd? Explain.

---

5. THE RATIONAL NUMBERS

Division is not always possible if one has only the integers. For example, try to divide 2 by 7. The answer is 2/7, which, of course, is not an integer. To overcome this difficulty, the rational numbers were invented. We define these as follows.

Definition 5.1   *A **rational number** is any number that can be written as*

$$\frac{a}{b},$$

where a and b are integers and b is not zero. Here a (the number on top) is called the numerator, and b (the number on the bottom) is called the denominator.

---

†* The dagger and asterisk together indicate, throughout, a difficult exercise involving algebra.

Notice that in this definition we stated that the denominator cannot be zero. Why?

Example 1   The following are examples of rational numbers:

$$\frac{2}{3}, \quad \frac{+5}{-3}, \quad \frac{-5}{+3}, \quad \frac{8}{2}, \quad \frac{-13}{1}.$$

The integer 4 is a rational number since it can be written as $\frac{4}{1}$.

Similarly -17 is a rational number. It can be written as $\frac{-17}{1}$.

0 is also a rational number. Why?

As a matter of fact, all the integers (and therefore all the natural numbers also) are rational numbers since we can write them as $\frac{\text{integer}}{1}$.

Comment   The rational numbers are so named, not because they are in better mental health than other numbers, but because they can be written as the **ratio** of two integers.

Equal Rational Numbers   If you have two quarters and your friend has a half-dollar, then clearly you both have the same amount of money. This means that $\frac{2}{4}$ represents the same quantity as $\frac{1}{2}$ (the half-dollar). Consequently,

$$\frac{2}{4} = \frac{1}{2}.$$

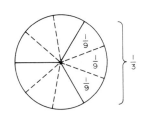

The rational numbers $\frac{3}{9}$ and $\frac{1}{3}$ are also equal. To see this, consider a pie divided into nine equal parts. Clearly $\frac{3}{9}$ of the pie is a third of the entire pie.

Consider the statement $\frac{2}{4} = \frac{1}{2}$. Notice that if we "cross-multiply"

$$\frac{2}{4} \diagup\!\!\!\!\diagdown \frac{1}{2},$$

we have $2 \cdot 2 = 4 \cdot 1$.

Similarly, in the statement $\frac{3}{9} = \frac{1}{3}$, if we cross-multiply

$$\frac{3}{9} \diagup\!\!\!\!\diagdown \frac{1}{3},$$

we get $3 \cdot 3 = 9 \cdot 1$.

In general, we have the following definition.

Definition 5.2   *The rational numbers $\dfrac{a}{b}$ and $\dfrac{c}{d}$ are equal, if when we cross-multiply*

$$\dfrac{a}{b} \diagup\!\!\!\!\diagdown \dfrac{c}{d},$$

we get $ad = bc$.

Example 2   The rational numbers $\dfrac{3}{8}$ and $\dfrac{12}{32}$ are equal since, when we cross-multiply

$$\dfrac{12}{32} \diagup\!\!\!\!\diagdown \dfrac{3}{8},$$

we get

$$12 \cdot 8 = 32 \cdot 3$$
$$96 = 96.$$

Example 3 $\dfrac{5}{11}$ is *not* equal to $\dfrac{4}{9}$ since, if we cross-multiply

$$\dfrac{5}{11} \diagup\!\!\!\!\diagdown \dfrac{4}{9},$$

we get $5 \cdot 9$ which is 45. This is not equal to $11 \cdot 4$ which is 44.

It is easy to see that $\dfrac{2}{3} = \dfrac{8}{12}$ since $2 \cdot 12 = 3 \cdot 8$. Let us look more closely at the rational number $\dfrac{8}{12}$. It can be written as $\dfrac{2 \cdot 4}{3 \cdot 4}$. Since 4 appears in the numerator and in the denominator, we can "**cancel**" **out** (divide the numerator and denominator by) the 4. As a result we have

$$\dfrac{2 \cdot \not{4}}{3 \cdot \not{4}} = \dfrac{2}{3}.$$

In general, if a, b, and c are any integers (with b and c not zero), then

$$\dfrac{ac}{bc} = \dfrac{a}{b} \qquad \text{and} \qquad \dfrac{ca}{cb} = \dfrac{a}{b}.$$

To justify this, apply the definition of equal rational numbers. We cross-multiply $\dfrac{ac}{bc} = \dfrac{a}{b}$, getting $(ac)b = (bc)a$. Using the commutative, associative, and closure laws for multiplication of integers, it can be shown that $(ac)b$ does equal $(bc)a$, so that the rational numbers $\frac{ac}{bc}$ and $\frac{a}{b}$ are equal. Similarly, we can show that $\frac{ca}{cb}$ and

$\frac{a}{b}$ are equal. Another justification of this cancellation principle will be given after we discuss multiplication of rational numbers.

The rational number $\frac{8}{12}$ can be written as $\frac{4 \cdot 2}{6 \cdot 2}$. If we cancel the 2, we are left

with $\frac{4}{6}$. So we have another rational number $\frac{4}{6}$ which is equal to $\frac{8}{12}$. Also, $\frac{8}{12}$ can

be written as $\frac{2}{3}$, as we saw above. The rational numbers $\frac{8}{12}, \frac{4}{6}$, and $\frac{2}{3}$ are all equal.

However, $\frac{2}{3}$ is different from the others because it is not possible to cancel any

more numbers from its numerator and denominator. When $\frac{8}{12}$ is written as $\frac{2}{3}$, we

say that it has been **reduced to lowest terms**.

Example 4 Reduce $\frac{15}{20}$ to lowest terms.

$$\frac{15}{20} = \frac{3 \cdot \cancel{5}}{4 \cdot \cancel{5}} = \frac{3}{4}$$

Example 5 Reduce $\frac{-18}{21}$ to lowest terms.

$$\frac{-18}{21} = \frac{(-6)(+\cancel{3})}{(+7)(+\cancel{3})} = \frac{-6}{7}$$

Example 6 A student was asked to reduce $\frac{360}{240}$ to lowest terms. He wrote $\frac{360}{240} = \frac{36 \cdot \cancel{10}}{24 \cdot \cancel{10}} =$

$\frac{36}{24}$. What is wrong? We see that this has not been reduced to lowest terms since

we can further reduce $\frac{36}{24}$ as

$$\frac{36}{24} = \frac{3 \cdot \cancel{12}}{2 \cdot \cancel{12}} = \frac{3}{2}.$$

Therefore, $\frac{360}{240}$ reduced to lowest terms is $\frac{3}{2}$.

Example 7 $\frac{7}{7} = \frac{1 \cdot \cancel{7}}{1 \cdot \cancel{7}} = \frac{1}{1}$. This is written as 1.

Addition and Subtraction

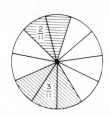

How do we add $\frac{3}{11}$ and $\frac{2}{11}$? Look at the pie on the left that has been divided into eleven equal parts. Clearly $\frac{2}{11} + \frac{3}{11}$ equals $\frac{5}{11}$. Both denominators are the same. Since $2 + 3 = 5$, then we have

$$\frac{2}{11} + \frac{3}{11} = \frac{2+3}{11} = \frac{5}{11}.$$

In general terms, we have the following rule.

Rule   *If $\frac{a}{b}$ and $\frac{c}{b}$ are rational numbers, then $\frac{a}{b} + \frac{c}{b} = \frac{a+c}{b}$.*

Now let us subtract $\frac{2}{11}$ from $\frac{3}{11}$, that is, $\frac{3}{11} - \frac{2}{11}$. Again referring back to the pie, we find that $\frac{3}{11} - \frac{2}{11} = \frac{1}{11}$. Since the denominators are the same and $3 - 2 = 1$, we have

$$\frac{3}{11} - \frac{2}{11} = \frac{3-2}{11} = \frac{1}{11},$$

which suggests the following rule.

Rule   *If $\frac{a}{b}$ and $\frac{c}{b}$ are rational numbers, then $\frac{a}{b} - \frac{c}{b} = \frac{a-c}{b}$.*

Example 8 $\dfrac{3}{10} + \dfrac{4}{10} = \dfrac{3+4}{10} = \dfrac{7}{10}.$

Example 9 $\dfrac{3}{14} + \dfrac{5}{14} = \dfrac{3+5}{14} = \dfrac{8}{14}.$ Since $\dfrac{8}{14}$ can be written as $\dfrac{4 \cdot \cancel{2}}{7 \cdot \cancel{2}}$, we can cancel the 2's, getting $\dfrac{4}{7}$. Thus $\dfrac{3}{14} + \dfrac{5}{14} = \dfrac{4}{7}.$

Example 10 $\dfrac{2}{5} - \dfrac{3}{5} = \dfrac{2-3}{5} = \dfrac{-1}{5}.$

Now let us add $\frac{2}{3}$ and $\frac{4}{5}$. Since they do not have the same denominators, the previous rule does not appear to apply. You may think (or wish) that these numbers cannot be added. However, we have a way out of this difficulty.

$\dfrac{2}{3}$ is the same as $\dfrac{2 \cdot 5}{3 \cdot 5} = \dfrac{10}{15}$ (Cancellation principle)

$\dfrac{4}{5}$ is the same as $\dfrac{4 \cdot 3}{5 \cdot 3} = \dfrac{12}{15}$ (Cancellation principle)

Therefore,

$\dfrac{2}{3} + \dfrac{4}{5}$ is the same as $\dfrac{10}{15} + \dfrac{12}{15}$.

Now we can use our rule:

$$\dfrac{10}{15} + \dfrac{12}{15} = \dfrac{10 + 12}{15} = \dfrac{22}{15}.$$

We conclude that $\dfrac{2}{3} + \dfrac{4}{5} = \dfrac{22}{15}$.

Our procedure is generalized in the following rule.

Rule   *To add two rational numbers $\dfrac{a}{b}$ and $\dfrac{c}{d}$ that have different denominators,*

1. *rewrite each number so that they have the same denominator, and*
2. *add the resulting rational numbers by the rule on p. 136.*

Subtraction is done by a similar procedure.

Example 11   Add $\dfrac{5}{6} + \dfrac{3}{4}$.

Solution $\dfrac{5}{6} = \dfrac{5 \cdot 4}{6 \cdot 4} = \dfrac{20}{24}$,   and $\dfrac{3}{4} = \dfrac{3 \cdot 6}{4 \cdot 6} = \dfrac{18}{24}$.

Therefore, $\dfrac{5}{6} + \dfrac{3}{4} = \dfrac{20}{24} + \dfrac{18}{24} = \dfrac{20 + 18}{24}$

$$= \dfrac{38}{24}$$ which can be reduced as $= \dfrac{19 \cdot \cancel{2}}{12 \cdot \cancel{2}} = \dfrac{19}{12}$.

We have $\dfrac{5}{6} + \dfrac{3}{4} = \dfrac{19}{12}$.

Example 12   Add $\dfrac{2}{3} + \dfrac{1}{9}$.

Solution $\dfrac{2}{3} = \dfrac{2 \cdot 3}{3 \cdot 3} = \dfrac{6}{9}$.

We don't have to do anything to $\dfrac{1}{9}$ since the denominator is already 9. Thus we have the following:

$$\frac{2}{3} + \frac{1}{9} = \frac{6}{9} + \frac{1}{9}$$

$$= \frac{6+1}{9} = \frac{7}{9}.$$

Our answer is $\dfrac{2}{3} + \dfrac{1}{9} = \dfrac{7}{9}$.

Example 13   Subtract $\dfrac{5}{7} - \dfrac{1}{2}$.

Solution $\dfrac{5}{7} = \dfrac{5 \cdot 2}{7 \cdot 2} = \dfrac{10}{14}$,   and $\dfrac{1}{2} = \dfrac{1 \cdot 7}{2 \cdot 7} = \dfrac{7}{14}$.

$$\frac{1}{2} = \frac{1 \cdot 7}{2 \cdot 7} = \frac{7}{14}.$$

Therefore, $\dfrac{5}{7} - \dfrac{1}{2} = \dfrac{10}{14} - \dfrac{7}{14}$

$$= \frac{10-7}{14} = \frac{3}{14}.$$

Thus, $\dfrac{5}{7} - \dfrac{1}{2} = \dfrac{3}{14}$.

Multiplication   The area of a rectangle may be calculated by multiplying the length times the width. The following two diagrams illustrate this.

3 | Area is $3 \cdot 5 = 15$.
5

4 | Area is $4 \cdot 6 = 24$.
6

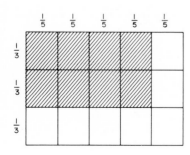

Now consider a rectangle measuring 1 by 1 that has been broken up into fifteen *equal* smaller rectangles, as shown on the left.

Each little rectangle measures $\dfrac{1}{3}$ by $\dfrac{1}{5}$. Suppose we wanted to find the area represented by the shaded portion of the diagram. From the area formula, this would be $\dfrac{2}{3} \cdot \dfrac{4}{5}$. On the other hand, the shaded portion contains 8 little rectangles. Since there are 15 altogether, the area is $\dfrac{8}{15}$ of the total area. Thus we see that

$$\frac{2}{3} \cdot \frac{4}{5} = \frac{8}{15}.$$

Notice that $2 \cdot 4 = 8$, and $3 \cdot 5 = 15$.

So we have

$$\frac{2}{3} \cdot \frac{4}{5} = \frac{2 \cdot 4}{3 \cdot 5} = \frac{8}{15}.$$

What we have done is multiply the numerators and also the denominators to get the product. This gives us the following rule.

Rule   *If $\dfrac{a}{b}$ and $\dfrac{c}{d}$ are rational numbers, then the product is $\dfrac{a}{b} \cdot \dfrac{c}{d} = \dfrac{ac}{bd}$.*

Example 14   Multiply $\dfrac{3}{7} \cdot \dfrac{5}{8}$.

Solution $\dfrac{3}{7} \cdot \dfrac{5}{8} = \dfrac{3 \cdot 5}{7 \cdot 8} = \dfrac{15}{56}$.

Example 15   Multiply $\dfrac{-10}{3} \cdot \dfrac{5}{9}$.

Solution $\dfrac{-10}{3} \cdot \dfrac{5}{9} = \dfrac{(-10) \cdot 5}{3 \cdot 9} = \dfrac{-50}{27}$.

Example 16   Multiply $4 \cdot \dfrac{6}{7}$.

Solution   Since 4 is the same as $\dfrac{4}{1}$, we have

$$\frac{4}{1} \cdot \frac{6}{7} = \frac{4 \cdot 6}{1 \cdot 7} = \frac{24}{7}.$$

Comment   We now have another way of justifying the cancellation principle introduced earlier.

$$\frac{ac}{bc} = \frac{a \cdot c}{b \cdot c} \qquad (b \text{ and } c \text{ are not zero})$$

$$= \frac{a}{b} \cdot \frac{c}{c} = \frac{a}{b} \cdot 1 = \frac{a}{b}.$$

Therefore,

$$\frac{ac}{bc} = \frac{a}{b} \qquad \text{which means that we can cancel the } c\text{'s.}$$

Division   In your previous mathematical studies, you learned a rather peculiar procedure for dividing one rational number by another. It can be stated as follows.

Rule   If $\dfrac{a}{b}$ and $\dfrac{c}{d}$ are rational numbers, then $\dfrac{a}{b} \div \dfrac{c}{d} = \dfrac{a}{b} \cdot \dfrac{d}{c}$.

In words, this says that you flip over $\dfrac{c}{d}$ to make $\dfrac{d}{c}$ and multiply $\dfrac{d}{c}$ by $\dfrac{a}{b}$.

Example 17   Divide $\dfrac{4}{7}$ by $\dfrac{3}{5}$.

Solution $\dfrac{4}{7} \div \dfrac{3}{5} = \dfrac{4}{7} \cdot \dfrac{5}{3} = \dfrac{4 \cdot 5}{7 \cdot 3} = \dfrac{20}{21}$.

Example 18   Divide $\dfrac{-2}{3}$ by $\dfrac{8}{9}$.

Solution $\dfrac{-2}{3} \div \dfrac{8}{9} = \dfrac{-2}{3} \cdot \dfrac{9}{8}$

$$= \dfrac{(-2) \cdot 9}{3 \cdot 8}$$

$$= \dfrac{-18}{24} \quad \text{which can be reduced as}$$

$$= \dfrac{(-3) \cdot \cancel{6}}{4 \cdot \cancel{6}} = \dfrac{-3}{4}.$$

Example 19   Divide $\dfrac{2}{5}$ by 3.

Solution $\dfrac{2}{5} \div 3 = \dfrac{2}{5} \div \dfrac{3}{1}$

$$= \dfrac{2}{5} \cdot \dfrac{1}{3} = \dfrac{2 \cdot 1}{5 \cdot 3} = \dfrac{2}{15}.$$

If you are like most students (including the authors), this rule probably seemed very mysterious when you first saw it. It was something you learned to humor the teacher, without really understanding why it works. The mystery will now be unraveled.

Remember that $6 \div 3 = 2$ means $6 = 3 \cdot 2$. This says that $6 \div 3$ is a number (namely 2) which when multiplied by 3 results in 6.

Similarly, $\dfrac{4}{7} \div \dfrac{3}{5}$ means a rational number which when multiplied by $\dfrac{3}{5}$ will give $\dfrac{4}{7}$. We claim that this number is $\dfrac{4}{7} \cdot \dfrac{5}{3}$, or $\dfrac{20}{21}$. So we want to multiply $\dfrac{3}{5}$ by some number and come up with an answer of $\dfrac{4}{7}$. This can be accomplished by first multiplying $\dfrac{3}{5}$ by $\dfrac{5}{3}$. This yields $\dfrac{3}{5} \cdot \dfrac{5}{3} = \dfrac{3 \cdot 5}{5 \cdot 3} = \dfrac{15}{15} = 1$.

Now if we multiply the 1 by $\frac{4}{7}$, we get

$$1 \cdot \frac{4}{7} = \frac{1}{1} \cdot \frac{4}{7} = \frac{1 \cdot 4}{1 \cdot 7} = \frac{4}{7}.$$

Therefore to get from $\frac{3}{5}$ to $\frac{4}{7}$, we multiply by $\frac{5}{3}$ and then by $\frac{4}{7}$ or by $\left(\frac{5}{3} \cdot \frac{4}{7}\right)$.

To check that we are right, we have

$$\frac{4}{7} = \frac{3}{5} \cdot \left(\frac{4}{7} \cdot \frac{5}{3}\right) = \frac{3}{5} \cdot \left(\frac{4 \cdot 5}{7 \cdot 3}\right)$$

$$= \frac{3}{5} \cdot \frac{20}{21} = \frac{3 \cdot 20}{5 \cdot 21} = \frac{60}{105}$$

$$= \frac{4 \cdot \cancel{15}}{7 \cdot \cancel{15}} = \frac{4}{7}.$$

Comment The commutative, associative, and distributive laws hold for addition and multiplication of rational numbers.

The number 1 is the identity for multiplication.
The number 0 is the identity for addition.

Decimals We know that the rational number $\frac{1}{2}$ can be written as 0.5 in **decimal** form.

Similarly 0.3333 . . . is the decimal representation of the rational number $\frac{1}{3}$. The

three dots indicate that there are infinitely many 3's appearing. Here the 3 forms a pattern that repeats forever.

The decimal 0.5 is called a **terminating** decimal since it ends after a specific number of places.

The decimal 0.3333 . . . is called a **repeating** decimal. It does not end after a specific number of places. The same number repeats itself endlessly.

The decimal 0.434343 . . . is also a repeating decimal. In this case, the repeating pattern consists of the two numbers 43 which repeat themselves endlessly.

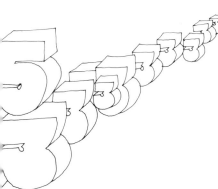

Definition 5.3 i) *A **terminating decimal** is one that ends after a specific number of places.*
ii) *A **repeating decimal** is one in which (after a certain point) the same pattern of numbers repeats itself endlessly,*
iii) *A **nonterminating and nonrepeating decimal** is one that goes on forever without the same group of numbers repeating over and over in some pattern.*

Example 20 0.25 is a terminating decimal.
0.76914 is a terminating decimal.

0.484848 . . . is a repeating decimal.
0.592592 . . . is a repeating decimal.
0.1434343 . . . is a repeating decimal.
0.12345678910111213 . . . is a nonterminating and nonrepeating decimal (What would the next few places be?).
0.43794162 . . . (where the numbers are chosen at random) is a nonterminating and nonrepeating decimal.

It can be shown that *every rational number can be written as either a terminating decimal or a repeating decimal.* It is also true that *repeating decimals and terminating decimals represent rational numbers.*

Example 21   The fraction $\frac{3}{5}$ can be converted to decimal form by the following procedure. We divide 5 into 3, getting

$$
\begin{array}{r}
0.6 \\
5\ \overline{\smash{)}\ 3.0} \\
3.0 \\
\hline
\end{array}
$$

Therefore $\frac{3}{5} = 0.6$.

Example 22   The fraction $\frac{1}{7}$ can be converted to decimal form by dividing 7 into 1:

$$
\begin{array}{r}
0.1428571\ldots \\
7\ \overline{\smash{)}\ 1.0000000} \\
7 \\
\hline
30 \\
28 \\
\hline
20 \\
14 \\
\hline
60 \\
56 \\
\hline
40 \\
35 \\
\hline
50 \\
49 \\
\hline
10 \\
7 \\
\hline
3
\end{array}
$$

Notice that as soon as we got a remainder of ① the pattern repeats. Thus, $\frac{1}{7} = 0.1428571\ldots$ is a repeating decimal.

Example 23   Convert 0.25 to a rational number.

Solution   The decimal 0.25 stands for $\dfrac{25}{100}$.

EXERCISES

1. Reduce the following rational numbers to lowest terms.

a) $\dfrac{63}{70}$ b) $\dfrac{-108}{180}$ c) $\dfrac{36}{144}$ d) $\dfrac{30}{-25}$

e) $\dfrac{-9}{9}$ f) $\dfrac{36}{9}$

2. Which of the following rational numbers are equal?

a) $\dfrac{5}{7}, \dfrac{15}{21}$ b) $\dfrac{10}{12}, \dfrac{6}{7}$ c) $\dfrac{11}{13}, \dfrac{77}{91}$ d) $\dfrac{8}{20}, \dfrac{6}{15}$

e) $\dfrac{9}{11}, \dfrac{8}{13}$ f) $\dfrac{3}{7}, \dfrac{4}{9}$

3. Write each of the given rational numbers in two different ways.

a) $\dfrac{5}{9}$ b) $\dfrac{7}{12}$ c) $\dfrac{8}{11}$ d) $\dfrac{6}{13}$

4. Perform the indicated operations and simplify the results.

a) $\dfrac{2}{3} + \dfrac{4}{3}$ b) $\dfrac{5}{9} - \dfrac{3}{9}$ c) $\dfrac{2}{5} + \dfrac{4}{10}$ d) $\dfrac{4}{7} + \dfrac{8}{9}$

e) $\dfrac{2}{3} - \dfrac{3}{2}$ f) $\dfrac{4}{11} \cdot \dfrac{5}{9}$ g) $\dfrac{3}{7} \cdot \dfrac{14}{6}$ h) $\dfrac{-5}{8} \cdot \dfrac{4}{12}$

i) $\left(\dfrac{-8}{9}\right) \cdot \left(\dfrac{-9}{8}\right)$ j) $\dfrac{4}{7} \div \dfrac{2}{3}$ k) $-\dfrac{3}{5} \div \dfrac{5}{3}$ l) $16 \div \dfrac{1}{4}$

m) $\dfrac{2}{3}\left(4 + \dfrac{1}{2}\right)$ n) $\dfrac{5}{7}\left(\dfrac{3}{8} - \dfrac{1}{4}\right)$

†*5. We stated on p. 120, that $(ac)b = (bc)a$ where a, b, and c are integers. Prove this, using the closure, commutative, and associative laws.

†6. What is wrong with the following?   Also:

$$\frac{3}{5} = \frac{3 \cdot 0}{5 \cdot 0} = \frac{0}{0}. \qquad\qquad \frac{4}{5} = \frac{4 \cdot 0}{5 \cdot 0} = \frac{0}{0}.$$

Since $\dfrac{3}{5}$ and $\dfrac{4}{5}$ are both equal to $\dfrac{0}{0}$, they are equal to each other. That is,

$$\frac{3}{5} = \frac{4}{5}.$$

*7. Let N represent the set of natural numbers, I the set of integers, and Q the set of rational numbers. Which of the following are true?

   a) $N \subset Q$ b) $I \subset Q$ c) $Q \subset I$ d) $N \cap I = I$

   e) $I \cup Q = Q$ f) $Q \cap I = I$

*8. Using the definition of multiplication of rational numbers, prove that

$$0 \cdot \frac{a}{b} = 0 \quad \text{where} \quad \frac{a}{b} \text{ is any rational number.}$$

†*9. Prove each of the following:

   i) $\dfrac{-a}{b} = \dfrac{a}{-b}.$

   ii) $\dfrac{-a}{b} = -\dfrac{a}{b}.$

   This shows that $-\dfrac{a}{b} = \dfrac{a}{-b} = -\dfrac{a}{b}.$

   (*Hint:* Use the definition of equal rational numbers.)

†*10. Show that $\dfrac{-a}{-b} = \dfrac{a}{b}.$

   (*Hint:* Use the definition of equal rational numbers.)

11. Identify each of the following decimals as being either terminating, repeating, or nonterminating and nonrepeating.

   a) 0.678   b) 0.121212 ...

   c) 0.191191   d) 0.21565656 ...

   e) 0.78654321 ...   f) 0.78964328

   g) 0.010010001 ...   h) 0.123125127129 ...

†12. If you add two repeating decimals, is the answer a repeating decimal? Explain your answer.

†*13. If you add two nonrepeating and nonterminating decimals, is the answer also a nonrepeating and nonterminating decimal? Explain your answer.

14. Convert each of the following rational numbers to decimals.

   a) $\dfrac{1}{9}$ b) $\dfrac{4}{7}$ c) $\dfrac{7}{25}$ d) $\dfrac{4}{11}$

15. Convert each of the following decimals to rational numbers.

   a) 0.86   b) 0.3   c) 5.8   d) 0.007

†16. The repeating decimal 0.4141 ... can be converted to a rational number as follows.

   Let N stand for the number.

   Then $N = 0.4141 ...$

Multiply by 100. We get

$$100N = 100(0.4141\ldots) \text{ or}$$
$$100N = 41.4141\ldots$$

$$
\begin{array}{r}
100N = 41.4141\ldots \\
\text{Subtract } N \quad -N = 0.4141\ldots \\
\hline
99N = 41.
\end{array}
$$

Dividing both sides by 99, we obtain

$$N = \frac{41}{99}.$$

By a similar procedure, convert each of the following repeating decimals to rational numbers.

a) $0.3131\ldots$ b) $0.9797\ldots$ c) $0.0808\ldots$ d) $0.567567\ldots$

---

6. THE IRRATIONAL NUMBERS

In the last section, we pointed out that there are three different kinds of decimals:

terminating decimals,
repeating decimals, and
nonrepeating and nonterminating decimals.

The first two kinds represent rational numbers. On the other hand, *numbers that can be written as nonrepeating and nonterminating decimals are called irrational numbers.* The fact that they are called irrational does not mean that they do not make sense. They are as meaningful as any other numbers. We do not introduce them merely to make our discussion of decimals complete. They are important and interesting in themselves, and mathematicians could not work without them.

Some examples of irrational numbers are π (which you will remember if you have studied geometry), $\sqrt{2}$, $\sqrt{3}$, etc.[3]

The story of the discovery of irrational numbers is one of the more interesting chapters in the history of mathematics. Irrational numbers were first discovered around the sixth century B.C. by the Pythagoreans, a school of Greek mathematicians. This school was named after its founder, Pythagoras, a philosopher, mathematician, and mystic. We know nothing certain about him except that he was born in Samos in Greece and is believed to have traveled widely, as far as Egypt, Babylon, and even India. When he returned from his travels he settled at Croton in southern Italy. There he founded a school for the study of religion, philosophy, mathematics, and science. An interesting story is told about Pythagoras's attempt to get students for his school. He found a poor workman and offered to pay him to learn geometry. Pythagoras promised to give him a coin for each theorem that he learned. The workman happily accepted the

---

3. Read $\sqrt{2}$ as "the square root of 2." It means a number which when multiplied by itself gives 2. For example, $\sqrt{4} = 2$ since $2 \cdot 2 = 4$, and $\sqrt{9} = 3$ since $3 \cdot 3 = 9$.

challenge and earned many coins. Gradually the workman became so interested in geometry that he wanted Pythagoras to teach him more and more. To persuade him, the workman now offered Pythagoras a coin for each theorem that he taught him. In the end, Pythagoras got back all his money.

The Pythagorean school was also a secret society. This society was in some ways like a modern commune. Men and women were equal, all property was common, and activities were communal. Even mathematical and scientific achievements were considered the work of the entire community.

Many of the Pythagoreans' religious beliefs seem somewhat strange to us. They believed in the transmigration of souls and would not eat meat, being afraid that they might be dining off some departed friend.

Among the things they considered sinful were

eating beans,
picking up anything that had fallen (a belief shared by many small boys),
eating from a whole loaf,
walking on highways (apparently Croton traffic was an earlier version of the Los Angeles Freeway),
letting swallows sit on one's roof.

When they were not worrying about the sacredness of beans, the Pythagoreans passed the time studying philosophy and mathematics. Pythagoras said "All is number," which meant that the form of all things in the world can be explained in terms of numbers. The mathematical basis of music was one of his important discoveries.

Probably the greatest accomplishment of this school was its investigation of what is today known as the Pythagorean theorem (refer back to p. 7). This theorem says: In a right triangle (labeled as shown on the left),

$a^2 + b^2 = c^2$.

It is said that Pythagoras was so pleased by this theorem that he sacrificed an ox to celebrate its discovery. Actually, various special cases of the theorem had been known for centuries before the Pythagoreans.

Unfortunately this achievement led to the downfall of the society and its philosophy. Remember that they believed that the universe could be explained entirely in terms of *numbers*, which for them meant rational numbers. Now consider the right triangle shown at the left.

The Pythagorean theorem says

$$a^2 + b^2 = c^2$$
$$1^2 + 1^2 = c^2$$
$$1 + 1 = c^2$$
$$2 = c^2$$
$$2 = c \cdot c.$$

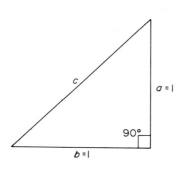

Thus c is a number which when multiplied by itself gives 2. Symbolically (see the footnote on p. 145) $c = \sqrt{2}$. It was discovered by a Pythagorean named Hippasus (at least so the story goes) that $\sqrt{2}$ *is not a rational number.* Thus we have a physical thing, that is, the longest side of our triangle (the hypotenuse), which is *not* a rational number. This, of course, upset the whole Pythagorean philosophy.

The discovery that $\sqrt{2}$ is not a rational number was thus a terrible shock to the Pythagoreans. According to one report, they drowned Hippasus in order to prevent him from spreading the "bad" news. (This is one way to solve a mathematical dispute.) Another version states that Hippasus was shipwrecked by the gods for his wickedness. However, nothing the Pythagoreans could do was able to change mathematical fact: $\sqrt{2}$ was not, and is not, rational. It is irrational. Let us see why. In order to prove this, we need a few simple definitions and facts.

Definition 6.1 *An integer is **even** if it can be divided exactly (no remainder) by 2. Thus every even number can be written as*

$2 \cdot \text{(integer)}.$

If you try to divide any number by 2, then either there is no remainder or the remainder is one. If there is no remainder, then the number is even. If the remainder is 1, then we have the following definition.

Definition 6.2 *An integer is said to be **odd** if it can be written as*

$2 \cdot \text{(integer)} + 1.$

Example 1 The integer 16 is even since we can write it as

$16 = 2 \cdot 8.$

Example 2 The integer 17 is odd since we can write it as

$17 = 2 \cdot 8 + 1.$

Now consider the following:

The number 9 is an odd number. If we multiply it by itself, we get 9×9, or 9^2, which is 81. Notice that 81 is also an odd number.

Similarly, 7 is an odd number. And 7 multiplied by itself, or 7^2, gives 49, which is also an odd number.

In general, we have this statement.

Statement 1 *If a number n is odd, then n multiplied by itself is odd. To put it another way, **if n is odd, then n^2 is odd**.*

Now look at the number 4, which is even; 4 is 2^2, and 2 is also even.

Similarly, 100 is also an even number; 100 is 10^2 and 10 is also even. This leads to the next statement.

Statement 2   *If the square of a number is even, then the number itself is even. Another way of saying this is,* **if n^2 is even, then n is even.**

We will now prove these two statements. The proof of Statement 1 requires a little algebra, so if you have had *no* algebra, go on to Example 3 below.

†*Proof of Statement 1*   We are given an odd number, n. We want to show that n^2 is also odd. Since n is odd, we can write n as

$2 \cdot$ (integer) $+ 1$.

Let us call the integer k. Then

$n = 2k + 1$.

Therefore,

$$n^2 = n \cdot n$$
$$= (2k + 1) \cdot (2k + 1)$$
$$= 4k^2 + 4k + 1 \qquad \text{[since the product } (2k + 1)(2k + 1) \text{ equals } 4k^2 + 4k + 1]$$
$$= (4k^2 + 4k) + 1 \qquad \text{(Here we use the associative law to group } 4k^2 + 4k \text{ together.)}$$
$$= 2(2k^2 + 2k) + 1 \qquad \text{(The distributive law is used here to factor out the 2.)}$$

Now by the closure laws for integers, $(2k^2 + 2k)$ is an integer also. Let us call it c. Then we have

$n^2 = 2c + 1$.

That is,

$n^2 = 2 \cdot$ (integer) $+ 1$.

This means that n^2 is an odd number, and our statement is proved.

Let us illustrate Statement 1 with some examples.

Example 3   3 is odd. Therefore, 3^2, which is 9, is also odd.

Example 4   5 is odd. Thus, 5^2, which equals 25, is also odd.

Before we prove Statement 2, we first need the idea of a **proof by contradiction**. Suppose you were invited to a party at a certain address. When you arrive there, you discover that the house has two apartments, A and B, and you do not know which of them is the right one. You would pick one apartment, say A, and ring the bell. If the occupants of apartment A said that the party was not there, you would then know that it was in apartment B.

The reasoning process you would use in the above situation is the same as the reasoning in a proof by contradiction. We can summarize it as follows:

1. You know that either possibility A or B must be true.
2. You try possibility A and find that it is wrong.
3. You conclude that possibility B is correct.

This method will be used on our proof of Statement 2.

†Proof of Statement 2   We are given an even number n^2 and asked to prove that n is also even. We know that *either n is odd* (possibility A) or n is even (possibility B).

Let us consider whether possibility A can be right. Possibility A says that n is odd. If n is odd, then Statement 1 tells us that n^2 is also odd. But we are given that n^2 is even. Thus, *n cannot be odd.*

We conclude that possibility A is wrong. It then follows that possibility B is right. This means that n must be even. Thus Statement 2 is proved.

Example 5   64, which is 8^2, is even. Therefore 8 is also even.

Example 6   36, which is the same as 6^2, is even. Thus 6 is also even.

Now we are ready to prove that $\sqrt{2}$ is not rational. The proof is again a proof by contradiction, and uses a little algebra. If you have *never* studied algebra, omit it.

†Proof that $\sqrt{2}$ is not rational   Either $\sqrt{2}$ is rational (possibility A) or $\sqrt{2}$ is not rational (possibility B).

Suppose A is correct, so that $\sqrt{2}$ is rational. Then, by our definition of rational number, $\sqrt{2}$ can be written as $\dfrac{a}{b}$, where a and b are integers and b is not 0.

Now we know that every rational number can be reduced to lowest terms. So we can assume that

$$\sqrt{2} = \frac{a}{b}.$$ where a and b have no common divisors (that is, a/b is reduced to lowest terms.)

Square both sides of this equation. We get

$$(\sqrt{2})^2 = \left(\frac{a}{b}\right)^2$$

$$2 = \frac{a^2}{b^2}.$$

Multiply both sides by b^2. We get

$$2b^2 = \frac{a^2}{b^2} \cdot b^2 \qquad \text{or upon simplifying,}$$

$$2b^2 = a^2.$$

We see that

$$a^2 = 2 \cdot b^2 = 2 \cdot \text{(integer)}, \tag{1}$$

which means that a^2 is even. Since a^2 is even, Statement 2 tells us that a must be even.

Now that we know a is even, we can write it as

$$a = 2 \cdot \text{(integer)}.$$

If we call this integer p, we have

$$a = 2p.$$

Squaring both sides we have

$$a^2 = (2p)^2 = 4p^2.$$

If we substitute $4p^2$ for the a^2 in equation (1) above, we get $4p^2 = 2b^2$. Dividing both sides by 2, we get

$$2p^2 = b^2$$

or

$$2 \cdot \text{(integer)} = b^2.$$

This means that b^2 is even. Since b^2 is even, Statement 2 tells us that b is even. Thus we can write b as

$$b = 2 \cdot \text{(integer)}.$$

Now we have

$$a = 2 \cdot \text{(integer)},$$
$$b = 2 \cdot \text{(integer)}.$$

So *a and b have a common divisor of 2*. But when we started out, *a and b had no common divisors*. This is obviously a **contradiction**. We must conclude that possibility A ($\sqrt{2}$ is rational) is wrong. Since A is wrong, B (which says that $\sqrt{2}$ is not rational) must be correct.

Finally, we conclude that $\sqrt{2}$ *is not rational*.

It turns out that $\sqrt{2}$ is not the only irrational number. There are infinitely many of them. In fact there are more irrational numbers than rational numbers. Another important irrational number that you might have come across in other mathematics courses is π (pronounced "pie"). In the Bible (I Kings 7:23 and II Chronicles 4:22) calculations indicate that 3 was used as the value of π. In recent years, π has been computed accurately to 500,000 places. The value of π correct to 14 places is 3.14159265358979. In the nineteenth century, the legislature of Indiana attempted to pass a law establishing a fixed finite decimal value for π. They gave up when the idea was ridiculed in the press.

EXERCISES

†*1. Two examples of irrational numbers are $5\sqrt{2}$ and $4\sqrt{2}$.

    a) Is their product irrational? Explain.

    b) Is their quotient irrational? Explain.

    c) Is their sum irrational? Explain.

†*2. Show, by examples, that the product of two irrational numbers may be rational or may be irrational.

†*3. $\sqrt{3}$ is another example of an irrational number. Construct a line segment that measures $\sqrt{3}$ inches long. (*Hint:* Use the Pythagorean theorem for triangles.)

7. THE REAL NUMBERS

We started this chapter with the natural numbers. However, subtraction was not always possible using only the natural numbers, so we needed the integers. Although this made subtraction a closed operation, we still found that division was not always possible. To remedy this situation, we introduced the rational numbers. Then we discovered that there exist certain numbers such as $\sqrt{2}$, $\sqrt{3}$, etc., that are not rational numbers, but do occur frequently in mathematics. These we called the irrational numbers. All of these numbers together make up what we call the **real numbers**. These are the numbers used in most of elementary mathematics and in everyday situations. They are defined as follows.

Definition 7.1

A real number is any number that is either a rational number or an irrational number. In the notation of sets, we have

real numbers = {rational numbers} ∪ {irrational numbers}.

We can draw a diagram illustrating the relationship among the different kinds of numbers.

Natural numbers
↓
Integers
↙
Rational numbers   Irrational numbers
↘   ↙
Real numbers

Another way of picturing the relationships between numbers is by means of a Venn diagram (see p. 65), as shown on the left.

The real numbers include everything in all the circles. The shaded portion represents the irrational numbers.

Example 1

Let R = {real numbers}, I = {integers}, Q = {rational numbers}, and W = {irrationals}.

a) Find $R \cap Q$.   b) Find $I \cap W$.

c) Is $W \subset I$ true?

Solution   a) $R \cap Q = Q$, since the diagram shows that the only elements that are in both sets are the rational numbers.

b) $I \cap W \doteq \varnothing$, since the diagram shows that there is no number that is both an integer and an irrational number.

c) If $W \subset I$ were true, then the circle for the irrationals would be inside the circle for the integers. Since it is not true, then the statement $W \subset I$ cannot be true.

The commutative, associative, distributive, and closure laws hold for addition and multiplication. Subtraction and division are closed for real numbers (with the exception of division by 0 which, of course, is not possible).

The identity for addition is 0.

The identity for multiplication is 1.

Thus we see that the real number system is a very complete system.

A very important property of the real numbers is that every real number is either positive, negative, or zero. A convenient way of picturing the real numbers is by means of *a real number line.* Such a line is shown below.

$$\sqrt{2} = 1.414\ldots$$

—3 —2 —1 —⅔ 0 +½ +1 +2   +3

We pick a point on the line and label it 0. Then we pick another point to the right of 0 and label it +1. Next, we take a point to the right of +1, which is the same distance from +1 as +1 is from 0. We label this point +2. Similarly, we mark off +3, +4, +5, etc., ... Points to the left of 0 are labeled —1, —2, —3, etc.

Fractions are points between these numbers. (For example, see $+\frac{1}{2}$ and $-\frac{2}{3}$ in the diagram).

Can we find a number like $\sqrt{2}$ on the number line? Look at the right triangle shown in Fig. 1.

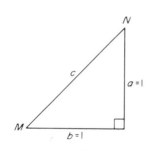

Figure 1

On pp. 146–147 we showed that side c has length equal to $\sqrt{2}$. Now suppose we take this triangle and place point M at 0 on the number line, and place side c along the number line as shown in Fig. 2. Then N will be exactly $\sqrt{2}$ units to the right of 0. Thus the point on the number line that N touches is $\sqrt{2}$. It can be shown that $\sqrt{2}$ is approximately equal to 1.414. The number $\sqrt{2}$ cannot be represented exactly by a repeating or terminating decimal (see pp. 149–150).

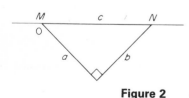

Figure 2

Every point on the line represents a real number. Similarly, every real number can be represented by a point on the line.

Pick any two numbers on the line. To be specific, let us take +3 and —2. The number +3 is to the right of —2. We say that +3 is larger than or greater than —2. This is symbolized by writing $(+3) > (-2)$. The symbol ">" stands for "is greater than."

We also see that −2 is to the left of +3. In this case we say that −2 is less than or smaller than +3. This is symbolized as (−2) < (+3). The symbol "<" stands for "is less than."

Sometimes we know that one number is either less than another number or equal to it. But we do not know which one is the case. In this situation we use the symbol "≤" which is read as "equal to or less than."

Similarly, the symbol "≥" means "equals to or greater than."

We can summarize our discussion in the following definition.

Definition 7.2   *If a and b are any real numbers, then*

$a < b$ means a is to the left of b,

$a > b$ means a is to the right of b,

$a \leq b$ means a is to the left of b or a is the same as b, and

$a \geq b$ means a is to the right of b or a is the same as b.

Comment   Given any two numbers a and b, then *one and only one* of the following must be true:

i) $a = b$,

ii) $a > b$, or

iii) $a < b$.

This is sometimes called the **law of trichotomy**.

Example 2 $2 < 3$ means 2 is to the left of 3, as shown below.

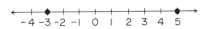

Example 3 $5 > -3$ means 5 is to the right of −3, as shown below.

Example 4 $10 \leq b$ means 10 is to the left of point b or the same as point b.

EXERCISES

1. Let R = {real numbers}, I = {integers}, Q = {rationals}, and W = {irrationals}. Find each of the following.

   a) $R \cap I$ b) $R \cup Q$ c) $I \cap Q$ d) $Q \cap W$ e) $Q \cup W$

   f) $W \cap I$ g) $W \cup R$ h) $W \cap R$ i) $I \cup Q$ j) $I \cap (Q \cup W)$

2. Using the notation of Exercise 1, which of the following are true?

   a) $Q \subset R$ b) $Q \subset I$ c) $I \subset Q$ d) $R \subset W$

   e) $W \subset R$ f) $I \subset (Q \cup W)$

3. Complete the following chart.

Number	Natural number	Integer	Rational number	Irrational number	Real number
-2	No	Yes	Yes	No	Yes
$\sqrt{7}$	No	No	No	Yes	Yes
4					
$-\dfrac{2}{3}$					
$3.1212\ldots$					
$-\sqrt{3}$					
$3.01001\ldots$					
0					
$2\sqrt{3}$					

4. Represent each of the following numbers on a number line:

   $5,\ -10,\ +\dfrac{1}{2},\ -0.6,\ 7.5,\ \sqrt{3}.$

5. Using a number line, determine which is the correct symbol ($>, =, <$) for each of the following pairs of numbers.

   a). $4.7, 4.9$ b) $+4, -5$ c) $-4, -5$ d) $\dfrac{1}{2}, \dfrac{1}{3}$

   e) $\dfrac{1}{2}, 0.5$ f) $\sqrt{2}, 2$

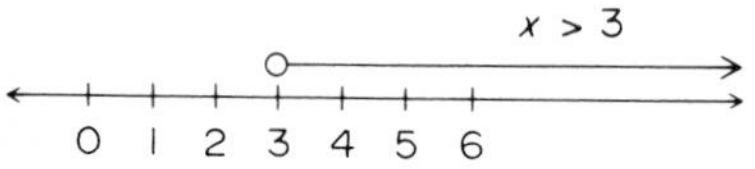

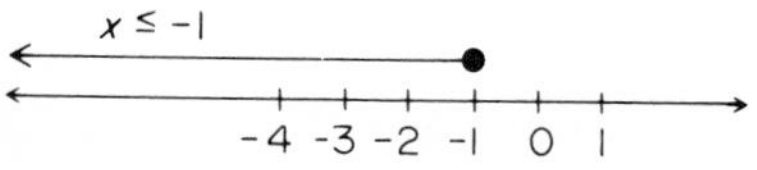

6. If x is any real number, then $x > 3$ means x is to the right of 3 on the number line. We can picture this as shown on the left. Similarly, $x \leq -1$ can be pictured as shown on the left.

   Picture each of the following, using a number line.

   a) $x \geq 2$ b) $x < -2$ c) $x > -1$ d) $x > \dfrac{1}{2}$

   e) $x \leq -3$ f) $x < -4$

†7. Let a, b, and x be any real numbers. Using the properties of real numbers, prove that if

   $a + x = b + x,$

   then

   $a = b.$

   This is called the cancellation law for addition.

†8. Let a, b, and x be any real numbers, with x not equal to zero. The cancellation law of multiplication says that if

$ax = bx,$

then

$a = b.$

a) Prove this using the properties of real numbers.

b) Why do we require that x not be equal to zero?

---

8. THE SQUARE ROOT OF −1

You may have the impression that there can be no numbers other than real numbers. This is not true. To see this, consider the innocent-looking $\sqrt{-1}$ (the square root of −1). What does this symbol really mean? The symbol $\sqrt{-1}$ means some number which, when multiplied by itself, gives −1. Let us call this number i. Then $i^2 = -1$. Now what kind of number is i? Is it negative, positive, or zero?

If i is negative, then i^2 would have to be positive (since a negative number times a negative number is a positive number). But i^2 is −1 which is negative. Therefore i cannot be negative.

If i is positive, then i^2 would have to be positive (since a positive number times a positive number is a positive number). Since i^2 is −1, then i cannot be positive.

If i is zero, then i^2 is 0 (since 0 times 0 is 0). This definitely is not −1.

So here we have a number, i (that is, $\sqrt{-1}$), that is neither negative, positive, nor zero. It follows that i cannot be a real number. (Remember that every real number is either negative, positive, or zero.)

Numbers like i are called **imaginary** or **complex numbers**. They play an important role in mathematics, physics, and technology. In particular, they are used in many branches of engineering, such as electrical engineering, heat conduction, elasticity, and aeronautical engineering.

Comment

Do not let the terms *real number* and *imaginary number* mislead you. Imaginary numbers exist just as much as real numbers do. The choice of names is just unfortunate.

STUDY GUIDE

In this chapter, we discussed the different types of numbers, the operations of addition, multiplication, subtraction, and division, and the rules that apply to these operations.

Different types of numbers
Natural or counting numbers (p. 115)
Zero (p. 124)
Integers (p. 128)
Negative numbers (p. 128)
Additive inverse (p. 130)

Rational numbers (p. 132)
Irrational numbers (p. 145)
Real numbers (p. 151)
Repeating decimals (p. 141)
Terminating decimals (p. 141)
Nonterminating, nonrepeating
    decimals (p. 141)
The number i (p. 155)
Complex numbers (p. 155)

Operations

Addition (p. 115)
Binary operation (p. 116)
Multiplication (p. 119)
Subtraction (p. 121)
Division (p. 122)

Rules

Commutative law (p. 116)
Associative law (p. 117)
Law of closure (p. 118)
Identity for addition (p. 125)
Identity for multiplication (p. 127)

Other important ideas

Division by 0 is impossible (p. 126)
Division into 0 is 0 (p. 126)
Numerator (p. 132)
Denominator (p. 132)
Cancellation principle (p. 134)
Reducing fractions to lowest terms (p. 135)
Proof by contradiction (p. 148)
Number line (p. 152)
Greater than or equal to ($\geq$) (p. 152)
Law of trichotomy (p. 153)

SUGGESTED FURTHER READINGS

Ablon, L., et al., *Series in Mathematics Modules.* Menlo Park, Calif.: Cummings, 1974. Modules I and II contain a clear basic treatment of algebraic operations.

Asimov, I., *The Realm of Numbers.* Greenwich, Conn.: Fawcett, 1967.

Beckmann, P., *A History of π.* Boulder, Colo.: The Golden Press, 1971. This book discusses how π has been calculated with the aid of the computer to 500,000 places.

Bergamini, D., et al., *Life, Mathematics* (Life Science Library). New York: Time-Life Books, 1970.

Boyer, C., *A History of Mathematics.* New York: Wiley, 1968. This book contains a complete history of mathematics from ancient to modern times.

Cooley, Hollis R., and H. E. Wahlert, *Introduction to Mathematics.* Boston: Houghton-Mifflin, 1968. Chapters 2, 3, and 4 discuss the real number system.

Gardner, M., *New Mathematical Diversions from the Scientific American.* New York: Simon and Schuster, 1966. See Chapter 8.

Gies, J., and F. Gies, *Leonardo of Pisa and the New Mathematics of the Middle Ages.* New York: T.Y. Cromwell, 1969.

Kline, M., *Mathematics for Liberal Arts.* Reading, Mass.: Addison-Wesley, 1967. Chapter 2 contains a historical introduction.

Niven, I., *Numbers Rational and Irrational.* New York: Random House, 1961.

Russell, B., *A History of Western Philosophy.* New York: Simon and Schuster, 1945. Chapter 3 contains an interesting account of Pythagorean philosophy and mathematics.

Smith, D. E., and J. Ginsburg, "From Numbers to Numerals and From Numerals to Computation." In James R. Newman, ed., *The World of Mathematics*, Vol. I, Part III, ch. 3. New York: Simon and Schuster, 1956.

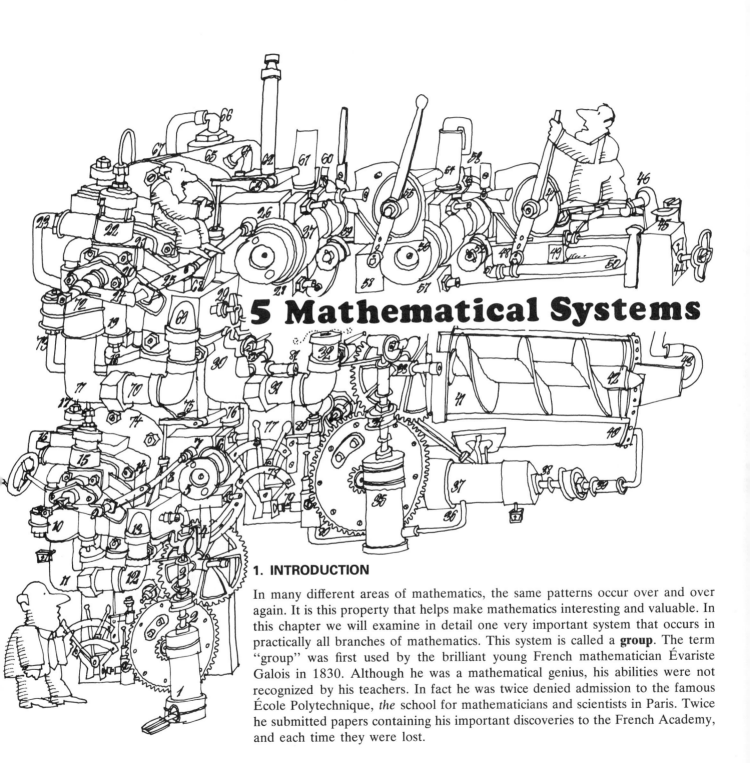

5 Mathematical Systems

1. INTRODUCTION

In many different areas of mathematics, the same patterns occur over and over again. It is this property that helps make mathematics interesting and valuable. In this chapter we will examine in detail one very important system that occurs in practically all branches of mathematics. This system is called a **group**. The term "group" was first used by the brilliant young French mathematician Évariste Galois in 1830. Although he was a mathematical genius, his abilities were not recognized by his teachers. In fact he was twice denied admission to the famous École Polytechnique, *the* school for mathematicians and scientists in Paris. Twice he submitted papers containing his important discoveries to the French Academy, and each time they were lost.

157

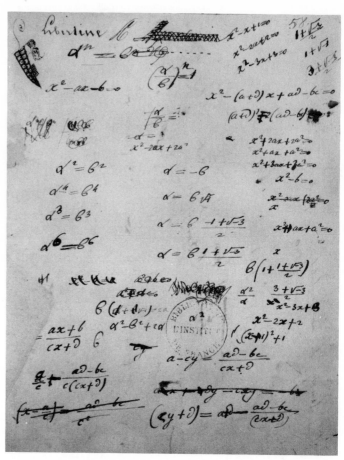

Évariste Galois (1811–1832).
At the right is a page from Galois' papers.

At the age of 20, he became involved in a quarrel over a woman and was challenged to a duel. He spent the entire night before the duel writing down his ideas. The next morning he was killed. It is interesting to think about what he might have contributed to mathematics had he lived a normal life span.

Galois studied groups in order to solve certain problems in algebra. His discoveries greatly expanded the field of algebra. Furthermore, his ideas have also been applied to physics. They have been particularly valuable in the study of quantum mechanics.

2. CLOCK ARITHMETIC

To introduce the idea of a group, let us consider a clock that has only four numbers on it, as shown on the following page.

Starting at 0, the clock hand will point to 1, 2, 3 in order, and then back to 0. This cycle repeats itself over and over. Let $\oplus$ represent the turning of the hand of the clock in the direction of the arrow. Then $2 \oplus 3$ means that the clock is at the

2 position and then moves through 3 positions. It stops at the 1 position, so we say that

$2 \oplus 3 = 1.$

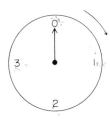

Similarly, $1 \oplus 2$ would mean that the clock is first in the 1 position and then moves 2 more places, ending up in the 3 position. Thus we have

$1 \oplus 2 = 3.$

We can make up a table showing all possible starting positions and all possible ending positions.

$$
\begin{array}{c|cccc}
\oplus & 0 & 1 & 2 & 3 \\
\hline
0 & 0 & 1 & 2 & 3 \\
1 & 1 & 2 & 3 & 0 \\
2 & 2 & 3 & 0 & 1 \\
\rightarrow 3 & 3 & 0 & ① & 2 \\
\end{array}
$$

This chart is read as follows: To find $3 \oplus 2$, for example, go to row 3 and then over to column 2. We have indicated this by means of arrows in the above table. We find the answer to be 1. Thus

$3 \oplus 2 = 1.$

In a similar manner, we read from the table that

$3 \oplus 3 = 2.$

We can think of the different positions of this clock as a set G, whose members are 0, 1, 2, and 3. So,

$G = \{0, 1, 2, 3\}.$

The operation of turning the hand, which we denoted by $\oplus$, can be considered a binary operation. (Recall that a binary operation involves combining any two elements of a set to get a third element.)

We can observe a number of interesting things about this system. First of all, the above table indicates that no matter where we start, we will always end up at one of the four positions 0, 1, 2, or 3. This means that the operation $\oplus$ is **closed**. (Remember, an operation is closed if, when we perform the operation, the result is always within the set we started with.)

Now notice the following:

$0 \oplus 0 = 0 \qquad 0 \oplus 0 = 0$
$0 \oplus 1 = 1 \qquad 1 \oplus 0 = 1$
$0 \oplus 2 = 2 \qquad 2 \oplus 0 = 2$
$0 \oplus 3 = 3 \qquad 3 \oplus 0 = 3$

Thus when we perform the operation $\oplus$ with 0 on any element, nothing happens. Here 0 is the **identity element** (see p. 125) for the operation $\oplus$.

Now suppose we are at any position on the clock. Can we always get back to 0 by the operation $\oplus$? The answer is clearly yes, as the following results indicate:

$$0 \oplus 0 = 0$$
$$1 \oplus 3 = 0$$
$$2 \oplus 2 = 0$$
$$3 \oplus 1 = 0$$

We call 3 the **inverse** of 1 for the operation $\oplus$, because when we perform $3 \oplus 1$, we get 0. Similarly, 2 is the inverse of 2 for the operation $\oplus$, 1 is the inverse of 3, and 0 is its own inverse.

Every element of this system has an inverse. In other words, we can always get back to 0. of a closed system

Consider the expression $2 \oplus 3 \oplus 1$. This can be interpreted in two different ways. One way is to first do $2 \oplus 3$, getting 1. Then do $1 \oplus 1$, getting 2. In other words,

$$(2 \oplus 3) \oplus 1$$
$$= \quad 1 \oplus 1$$
$$= \quad 2.$$

Another way is to first do $3 \oplus 1$, getting 0. Then do $2 \oplus 0$, which gives 2. That is,

$$2 \oplus (3 \oplus 1)$$
$$= \quad 2 \oplus 0$$
$$= \quad 2.$$

In both cases, our answer is 2. We conclude that $(2 \oplus 3) \oplus 1 = 2 \oplus (3 \oplus 1)$. This shows that the associative law holds for these three numbers. In a similar manner, we can show that *the associative law holds for any three numbers in this system.*

This clock is called a **mod 4 clock** because it has four positions on it.

Next, let us look at a familiar object, a three-way switch on a table lamp. Such a switch can be pictured as shown on the left.

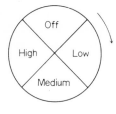

We turn the switch in a clockwise direction as indicated by the arrow. If the switch is in the "off" position, turning it once puts it in "low." Another turn moves it to "medium." Still another turn moves it to "high." The next turn moves it to the off position again. If the switch is in the low position, then turning it twice moves it to the high position. Turning it three times puts it in the off position. This cycle can be repeated as often as we like.

If we start at the off position, then we need 0 turns to get to the off position, 1 turn to get to low, 2 turns to medium, and 3 turns to high. Thus we let

⊕	0	1	2	3
0	0	1	2	3
1	1	2	3	0
2	2	3	0	1
3	3	0	1	2

0 stand for the off position,
1 stand for the low position,
2 stand for the medium position,
3 stand for the high position, and
⊕ stand for turning the switch clockwise.

The chart on the left shows the final position of the switch, depending on where you start and the number of turns you make.

Compare this table with the table for the mod 4 clock (p. 159). It is exactly the same. What this means is that both situations have the same mathematical structure although they appear to be completely different.

Example 1

Let us construct a table for a mod 3 clock (that is, a clock with 3 positions). The clock is pictured on the left. As before let ⊕ represent turning the clock. We then have the following:

⊕	0	1	2
0	0	1	2
1	1	2	0
2	2	0	1

Example 2   What would 6 ⊕ 5 be in a mod 10 clock? in a mod 8 clock?

Solution   We draw both clocks.

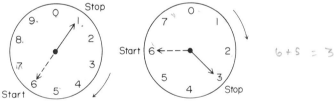

6 + 5 = 1

6 + 5 = 3

Mod 10 clock   Mod 8 clock

Start at 6 and move through 5 additional positions. In the mod 10 clock, we stop at 1 as indicated in the diagram. Thus,

6 ⊕ 5 = 1 in mod 10.

In the mod 8 clock, we stop at 3, again as indicated in the diagram. Thus,

6 ⊕ 5 = 3 in mod 8.

Example 3   In what clock is 4 ⊕ 3 = 1?

Solution

The number 4 shows us that the clock must be *at least* a mod 5 clock. (Why?) Let us try mod 5. A quick check (which the reader should verify by actually drawing such a clock) will show that

4 ⊕ 3 = 2 in mod 5.

So mod 5 is wrong. Next we try mod 6. This works. (Try it, to convince yourself.)

$4 \oplus 3 = 1$ in mod 6.

Example 4   Let us define "subtraction" on a clock as "turning the clock backwards." We will denote this as $\ominus$. Let us find $3 \ominus 5$ in a mod 8 clock. (See the mod 8 clock above.) We start at 3 and move back 5 positions, stopping at 6. Thus,

$3 \ominus 5 = 6$ in mod 8.

Comment   Some readers may say that $3 \ominus 5 = -2$. This is incorrect, since there is no -2 on the clock. However, if we interpret -2 to mean "start at 0 and go back 2," then, of course, we stop at 6. We will always give our answer as 6 rather than as -2.

Example 5   In a mod 7 clock, $4 \ominus x = 6$. Find x.

Solution   We start at 4 and move back x places until we get to 6.

From the clock on the left, we find that

$4 \ominus 5 = 6$.

Thus $x = 5$.

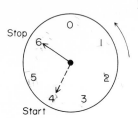

EXERCISES

1. Make up a table for the operation $\oplus$ on the following clocks.

   a) mod 6   b) mod 9   c) mod 3   d) mod 5

2. Calculate the following.

   a) $2 \oplus 5$ in mod 6   b) $3 \oplus 4$ in mod 5   c) $3 \ominus 8$ in mod 9

   d) $5 \ominus 6$ in mod 11   e) $2 \oplus 2$ in mod 5   f) $2 \oplus 3$ in mod 5

   g) $7 \ominus 7$ in mod 39   h) $6 \ominus 10$ in mod 12   i) $1 \oplus 1$ in mod 13

   j) $1 \oplus 1$ in mod 2   k) $6 \ominus 0$ in mod 8   l) $3 \oplus 6$ in mod 9

   m) $2 \ominus 3$ in mod 7   n) $2 \ominus 8$ in mod 10

3. Find x in each of the following.

   a) $2 \oplus x = 6$ in mod 8   b) $7 \ominus x = 8$ in mod 10   c) $4 \oplus x = 2$ in mod 5

   d) $x \ominus 3 = 5$ in mod 8   e) $3 \oplus x = 3$ in mod 7   f) $1 \ominus x = 0$ in mod 4

   g) $1 \ominus x = 3$ in mod 5   h) $x \ominus 2 = 3$ in mod 5

4. In what clock is each of the following true?

   a) $4 \oplus 4 = 3$ b) $4 \ominus 6 = 3$ c) $2 \oplus 2 = 1$

   d) $4 \ominus 5 = 7$ e) $7 \oplus 5 = 3$ f) $5 \ominus 8 = 3$

   g) $1 \oplus 1 = 0$ h) $6 \ominus 7 = 6$

5. We mentioned on p. 160 that the associative law holds for all numbers in the mod 4 clock. We verified this only for the case, $(2 \oplus 3) \oplus 1 = 2 \oplus (3 \oplus 1)$. Verify it for at least two other cases.

6. In a mod 6 clock, name the inverse of each element.

7. In a mod 9 clock, name the inverse of each element.

8. In designing tiles, we can often use the addition or multiplication tables of the various clock arithmetics to obtain different patterns. Thus in mod 3 we have the following addition and multiplication tables.

$\oplus$	0	1	2		$\ominus$	0	1	2
0	0	1	2		0	0	0	0
1	1	2	0		1	0	1	2
2	2	0	1		2	0	2	1

If we represent the number 0 as ✦, the number 1 as a box with three lines in it ▤, and the number 2 as a box with a little square in it, then we get the following tile design based upon the addition table.

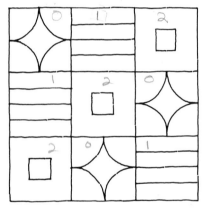

A design based upon the multiplication table might be pictured as follows (using the same representations).

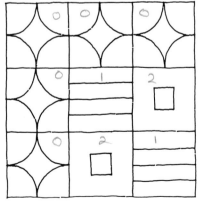

Using the addition and multiplication tables for the mod 5 and 6 clocks, construct tiles with appropriate design based upon these tables.

3. MODULAR ARITHMETIC

Let us look again at the mod 4 clock of the last section.

Suppose we were to start at 0 and move through 11 places. We would make 2 complete turns and then stop at 3. Notice that if you divide 11 by 4 you get a remainder of 3. This is no coincidence. Every time we make a complete turn on the clock, we go through 4 positions. Thus the number of complete turns is the number of times that 4 goes into 11 (that is, 2). We then have to go through the remaining 3 positions. This is the remainder when 11 is divided by 4.

Similarly, if we start at 0 and move through 18 positions, we would stop at 2. If we divide 18 by 4, the remainder would also be 2.

In other words, each position on the clock represents the remainder that we get when we divide a number by 4. This gives us a new way of looking at this clock. We can think of it as just the remainders we get when we divide by 4. From this point of view, this system is called a **modulo 4 arithmetic**. The number 4 is called the **modulus**.

Example 1   In a modulo 5 arithmetic, the numbers would be the remainders we get when we divide by 5. These are 0, 1, 2, 3, 4. In this system, the modulus is 5. In a mod 5 clock, moving through 17 positions, starting at 0, leaves us at 2. We get the same result if we divide 17 by 5.

In a modulo 4 arithmetic, suppose we were to add $3 + 2$. We know that $3 + 2 = 5$. However, we are only interested in the remainder when we divide by 4, so our answer would be 1. We write this answer as $3 + 2 \equiv 1$ in modulo 4 arithmetic. The three-lined equal sign indicates that we are working only with remainders. This corresponds to $3 \oplus 2 = 1$ of the table for the mod 4 clock given on p. 159.

Notation   Instead of writing out the words "in modulo 4 arithmetic" for the previous example, we abbreviate this as "(mod 4)." We write the answer as $3 + 2 \equiv 1$ (mod 4). This is read as

3 plus 2 is **congruent** to 1 modulo 4.

Example 2   Add $4 + 5$ (mod 6).

Solution $4 + 5$ is 9. If we divide 9 by 6, our remainder will be 3. Therefore, $4 + 5 \equiv 3$ (mod 6).

Example 3   Add $7 + 9$ (mod 10).

Solution $7 + 9 = 16$. Dividing 16 by 10, we get a remainder of 6. Therefore, $7 + 9 \equiv 6$ (mod 10).

Example 4   Add $5 + 5$ (mod 7).

Solution $5 + 5 = 10$. Dividing 10 by 7, the remainder is 3. Our answer is $5 + 5 \equiv 3$ (mod 7).

Example 5   Multiply $7 \cdot 3$ (mod 5).

Solution $7 \cdot 3 = 21$. If we divide 21 by 5 we get a remainder of 1. Thus,

$7 \cdot 3 \equiv 1$ (mod 5).

On a mod 5 clock, this means that we move through 21 positions, stopping at the 1 position.

Example 6   Multiply $5 \cdot 9$ (mod 15).

Solution $5 \cdot 9 = 45$. Dividing 45 by 15, we get a remainder of 0. Thus,

$5 \cdot 9 \equiv 0$ (mod 15).

Example 7   Subtract $5 - 2$ (mod 6).

Solution $5 - 2 = 3$. Dividing 3 by 6, we get a remainder of 3. Therefore,

$5 - 2 \equiv 3$ (mod 6).

Example 8   Subtract $3 - 6$ (mod 9).

Solution   To do this problem, we look at a mod 9 clock. We start at 3 and go back 6, stopping at 6. Thus,

$3 - 6 \equiv 6$ (mod 9).

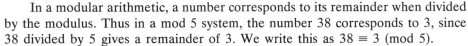

In a modular arithmetic, a number corresponds to its remainder when divided by the modulus. Thus in a mod 5 system, the number 38 corresponds to 3, since 38 divided by 5 gives a remainder of 3. We write this as $38 \equiv 3$ (mod 5).

Notice that $38 - 3$ is exactly divisible by 5. This leads us to a useful way of rephrasing the idea of modular arithmetics.

Definition 3.1   *We say that two integers a and b **are congruent modulo** m, if a − b can be divided exactly by m (m is a natural number). We write this as*

$a \equiv b$ (mod m).

Example 9 $7 \equiv 3$ (mod 2) means 7 is congruent to 3 modulo 2. This is true because $7 - 3$ can be divided exactly by 2.

Notice that if we divide 7 by 2, the remainder is 1, and if we divide 3 by 2, the remainder is also 1. *If any two numbers are congruent mod m, then their remainders when divided by m are the same.*

Example 10 $10 \equiv 15$ (mod 5) because $10 - 15$, which is -5, is exactly divisible by 5.

Example 11 $12 \equiv 7$ (mod 3) is *not* true because $12 - 7$, which is 5, cannot be exactly divided by 3.

Example 12 $25 \equiv 7$ (mod 9) is true since $25 - 7$, which equals 18, is divisible by 9.

Example 13 $38 \equiv 5$ (mod 11) is true because $38 - 5 = 33$, and 33 is exactly divisible by 11.

Example 14 $-5 \equiv 6 \pmod{11}$ is true since $-5 - 6$, which is -11, can be divided exactly by 11.

Example 15 $5 \equiv -13 \pmod 9$ because $5 - (-13)$, which equals $5 + 13$ or 18, is exactly divisible by 9.

We wish to emphasize a comment made in Example 9. We pointed out that if $a \equiv b \pmod m$, then when we divide a by m or b by m, the remainder is the same in both cases. Thus, whether two numbers are congruent modulo m depends upon whether they have the same remainders when we divide them by m. To put it another way, if $a \equiv b \pmod m$, the positions of a and b on the mod m clock are the same.

EXERCISES

1. Perform the following operations in the indicated modular arithmetics.

   a) $5 + 7 \pmod 3$ b) $7 + 10 \pmod 8$ c) $4 \cdot 6 \pmod 9$

   d) $6 \cdot 3 \pmod 4$ e) $3 - 8 \pmod 3$ f) $1 - 5 \pmod 6$

   g) $0 - 2 \pmod 5$ h) $-2 \cdot 5 \pmod 7$ i) $-4 \cdot 5 \pmod{11}$

   j) $-3 \cdot 4 \pmod 2$

2. Which of the following are true?

   a) $17 \equiv 13 \pmod 2$ b) $-6 \equiv -2 \pmod 4$ c) $12 \equiv -8 \pmod{10}$

   d) $12 \equiv 5 \pmod 7$ e) $2 \cdot 9 \equiv 3 \pmod 5$ f) $-7 \equiv 4 \pmod 3$

   g) $7 \equiv 7 \pmod 6$ h) $2 \equiv 1 \pmod 2$ i) $18 \equiv 1 \pmod 2$

   j) $5 \cdot 3 \equiv -1 \pmod 8$

3. Find one possible replacement for x in each of the following, so that the resulting statement will be true.

   a) $x \equiv 3 \pmod 5$ b) $2x \equiv 4 \pmod{10}$ c) $x \equiv 3 \pmod 8$

   d) $x + 3 \equiv 3 \pmod 4$ e) $4 - x \equiv 9 \pmod 9$ f) $x - 7 \equiv 4 \pmod 5$

   g) $2x \equiv 4 \pmod 6$ h) $x + 7 \equiv 1 \pmod 6$ i) $x - 1 \equiv 4 \pmod 7$

   j) $3x + 1 \equiv 1 \pmod 2$

4. Find a number m for which each of the following are true (more than one answer is possible).

   a) $7 \equiv 3 \pmod m$ b) $-3 \equiv -6 \pmod m$ c) $3 \equiv -8 \pmod m$

   d) $-5 \equiv -5 \pmod m$ e) $13 \equiv 3 \pmod m$ f) $4 \equiv -10 \pmod m$

   g) $2 \equiv -6 \pmod m$ h) $10 \equiv -4 \pmod m$

The following example illustrates the technique needed to do Exercises 5 through 8.

Example   If New Year's Day is Thursday, on what day of the week will February 1 fall?

Solution   January has 31 days. Since there are 7 days in a week, we can consider this as a modulo 7 system, where Sunday is 0, Monday is 1, and so forth. If we now divide 31 by 7, our remainder is 3. Thus from January 1 until February 1 there are exactly 4 weeks and 3 days. Therefore February 1 will occur 3 days after a Thursday. It will occur on a Sunday.

*5. In a certain year April 1 falls on a Friday. In that same year, on what day of the week does July 4 fall?

*6. If this year is a leap year and my birthday, which is January 10, is on Tuesday, on what day of the week will my birthday be next year?

*7. If this year Christmas falls on a Wednesday, on what day of the week will April Fool's Day fall next year? (Assume that next year is not a leap year.)

*8. January 1, 1976 fell on a Thursday. On what day of the week will January 1, 1986 fall?

---

4. CASTING OUT 9's

Whenever we perform a computation, we like to check our answer. One way of doing this is by **casting out 9's**. It works this way.

Add: 476
     + 237
     ——
      713

We check this by adding the digits in each number as shown. Thus we have

$$476 \rightarrow 4 + 7 + 6 = 17 \rightarrow 1 + 7 = \quad 8$$
$$+\ 237 \rightarrow 2 + 3 + 7 = 12 \rightarrow 1 + 2 = \quad \underline{3}$$
$$11 \rightarrow 1 + 1 = ②$$
$$713 \rightarrow 7 + 1 + 3 = 11 \rightarrow 1 + 1 = ②$$

Let us examine this procedure. We first add the digits in 476. This gives 17. Now we add the digits in 17. This gives 8.

We repeat the process on 237. First we get 12. Then adding the digits of 12, we get 3.

Next we *add* the results 8 and 3, getting 11. Finally adding the digits of 11, we get 2.

Now we do the same thing to the total 713. Again we end up with 2.

Since in both cases the result is 2, our answer is *probably* correct. There is a *slight* possibility that the answer is wrong.

If the results are not the same, then we know definitely that we have made a computational error. To illustrate this, suppose we add 2437 and 5617 and obtain the incorrect answer 8044. Let us check.

$$2437 \rightarrow 2 + 4 + 3 + 7 = 16 \rightarrow 1 + 6 = \quad\quad\quad 7$$
$$+\ 5617 \rightarrow 5 + 6 + 1 + 7 = 19 \rightarrow 1 + 9 = 10 \rightarrow 1 + 0 = \underline{1}$$
$$⑧$$
$$8044 \rightarrow 8 + 0 + 4 + 4 = 16 \rightarrow 1 + 6 = ⑦$$

The fact that one result is 8 and the other is 7 tells us that we made a mistake. Find the error.

Example 1   Add 368, 47, and 5928, and check by casting out 9's.

Solution

$$368 \rightarrow 3 + 6 + 8 \quad = 17 \rightarrow 1 + 7 = \; 8$$
$$47 \rightarrow 4 + 7 \qquad = 11 \rightarrow 1 + 1 = \; 2 \quad \text{We add these.}$$
$$+ \; 5928 \rightarrow 5 + 9 + 2 + 8 = 24 \rightarrow 2 + 4 = \; 6$$
$$16 \rightarrow 1 + 6 = \;⑦$$
$$6343 \rightarrow 6 + 3 + 4 + 3 = 16 \rightarrow 1 + 6 = ⑦$$

Since we get 7 in both cases, the addition is *probably* correct.

Example 2   Multiply 731 by 26 and check by casting out 9's.

Solution

$$731 \rightarrow 7 + 3 + 1 = 11 \rightarrow 1 + 1 = \; 2$$
$$\times \, 26 \rightarrow 2 + 6 \qquad\qquad = \; 8 \quad \text{We multiply these.}$$
$$\overline{4386} \qquad\qquad\qquad\qquad 16 \rightarrow 1 + 6 = ⑦$$
$$1462$$
$$\overline{19006} \rightarrow 1 + 9 + 0 + 0 + 6 = 16 \rightarrow 1 + 6 = ⑦$$

Since the result in both cases is 7, our answer is *probably* correct.

This time, because it is a multiplication problem, we multiply the 2 and the 8, as we indicated.

Example 3   Multiply 68 by 281 and check by casting out 9's.

Solution

$$68 \rightarrow 6 + 8 \qquad = 14 \rightarrow 1 + 4 = \; 5$$
$$\times \, 281 \rightarrow 2 + 8 + 1 = 11 \rightarrow 1 + 1 = \; 2 \quad \text{We multiply these.}$$
$$\overline{68} \qquad\qquad\qquad\qquad 10 \rightarrow 1 + 0 = ①$$
$$564$$
$$136$$
$$\overline{19308} \rightarrow 1 + 9 + 3 + 0 + 8 = 21 \rightarrow 2 + 1 = ③$$

The difference in the results tells us that we have made a mistake. Can you find it?

Example 4   Subtract 256 from 468 and check by casting out 9's.

Solution

$$468 \rightarrow 4 + 6 + 8 = 18 \rightarrow 1 + 8 = 9$$
$$- \, 256 \rightarrow 2 + 5 + 6 = 13 \rightarrow 1 + 3 = 4 \quad \text{We subtract.}$$
$$⑤$$
$$212 \rightarrow 2 + 1 + 2 = ⑤$$

Example 5   Subtract 273 from 465 and check by casting out 9's.

Solution

$$465 \rightarrow 4 + 6 + 5 = 15 \rightarrow 1 + 5 = 6$$
$$- \, 273 \rightarrow 2 + 7 + 3 = 12 \rightarrow 1 + 2 = 3 \quad \text{We subtract.}$$
$$③$$
$$192 \rightarrow 1 + 9 + 2 = 12 \rightarrow 1 + 2 = ③$$

Example 6   Subtract 213 from 778 and check by casting out 9's.

Solution

$$778 \rightarrow 7 + 7 + 8 = 22 \rightarrow 2 + 2 = \quad 4 \atop - 213 \rightarrow 2 + 1 + 3 \qquad\qquad = \quad 6 \Big\} \text{We subtract.}$$

$$\boxed{-2}$$

$$565 \rightarrow 5 + 6 + 5 = 16 \rightarrow 1 + 6 = ⑦$$

On first thought it would appear that our answer is incorrect. However, note that 7 is congruent to $-2 \pmod 9$ because $7 - (-2) = 7 + 2 = 9$ is exactly divisible by 9. Thus in mod 9 arithmetic, -2 and 7 are really the same, so that our answer is *probably* right.

You are probably wondering why this mysterious procedure works. The secret will now be revealed.

Consider the number 3221 which can be written as

3(thousands) + 2(hundreds) + 2(tens) + 1

3(1000) + 2(100) + 2(10) + 1.

If you divide 1000 by 9, the remainder is 1. Thus if you divide 3(1000) by 9, the remainder is 3(1), or 3.

Similarly if we divide 100 by 9, the remainder is 1. Therefore if we divide 2(100) by 9, the remainder is 2(1) = 2.

Also if we divide 10 by 9, the remainder is 1. So if you divide 2(10) by 9, the remainder is 2(1), which is 2.

But 1 divided by 9 doesn't go. We just have a remainder of 1.

Putting this all together, if we divide 3221 by 9, we will get a total remainder of $3 + 2 + 2 + 1$, or 8. (You should verify this by actually dividing 3221 by 9.)

Notice that the remainder is exactly the sum of the digits.

In general, if we divide any number by 9, the remainder will be the sum of the digits. (If the sum of the digits is greater than 9, repeat the procedure.)

Now let us look at a casting out 9's example.

$$436 \rightarrow 4 + 3 + 6 = 13 \rightarrow 1 + 3 = 4 \atop + 237 \rightarrow 2 + 3 + 7 = 12 \rightarrow 1 + 2 = 3 \Big\} \text{We add.}$$

$$⑦$$

$$673 \rightarrow 6 + 7 + 3 = 16 \rightarrow 1 + 6 = ⑦$$

The numbers 4 and 3 are the remainders we get when we divide 436 and 237, respectively, by 9. If the addition is correct, then the sum of these remainders should equal the remainder we get when we divide 673 by 9. It is the same in this case, so we are *probably* correct.

Comment   We stated that if the remainders check, then our answer is *probably* correct. You cannot be 100% sure that your answer is correct, because if you make a mistake that is a multiple of 9, then the remainders will not be affected by the error. The remainders will still check.

EXERCISES   1. Perform the indicated operations and check your result by casting out 9's.

a)   543
     + 815

b)   2976
        3301
     + 5029

c)   361
     − 163

d)   84190
     −36221

e)   4311
     − 3567

f)   215
     × 314

g)   86
     × 48

h)   3144
     × 234

i)   1234
       5678
     + 9012

j)   9876
     − 5432

k)   2233
     × 4455

l)   1234
     × 1234

2. A teacher making up an arithmetic exam accidently wrote the problem

5463   as   5643
+ 1229   + 1229

She has prepared an answer sheet in advance and decides to check by casting out 9's. She does not find her error. Why?

*3. What does casting out 9's have to do with congruence and modular arithmetic?

*4. In base 3, the method of checking by casting out 9's will not work since there are no 9's. Can you think of a method similar to that of casting out 9's that would work for base 3?

*5. Is there any way to tell if a number is divisible by 9 (without actually dividing by 9)?

*6. By looking at the last digit, can we tell if a number is divisible by 9?

---

5. CASTING OUT 11's   Another method commonly used to check computations is **casting out 11's**. In this method, for each number, *we start at the right* and move to the left. We put a + in front of the first digit, then a − in front of the second digit, and continue alternating the signs. We add the results. This is illustrated below.

$2741 \rightarrow -2 + 7 - 4 + 1 = 2$

$5673 \rightarrow -5 + 6 - 7 + 3 = -3$

$781 \rightarrow +7 - 8 + 1 \quad = 0$

From here on, the procedure is the same as that for casting out 9's.

Example 1   Add 2741 and 781 and check the result by casting out 11's.

Solution $2741 \rightarrow -2 + 7 - 4 + 1 = 2$ ⎫ We add.
     $+ \ \ 781 \rightarrow +7 - 8 + 1 \quad = 0$ ⎭

     $3522 \rightarrow -3 + 5 - 2 + 2 = ②$

Since in both cases we get 2, our answer is probably correct.

Example 2   Add 142936 and 782225 and check the result by casting out 11's.

Solution

$$142936 \rightarrow -1 + 4 - 2 + 9 - 3 + 6 = 13 \rightarrow -1 + 3 = 2$$
$$+ \: 782225 \rightarrow -7 + 8 - 2 + 2 - 2 + 5 = \qquad\qquad 4$$

We add.

$$925161 \rightarrow -9 + 2 - 5 + 1 - 6 + 1 = \underbrace{-16}$$

$$\overline{6}$$

−16 is congruent to 6 (mod 11) because −16 − 6, which is −22, is exactly divisible by 11. Hence, our answer is probably correct.

Example 3 Multiply 321 and 68 and check by casting out 11's.

Solution

$$321 \rightarrow +3 - 2 + 1 = 2$$
$$\times \: 68 \rightarrow -6 + 8 \qquad = 2$$

We multiply.

$$\textcircled{4}$$

$$\underline{}$$
$$2568$$
$$1926$$
$$\overline{21828} \rightarrow 2 - 1 + 8 - 2 + 8 = 15 \rightarrow -1 + 5 = \textcircled{4}$$

Hence our answer is probably correct.

EXERCISES

1. Check all the problems of Exercise 1 from the previous section (p. 170) by casting out 11's.

*2. Can you explain why the method of casting out 11's works?

---

6. GROUPS   In Section 2 of this chapter, we pointed out that the mod 4 clock (or modulo 4 arithmetic, which is the same thing) has the following important properties.

1. The binary operation $\oplus$ is closed.
2. $\oplus$ is associative.
3. 0 is an identity for $\oplus$.
4. Every number on the clock has an inverse for $\oplus$.

Clearly any other clock besides the mod 4 clock has these four properties also. Many other systems found in widely varying branches of mathematics also have these properties. Such systems are called **groups**. Because they occur so often, mathematicians devote a good deal of attention to them. We will first give a formal definition of a group and then illustrate it with several examples.

Definition 6.1   *A **group** is a set of elements, G, together with a binary operation ∘, with the following properties*:

Property 1.   ∘ is closed.

Property 2.   ∘ is associative.

Property 3.   There is an identity element in G for ∘. If we call this identity element i, then this means that for any element a in G we have a ∘ i = a and i ∘ a = a. In other words, i is the "do-nothing" element for ∘. In the mod 4 clock, 0 was the identity for the operation ⊕.

Property 4.   Every element in G has an inverse for ∘. This means that for any element a, we can find another element which operating on a with ∘, gives i. For the mod 4 clock we actually listed the inverse for each element on p. 160.

Comment   The operation ∘ may vary from one situation to another, as we shall see. It may sometimes be ordinary addition or multiplication. Sometimes it will be the turning of a clock hand as in clock arithmetic. In other situations, it will be a totally different operation.

The best way to understand groups is to look at some examples.

Example 1   Let G = {integers} and let ∘ stand for ordinary addition. This system is a group. Let us see why.

In Chapter 4, we saw that addition of integers is closed and associative. We also saw that 0 is the identity element. So the first three properties are satisfied.

Let us verify Property 4. Consider the integer 3. Its inverse is -3, since $3 + (-3) = 0$, which is the identity. Similarly the inverse of -11 is $+11$, since $(-11) + (+11) = 0$. It is clear that *every* element has an inverse. Thus Property 4 is satisfied.

Since all four properties are satisfied, it follows that this system is a group.

Example 2   Let G = {integers} and let ∘ be ordinary multiplication. Let us see if this system is a group.

Again we know from our work in Chapter 4 that multiplication of integers is closed and associative. We also know that 1 is the identity for multiplication. So Properties 1, 2, and 3 are satisfied. What about Property 4? Consider the integer 3. Does it have an inverse for multiplication? We are asking if there is an **integer** such that

$3 \cdot (\text{integer}) = 1.$

The only number which when multiplied by 3 gives 1 is $\frac{1}{3}$. However, $\frac{1}{3}$ is not an integer. Thus *there is no element in G that is an inverse for 3*. Property 4, which says that *every* element must have an inverse, is *not* true in this case. So this system is *not* a group.

Example 3   In Example 2 we saw that the set of integers with the operation of multiplication did not form a group. The problem was that not every element had an inverse. The only numbers that could be inverses were rational numbers like $\frac{1}{3}$.

If we change G from the integers to the rational numbers, then we *do* have numbers like $\frac{1}{3}$. Thus, if we let G = {rational numbers} and let ∘ be ordinary multiplication, it looks as if we have a group. Unfortunately this is not the case; 0 has no inverse. If you doubt this, try to find a number which when multiplied by 0 will give you 1. It can't be done! However, if we let G = {rational numbers, without 0} and let ∘ be multiplication, then we *do* have a group.

Example 4 Let $G = \{-1, 1\}$ and let $\circ$ be ordinary multiplication. We will verify that this is a group.

Property 1.
$$1 \cdot 1 = 1$$
$$1 \cdot (-1) = -1$$
$$(-1) \cdot 1 = -1$$
$$(-1) \cdot (-1) = +1$$

These results are summarized in the following table:

$\circ$	-1	1
-1	1	-1
1	-1	1

All possible ways of multiplying elements of G result in some element of G. Thus $\circ$ is a closed operation.

Property 2. Since $\circ$ is just ordinary multiplication, we already know that the associative law holds for $\circ$.

Property 3. 1 is the identity.

Property 4. The inverse of 1 is 1, since $1 \cdot 1 = 1$. The inverse of -1 is -1, since $(-1) \cdot (-1) = 1$.

Therefore this system is a group.

Example 5 Let $G = \{1, -1\}$ and let $\circ$ be ordinary addition. Then this system is *not* a group. For one thing $\circ$ is not a closed operation, since $1 + 1$ is 2, and 2 is not an element of G. Moreover, there is no identity Why?

Example 6 Consider a set G whose elements are a, b, and c. Let the operation $\circ$ be given by the table shown on the left. Does this system form a group?

$\circ$	a	b	c
a	a	b	c
b	b	c	a
c	c	a	b

Solution We must verify that the four properties of a group hold.

Property 1. Since all the elements in the table are elements of G, $\circ$ is a closed operation.

Property 2. The associative law holds. We will verify it for one case. Readers should verify it for a few others to convince themselves.

$$
\begin{array}{c|c}
a \circ (b \circ c) & (a \circ b) \circ c \\
= \quad a \circ a & = \quad b \circ c \\
= \quad a & = \quad a
\end{array}
$$

Property 3. We see from the table that a is the identity since a applied to any element results in the same element.

Property 4. Every element has an inverse.

The inverse of a is a because $a \circ a = $ the identity which is a.

The inverse of b is c because $b \circ c = $ the identity which is a.

The inverse of c is b because $b \circ c = $ the identity which is a.

Since all four properties are satisfied, this system forms a group.

Some groups have an additional property called the *commutative* property. This is defined as follows.

Definition 6.2   *A group is called an **abelian group** if the commutative property holds. This means that if a and b are any elements of G, then a ∘ b = b ∘ a.*

All of the groups we have discussed so far are commutative.

Abelian groups are named after the Norwegian mathematician Niels Henrik Abel. A gifted mathematician, Abel unfortunately died of consumption before he was 27 years old. Despite his illness, personal hardships, and many disappointments in his career, he managed to make a very solid contribution to mathematics.

EXERCISES   1. Determine whether each of the following are groups.

a) The even integers with the operation of addition

b) The odd integers with the operation of addition

c) The positive integers with the operation of addition

d) The negative integers with the operation of addition

e) Mod 5 arithmetic, that is, 0, 1, 2, 3, 4, mod 5 with the operation of multiplication

f) Mod 5 arithmetic, that is, 0, 1, 2, 3, 4, mod 5 without zero, with the operation of multiplication

g) Mod 4 arithmetic, that is, 0, 1, 2, 3, mod 4 with the operation of multiplication

h) Mod 4 arithmetic, that is, 0, 1, 2, 3, mod 4 without 0, with the operation of multiplication

i) The real numbers with the operation of addition

j) The real numbers with the operation of multiplication

2. In each of the following, an operation $\circ$ is given by the table shown. Determine whether each is a group.

a)

$\circ$	a	b
a	a	b
b	b	a

b)

$\circ$	x	y
x	x	y
y	y	y

c)

$\circ$	x	y	z
x	y	z	x
y	z	x	y
z	x	y	z

d)

$\circ$	x	y	z
x	x	x	x
y	x	y	z
z	x	y	z

Niels Henrik Abel (1802–1829)

∘	1	2	3	4	5	6
1	1	1	1	1	1	1
2	1	2	3	4	5	6
3	1	3	5	1	3	5
4	1	4	1	4	1	4
5	1	5	3	1	5	3
6	1	6	5	4	3	2

∘	0	3	6
0	0	3	6
3	3	6	0
6	6	0	3

3. Which of the groups of Exercises 1 and 2 are abelian (or commutative) groups?

*4. Let $G = \{$all integers$\}$ and let ∘ be defined as follows: If a and b are in G, then $a \circ b = a + b + 2$. For example, if a and b are 5 and 4, then $a \circ b = 5 + 4 + 2 = 11$. Does this system form a group?

5. Let $G = \{x, y, z\}$ and let ∘ be the operation given by the following table:

∘	x	y	z
x	x	y	z
y	y	x	z
z	z	y	x

a) Find the identity element.

b) Find the inverse of each element.

c) Is G an abelian group?

†*6. Let $G = \{1, -1, i, -i\}$. (Remember $i = \sqrt{-1}$.) Let the operation ∘ be multiplication. Does this system form a group? Explain.

7. Let $G = \{0, 2, 4, 6\}$ and let ∘ be defined by the following table:

∘	0	2	4	6
0	0	0	0	0
2	0	2	0	4
4	0	4	2	6
6	0	4	6	4

Does this system form a group? Explain.

8. Consider the set $G = \{1, 2, 3, 4, 5, 6\}$ and an operation ∘ given by the table on the left.

a) Is ∘ a closed operation?

b) Is there an identity? If yes, find it.

c) Is there an inverse for 3? If yes, find it.

d) Is there an inverse for 6? If yes, find it.

e) Is ∘ commutative?

f) Does this system form a group?

9. Let $G = \{0, 3, 6\}$ and let ∘ be defined by the table on the left. Does this system form a group?

10. Let $G = \{1, 3, 5, 7\}$ and let ∘ be given by the table shown below.

∘	1	3	5	7
1	1	3	5	7
3	3	1	7	5
5	5	7	1	3
7	7	5	3	1

a) Is ∘ a closed operation?

b) Is ∘ a commutative operation?

c) Is there an identity element? If yes, find it.

d) Find the inverse of each element.

e) Is this system a group?

*7. MORE EXAMPLES OF GROUPS

Figure 1

In this section we will examine some interesting groups that arise when we move around geometric shapes such as triangles, squares, etc. (A background in geometry is not needed. You only have to know what triangles, squares, and rectangles are.)

Example 1   Let us look at an **equilateral triangle**. This is just a triangle with three equal sides and three equal angles as shown (Fig. 1).

Suppose we cut such a triangle out of a wooden block. We ask the following question: How many ways can we put the triangle back into the wooden block without turning the triangle over? (We suggest that the reader cut such a triangle out of a piece of paper or cardboard and refer to it as he or she reads along.) The triangle and block are shown in Fig. 2.

Figure 2

One way of putting the triangle back into the block is to put it back exactly as it was taken out. This we will call rotation X.

Another way of putting the triangle back into the block is to rotate it one-third of the way around (120°). This we call rotation Y.

The only other way of putting the triangle back into the block is to rotate it two-thirds of the way around (240°). This we will call rotation Z.

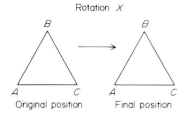

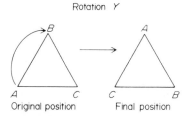

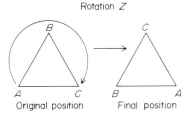

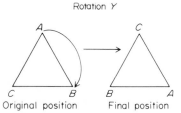

These illustrations show the rotation performed on the triangles starting in the original positions only. However, we may perform the rotation from any starting position that fits into the block. One such possibility is shown in the next diagram of rotation Y at the left.

Let G be the set containing these three rotations, that is, $G = \{X, Y, Z\}$. Now we define an operation $\circ$ on G in the following way: $Y \circ Z$ means "first apply rotation Y and then apply rotation Z to the result." If we start with the triangle in the original position, then $Y \circ Z$ is as illustrated below. So $Y \circ Z$ leaves the triangle in the original position. This is the same as rotation X. Thus $Y \circ Z = X$.

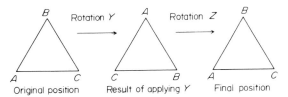

Similarly, we define $X \circ Z$ as first applying X and then Z. If we start in the original position, then the result is as shown below. Thus $X \circ Z = Z$.

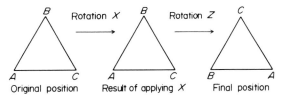

In the same way, we define $\circ$ for other elements of G. We can summarize all possibilities by the following table:

$\circ$	X	Y	Z
X	X	Y	Z
Y	Y	Z	X
Z	Z	X	Y

Now we ask: Is this system a group? We must verify the four group properties.

Property 1.   From the table we see that $\circ$ is a closed operation.

Property 2.   It is easy to verify that the associative law holds. Try it.

Property 3.   The identity is X.

Property 4.   The inverse of X is X.
　　　　　　　　The inverse of Y is Z.
　　　　　　　　The inverse of Z is Y.

Hence this system is a group. As a matter of fact, it is even a commutative or abelian group.

Example 2   We consider the same situation as in the last example. However, we now allow the triangle to be turned over as well. We still have the three rotations as in the previous example, but we rename them *symmetries X, Y,* and *Z*. In addition, we have three new ways of putting the triangle back into the block. These are shown in the figure below and are called symmetries P, Q, and R. In each case, we flip the triangle over around the dotted line.

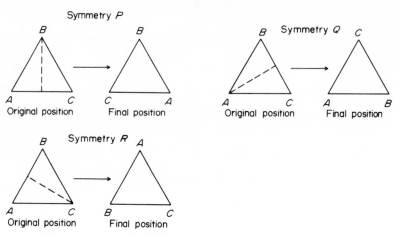

The operation ∘ is defined as in Example 1. Thus $X \circ P$ means first apply X and then apply P.

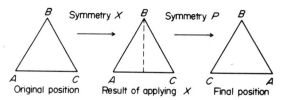

Similarly, $Q \circ Y$ means first apply Q and then apply Y.

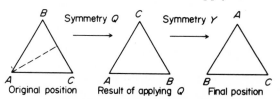

All possible results are summarized in the following table:

∘	X	Y	Z	P	Q	R
X	X	Y	Z	P	Q	R
Y	Y	Z	X	R	P	Q
Z	Z	X	Y	Q	R	P
P	P	Q	R	X	Y	Z
Q	Q	R	P	Z	X	Y
R	R	P	Q	Y	Z	X

This system also forms a group called the **symmetries of a triangle**. It is *not* a commutative group. (See page 174.)[1]

Example 3   Suppose we now have a square cut out of a block of wood, as in the previous two examples. Let us find all possible ways of replacing the square in the block of wood. Flipping over is allowed. There are eight possible ways of doing this, as shown below. These are called **symmetries**.

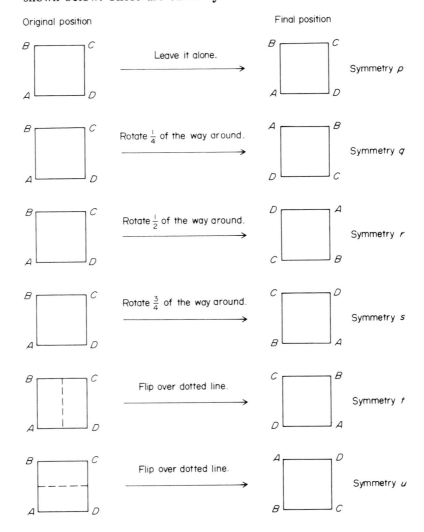

Original position   Final position

Leave it alone.   Symmetry p

Rotate $\frac{1}{4}$ of the way around.   Symmetry q

Rotate $\frac{1}{2}$ of the way around.   Symmetry r

Rotate $\frac{3}{4}$ of the way around.   Symmetry s

Flip over dotted line.   Symmetry t

Flip over dotted line.   Symmetry u

---

1. The smallest group that is not commutative has six elements; that is, a group with less than six elements *must* be commutative.

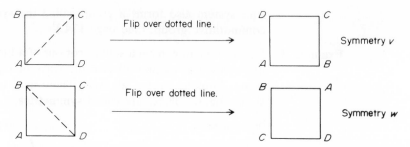

We define ∘ as we did for the triangle. For example, $r \circ v$ means "first apply r and then apply v to the result." The table for ∘ is shown below.

∘	p	q	r	s	t	u	v	w
p	p	q	r	s	t	u	v	w
q	q	r	s	p	w	v	t	u
r	r	s	p	q	u	t	w	v
s	s	p	q	r	v	w	u	t
t	t	v	u	w	p	r	q	s
u	u	w	t	v	r	p	s	q
v	v	u	w	t	s	q	p	r
w	w	t	v	u	q	s	r	p

This system forms a group called the **symmetries of the square**. It is clear from the table that ∘ is a closed operation.

The associative law holds. Since there are 512 possible combinations, we will check just one of them:

$$t \circ (q \circ r) \qquad | \qquad (t \circ q) \circ r$$
$$= \quad t \circ s \qquad \qquad = \quad v \circ r$$
$$= \quad w \qquad \qquad \quad = \quad w$$

The identity element is obviously p. Every element has an inverse. For example, the inverse of s is q.

EXERCISES

*1. Verify all cases of the associative law for Example 1.

*2. In Example 3, verify the associative law for three different cases.

*3. In Example 2, find the inverse of each element.

*4. a) Find all the symmetries (possible ways of putting it back into the wood block) for any rectangle that is not a square. (*Hint:* There are four of them.)

   b) Show that this system is a group. It is called the **Klein 4 group**.

Side a = side b

*5. An isosceles triangle is a triangle with two equal sides as shown on the left. (The third side does not necessarily have to equal the other two sides.)

    a) Find all the symmetries of this isosceles triangle.

    b) Do these symmetries form a group?

*6. A regular pentagon has five equal sides as shown on the left.

    a) Find all the symmetries of a regular pentagon.

    b) Do these symmetries form a group?

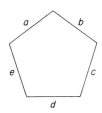

$a = b = c = d = e$

*7. Private Bond is standing at attention. Sergeant Stone comes along and gives him orders such as, "right turn," "left turn," "about face," or "as you were." Let

    r stand for right turn,
    l stand for left turn,
    a stand for about face, and
    s stand for as you were.

Let ∘ represent any combination of these orders. For example, $l \circ r$ would mean first make a left turn and then a right turn. Your result is, of course, the starting position. So $l \circ r = s$.

    a) Make up a table for ∘.

    b) Is ∘ closed?

    c) Is ∘ commutative?

    d) Is there an identity element? If yes, find it.

    e) Find the inverse of each element.

    f) Find $l \circ (s \circ a)$ and $(l \circ s) \circ a$.

    g) Does this system form a group?

*8. Show by example that the group of Example 2 (p. 177) is not commutative.

---

8. CONCLUDING REMARKS

As we have already pointed out, groups occur in many areas of mathematics. By studying groups in general, we can discover the properties that *all* groups have. Then, when we work with a specific group, we know that *this* group will have all those properties. We do not have to reestablish these properties for this specific group.

    For example, it can be shown that the identity for a group is unique. This means that there cannot be more than one identity. As an illustration, if we look at the group of symmetries of the square (Example 3, p. 179), we notice that the identity is p. There is no other identity. Now if we were to study *any* other group, we would immediately know that there is only one identity. We would not have to prove it.

This process is typical of mathematical activity. Mathematicians try to discover general properties and then apply them to specific situations as they arise. Of course, in establishing general concepts, they are usually motivated by specific examples. (Refer back to the discussion of inductive and deductive reasoning in Chapter 1, pp. 2–6).

STUDY GUIDE   In this chapter we discussed mathematical systems and the concept of a group. The following ideas were covered.

Systems

Clock arithmetic (p. 158)
Modular arithmetic (p. 164)
Group (p. 171)
Abelian groups (p. 174)
Symmetries of a triangle (p. 178)
Symmetries of a square (p. 180)
Klein 4 group (p. 180)

Properties of these systems

Closure property (p. 171)
Associative law (p. 171)
Identities (p. 171)
Inverses (p. 172)
Commutative law (p. 174)

Other ideas discussed

Modulus (p. 164)
Congruence modulo a natural number
   (p. 165)
Binary operation (p. 171)

Applications were given for:

a) casting out 9's (p. 167),
b) casting out 11's (p. 170).

SUGGESTED FURTHER READINGS

Bell, E. T., *Men of Mathematics*. New York: Simon and Schuster, 1961. Chapters 17 and 20 contain biographies of Abel and Galois.

Courant, R., and H. Robbins, *What is Mathematics?* New York: Oxford University Press, 1960. Pages 31–40 discuss congruences.

Dinkines, F., *Abstract Mathematical Systems*. New York: Appleton-Century-Crofts, 1964. A general basic discussion of groups.

Eddington, A. S., "The Theory of Groups." In James R. Newman, ed., *The World of Mathematics*, Vol. III, Part IX. New York: Simon and Schuster, 1956. Pages 1534–1574 discuss groups.

Gardner, M., *New Mathematical Diversions from the Scientific American*. New York: Simon and Schuster, 1966. Chapter 2.

Keyser, C., "The Group Concept." In James R. Newman, ed., *The World of Mathematics*. New York: Simon and Schuster, 1956.

Mathematics in the Modern World (Readings from *Scientific American*). San Francisco: W. H. Freeman, 1968. Article 15 discusses groups and Article 33 discusses some applications of groups to science.

Sawyer, W. W., *Prelude to Mathematics*. Baltimore, Md.: Penguin Books, 1959. Chapter 14 discusses groups.

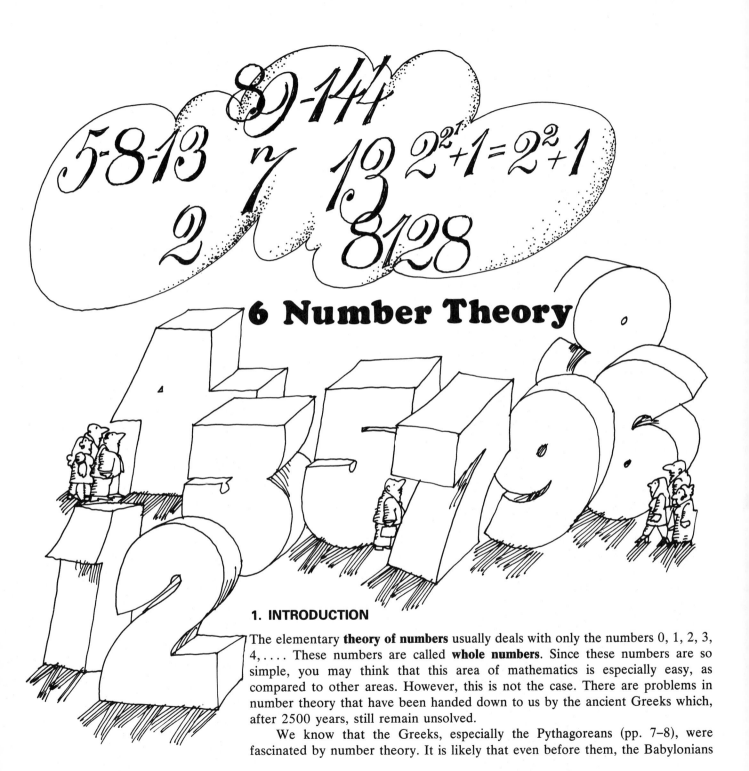

$5 \cdot 8 \cdot 13$ $89 \cdot 144$

2 π 13 $2^{2^{2^{21}}}+1=2^{2^2}+1$

8128

6 Number Theory

1. INTRODUCTION

The elementary **theory of numbers** usually deals with only the numbers 0, 1, 2, 3, 4, ... These numbers are called **whole numbers**. Since these numbers are so simple, you may think that this area of mathematics is especially easy, as compared to other areas. However, this is not the case. There are problems in number theory that have been handed down to us by the ancient Greeks which, after 2500 years, still remain unsolved.

We know that the Greeks, especially the Pythagoreans (pp. 7–8), were fascinated by number theory. It is likely that even before them, the Babylonians

183

studied it. Why should this subject arouse so much interest? There are several answers.

In the first place, as we have already pointed out, this branch of mathematics deals with rather simple numbers. This makes it easy, even for the nonmathematician, to state and understand many number theory problems. (Solving them is quite a different story.)

Another reason for the Greeks' interest in number theory was their belief that numbers had certain magical powers. Some of these superstitions survive even today. We consider 7 lucky, except if you break a mirror, in which case you get seven years of bad luck. We also think of 11 as being a lucky number. The number 13 is unlucky to many people, and some tall buildings have no thirteenth floor because it cannot be rented easily. How do you feel about taking an exam on Friday, the thirteenth? There are some smokers who will not light 3 cigarettes with one match.

The Pythagoreans went even further. They believed that the entire universe could be explained in terms of the numbers 0, 1, 2, 3, ... According to their beliefs, the number 1 gave rise to all other numbers, and they considered it the "number of reason." They called 2 the first female number (all other even numbers were also female). On the other hand, 3 was the first male number (all other odd numbers were also male). We do not care to comment on their belief that 2 was the "number of opinion" and 3 the "number of harmony." They believed that 5 was the "number of marriage" (can you see why?), and that 6 was the "number of creation" (remember, the Bible tells us that God created the world in six days).

The Pythagoreans believed that 10 was the holiest number. It was the number of the universe, and the symbol of health and harmony. Because they believed that 10 was the perfect number, they believed that there were 10 heavenly bodies. At the center of the universe there was a fire. Around this fire, the sun, earth, moon, and five planets revolved. Since this totaled only 9 heavenly bodies, they invented a tenth one called a "counterearth."

Finally, some of the ideas of number theory are closely related to geometry. This may not seem like much of an attraction to you, but the study of geometry was itself of major concern to both the Babylonians and Greeks.

Whatever the reason, number theory has attracted, and still attracts, the attention of many first-rate mathematicians. The great German mathematician, Gauss, once said, "Mathematics is the queen of the sciences, and number theory is the queen of mathematics."

Remark   For the rest of this chapter, when we say "number," we will mean one of the *numbers* 0, 1, 2, 3, ...

2. NUMBER PATTERNS

One of the reasons that numbers are so fascinating is that in working with them, we discover many surprising and interesting patterns.

Example 1   Let us examine the following number facts carefully to discover a method for finding the square of a number.

$$1^2 = 1$$
$$2^2 = 1 + 2 + 1 = 4$$
$$3^2 = 1 + 2 + 3 + 2 + 1 = 9$$
$$4^2 = 1 + 2 + 3 + 4 + 3 + 2 + 1 = 16$$
$$5^2 = 1 + 2 + 3 + 4 + 5 + 4 + 3 + 2 + 1 = 25$$
$$6^2 = 1 + 2 + 3 + 4 + 5 + 6 + 5 + 4 + 3 + 2 + 1 = 36$$
$$7^2 = 1 + 2 + 3 + 4 + 5 + 6 + 7 + 6 + 5 + 4 + 3 + 2 + 1 = 49$$

From the above patterns, it is obvious that to find the square of a number, we start with 1 and keep adding 1 until we get to our number. Then we go down by ones until we get back to 1 again. Thus to find 8^2, we have

$$1 + 2 + 3 + 4 + 5 + 6 + 7 + 8 + 7 + 6 + 5 + 4 + 3 + 2 + 1$$

which equals 64. Therefore, $8^2 = 64$. We can generalize this to obtain the square of any number n. We have

$$1 + 2 + 3 + \cdots + (n - 1) + n + (n - 1) + \cdots + 3 + 2 + 1.$$

When we add these we get n^2.

Example 2   The mathematician Nichomachus of Gerasa (about A.D. 100) noticed that if you write the odd numbers in the pattern

$$1 \qquad 3, 5 \qquad 7, 9, 11 \qquad 13, 15, 17, 19 \qquad 21, 23, 25, 27, 29, \ldots,$$

then the sums of the above groupings in order are

$$1 = 1 = 1^3$$
$$3 + 5 = 8 = 2^3$$
$$7 + 9 + 11 = 27 = 3^3$$
$$13 + 15 + 17 + 19 = 64 = 4^3$$
$$\vdots \qquad\qquad \vdots \quad \vdots$$
$$\vdots \qquad\qquad \vdots \quad \vdots$$
$$\vdots \qquad\qquad \vdots \quad \vdots$$

Thus we see that these successive sums are the cubes of the integers.

Example 3   The odd numbers are the numbers 1, 3, 5, 7, 9, 11, ... We see that

$$1 = 1 \text{ which is the same as } 1^2$$
$$1 + 3 = 4 \text{ which is the same as } 2^2$$
$$1 + 3 + 5 = 9 \text{ which is the same as } 3^2$$
$$1 + 3 + 5 + 7 = 16 \text{ which is the same as } 4^2$$
$$1 + 3 + 5 + 7 + 9 = 25 \text{ which is the same as } 5^2$$
$$1 + 3 + 5 + 7 + 9 + 11 = 36 \text{ which is the same as } 6^2$$

There is a relationship between the number of odd numbers we are adding and the sum. The relationship can be seen when we rewrite the above list as follows:

Sum of first odd number equals 1, which is 1^2.

Sum of first two odd numbers equals 4, which is 2^2.

Sum of first three odd numbers equals 9, which is 3^2.

In general, the sum of first n odd numbers equals n^2.

Thus, for example, the sum of the first ten odd numbers ($1 + 3 + 5 + 7 + 9 + 11 + 13 + 15 + 17 + 19$) is 100 or 10^2.

Compare this result with Example 1. What do you notice?

Example 4   Consider the sequence of numbers 1, 4, 7, 10, 13, ... Each number is 3 more than the one before it. Such a sequence of numbers is called an **arithmetic progression**. Can we find the 100th term in this sequence without actually listing all the numbers?

Solution   Let us rewrite the above numbers as follows.

Number of term in sequence	Number	Rewritten form
1	1	1
2	4	$1 + (1 \cdot 3)$
3	7	$1 + (2 \cdot 3)$
4	10	$1 + (3 \cdot 3)$
5	13	$1 + (4 \cdot 3)$
6	16	$1 + (5 \cdot 3)$
7	19	$1 + (6 \cdot 3)$
$\vdots$	$\vdots$	$\vdots$

In general, we have n | ? | $1 + (n - 1)3$

Thus to find the 100th term, we take $n = 100$ and get

$$1 + (n - 1)3$$
$$= 1 + (100 - 1)3$$
$$= 1 + (99 \cdot 3)$$
$$= 298.$$

Therefore the 100th term in the above sequence is 298. To convince yourself that this is true, write out the first one hundred numbers in the sequence.

Example 5   Consider the sequence 1, 2, 4, 8, 16, 32, ... In this sequence, each number is 2 times the one before it. Such a sequence is called a **geometric progression**. Find the 10th term in this progression.

Solution   We write the above numbers as follows.

Number of term in sequence	Number	Rewritten form
1	1	1 (also written as 2^0)
2	2	2^1
3	4	2^2
4	8	2^3
5	16	2^4
.	.	.
.	.	.
.	.	.
In general we have n	?	2^{n-1}

Thus to find the 10th term we take $n = 10$ and get

2^{n-1}

$= 2^{10-1} = 2^9$

$= 512.$

Therefore, the 10th term in the sequence is 512.

EXERCISES

1. Using the results of Example 2, find the following sums.

   a) $1 + 3 + 5 + \cdots + 11$ b) $1 + 3 + 5 + \cdots + 99$ c) $13 + 15 + \cdots + 99$

2. Using a similar method to that of Example 4, find the indicated term of the following arithmetic progressions.

   a) $2, 4, 6, \ldots,$ the 30th term

   b) $1, 6, 11, \ldots,$ the 12th term

   c) $1, 11, 21, \ldots,$ the 100th term

   d) $100, 93, 86, 79, \ldots,$ the 15th term

   e) $10, 20, 30, 40, \ldots,$ the nth term

3. Find the 10th term in the following sequence:

   $1, 3, 9, 27, \ldots.$

4. Complete each of the following. Can you generalize?

   i) $1 \cdot 9 + 2 = ?$
   $12 \cdot 9 + 3 = ?$
   $123 \cdot 9 + 4 = ?$
   etc.

   ii) $9 \cdot 9 + 7 = ?$
   $98 \cdot 9 + 6 = ?$
   $987 \cdot 9 + 5 = ?$
   etc.

   iii) $1 \cdot 8 + 1 = ?$
   $12 \cdot 8 + 2 = ?$
   $123 \cdot 8 + 3 = ?$
   etc.

   iv) $3 \cdot 37 = 111$ and $1 + 1 + 1 = 3$
   $6 \cdot 37 = $? and ?
   $9 \cdot 37 = $? and ?
   etc.

   v) $1 \cdot 1 \ = ?$
   $11 \cdot 11 \ = ?$
   $111 \cdot 111 = ?$
   etc.

vi) $7 \cdot 7$ = ?   vii) $7 \cdot 15873$ = ?
      $67 \cdot 67$ = ? $14 \cdot 15873$ = ?
      $667 \cdot 667$ = ? $21 \cdot 15873$ = ?
      etc.   etc.

5. Pick a 3-digit number, such as 273. Write the digits again in the same order to make the 6-digit number 273273. Divide 273273 by 13 and you will see that there is no remainder. Now pick another 3-digit number, rewrite the digits again to make a 6-digit number, and divide the result by 13. Again you will get no remainder.

   a) Generalize this result.

   *b) Try to explain why this happens.

6. Follow the same procedure as in Example 5, except divide by 7 instead of 13. What happens? Can you generalize this result? Can you explain why this happens?

7. The president of Geometric Progressions of America Inc. is paid a monthly salary (31 days) of $25,000. The stockholders believe that he is overpaid. Being familiar with geometric progressions, he agrees to take an immediate drastic salary cut. He offers to be paid according to the following schedule:

   1¢ first day of month,
   2¢ second day of month,
   4¢ third day of month,
   8¢ fourth day of month,

   that is, each day's salary is double the previous day's salary. The stockholders eagerly agree to this proposal. Is the stockholder's decision a wise one? Explain your answer.

---

3. SOME INTERESTING NUMBERS

In this section, we will discuss some special types of numbers that have fascinated mathematicians over the years because of their unusual properties. We will start with the Fibonacci numbers.

Fibonacci Numbers

Suppose we have decided to breed rabbits. Obviously we must start with a pair, whom we shall call Jack Rabbit and Bunny Rabbit. We will assume the following:

1. Jack and Bunny are newborn when we start.

2. Rabbits begin to reproduce exactly two months after their own birth.

3. Thereafter, every month a pair of rabbits will produce exactly *one* other pair (a male and a female).

4. None of the rabbits die.

Let us see how fast the number of pairs of rabbits increases.

   The first pair, Jack and Bunny Rabbit, will have their first pair of children after two months. This gives us 2 pairs of rabbits.

   After three months, Jack and Bunny will have another pair. We now have 3 pairs.

   After four months, Jack and Bunny will again have another pair. Also, their first pair of children will have *their* first pair. This brings the total to 5 pairs.

Number of months passed	Total number of pairs of rabbits
0	1
1	1
2	2
3	3
4	5
5	8
6	13
7	21
8	34
9	55
10	89
11	144
12	233

This process will continue. The following diagram shows the number of rabbits at the end of each month through the first six months.

End of month		Total pairs
0		1
1		1
2		2
3		3
4		5
5		8
6		13

The total number of pairs of rabbits for the first twelve months is shown in the table at the left.

We see that at the end of the first year, there will be 233 pairs of rabbits! (Maybe we should have started with mink!) If we were to continue this process beyond 12 months, we would get more and more pairs of rabbits.

Now let us examine the numbers in the second column of the table. They are 1, 1, 2, 3, 5, 8, 13, 21, 34, 55, 89, 144, 233. There is something especially interesting about these numbers (even to mathematicians who don't like rabbits). If we look at them carefully, we see that when we add the first two numbers, we get $1 + 1 = 2$. This is the next number. Then $1 + 2 = 3$. This is the next number. And $2 + 3 = 5$. This again is the next number.

In general, we see that if we add any 2 consecutive numbers, we always get the next number in the list. The list of numbers 1, 1, 2, 3, 5, 8, 13, 21, 34, 55, 89, 144, 233, . . . is called a **Fibonacci sequence** in honor of its discoverer. He was Leonardo of Pisa, a remarkable Italian mathematician who lived in the last part of the twelfth century and the early thirteenth century. His father's name was Bonaccus, and Leonardo was nicknamed Filius Bonacci (which is Latin for "son of Bonaccus"). This nickname was shortened to Fibonacci.

Another entertaining application of Fibonacci numbers is to the game of Fibonacci Nim. This game is played by two players. There is one pile of matches. The first player takes one or more matches, but cannot take the whole pile. The next player can take up to twice the number of matches his opponent took, but no more. The first player can now also take up to twice the number of matches his opponent took on the last play, but no more. The play continues in this way. For example, if one player takes 4 matches on any play, his opponent may take as many as 8 matches on the next play, but no more. The game continues until all the matches are gone. The player who takes the last match wins.

It can be shown that if the number of matches in the original pile is a Fibonacci number, then the second player can *always* win if he or she plays

What connection does this picture have with Fibonacci numbers?

correctly. If the number of matches in the original pile is not a Fibonacci number, then the first player can *always* win if he or she plays correctly.

Play this game with a friend and figure out the winning strategy.

Fibonacci numbers can also be applied to the situation shown in the figure on the left.

Suppose the bee wants to go to cell 3. The bee must *always move to the right* and always to a cell right next to it. In how many different ways can he get to cell 3?

One possible way is for him first to go to cell 1 and then to cell 3. We write this as

$1 \rightarrow 3$.

Another possible path is

$0 \rightarrow 2 \rightarrow 3$.

The other possible paths are

$0 \rightarrow 1 \rightarrow 2 \rightarrow 3$,
$1 \rightarrow 2 \rightarrow 3$,
$0 \rightarrow 1 \rightarrow 3$.

Thus, the bee has 5 possible paths by which he can get to cell 3.

If you also calculate the number of possible paths by which the bee can get to cells 0, 1, 2, and 4, you get the results shown at the left.

Notice that the number of possible paths form a Fibonacci sequence (except for the first term).

Cell	Number of possible paths
0	1
1	2
2	3
3	5
4	8

The Fibonacci sequence has many other interesting properties and applications. If you are interested, you can find some of these in the suggested further readings at the end of the chapter.

Perfect Numbers The numbers that divide 6 evenly, excluding 6 itself, are 1, 2, and 3. Notice that $1 + 2 + 3 = 6$. We call a number **perfect** if it equals the sum of all the numbers that divide it and if these numbers are smaller than the number itself.

The diagram above shows two opposite sets of spirals in a daisy—21 spirals in a clockwise direction and 34 counterclockwise. This ratio corresponds to the 21:34 sequence in the Fibonacci series. In pine cones, opposing spirals go 5 one way and 8 the other; in pineapples 8 one way and 13 the other.

Bruce Anderson

Example 1 The number 12 is *not* a perfect number, since the numbers that divide it (that are less than 12) are 1, 2, 3, 4, and 6. We find that

$$1 + 2 + 3 + 4 + 6 = 16.$$

So 12 is *not* equal to the sum of these numbers.

The number 15 is *not* perfect since the numbers that divide it (that are less than 15) are 1, 3, and 5. If we add these numbers, we get $1 + 3 + 5$ which equals 9. Thus 15 is *not* equal to the sum of these numbers.

The number 28 *is* perfect since the numbers less than 28 that divide it are 1, 2, 4, 7, and 14. Adding these, we get $1 + 2 + 7 + 14 = 28$.

It was once believed that perfect numbers had magical properties. In fact, it was believed that God created the world in 6 days because 6 is the smallest perfect number.

The first four perfect numbers are 6, 28, 496, and 8128. No one knows how many perfect numbers there are, although it is likely that there are infinitely many. All the known perfect numbers are even, for no one has yet discovered an odd one. However, there may be odd perfect numbers. It *is* known that every even perfect number ends in either 6 or 8.

There is a formula that will give all the even perfect numbers. It is

$$2^{P-1}(2^P - 1),$$

where P is a prime number (see p. 8 or p. 197), and where $2^P - 1$ is also a prime number.

EXERCISES

1. We have given the first twelve numbers in the Fibonacci sequence. Write the next three.

2. a) Add the first three numbers of the Fibonacci sequence. You get $1 + 1 + 2 = 4$. This is one less than the fifth number, which is 5. Now add the first four numbers. How does this *sum* compare to the sixth number?

   b) Add the first five numbers of the Fibonacci sequence. How does this *sum* compare to the seventh number?

   c) Without actually adding them, can you say what would be the sum of the first six numbers in the Fibonacci sequence?

3. Divide each number of the Fibonacci sequence by 4, and write down the remainders. Do this for the first twenty numbers. Do you see any pattern?

4. In the application of Fibonacci numbers to the bee's path, find all the possible paths by which it can get to cells 5, 6, and 7.

5. Show that 10 is not a perfect number.

6. Write each of the four perfect numbers 6, 28, 496, and 8128 in binary notation. (See Chapter 3, if you have forgotten how to do this.) Do you see a pattern?

---

4. DIVISIBILITY   A basic idea of number theory is that of one number dividing another number evenly. For example, 4 divides 12 evenly, but 4 does *not* divide 11 evenly. We therefore begin with a discussion of what we mean by **divisibility**.

We say that 4 divides 12 evenly because 12 consists of exactly three 4's. To put it another way, $4 \cdot 3 = 12$.

Similarly, we know that 4 does not divide 11, because 11 does not consist exactly of a whole number of 4's. In other words, there is no whole number, m, such that $4 \cdot m = 11$. This leads us to the following definition.

Definition 4.1   *We say that x **divides** y whenever there is a whole number, m, such that $x \cdot m = y$. If no such number m exists, then we say that x does not divide y. If x divides y, then we call x a **divisor** of y.*

Comment   In this definition, we *never* let $x = 0$. Remember, you cannot divide by 0. (See p. 126.) However, y *can* equal 0.

Example 1   7 divides 21 because $7 \cdot 3 = 21$.

9 does *not* divide 21, because there is no whole number, m, such that $9 \cdot m = 21$.

8 divides 200, because $8 \cdot 25 = 200$.

8 does *not* divide 100, because there is no whole number, m, such that $8 \cdot m = 100$.

Example 2 The divisors of 18 are 1, 2, 3, 6, 9, and 18 itself, since each of these divides 18.

Example 3 5 divides 50 and 5 also divides 15. Let us now add 50 and 15. We get 65. How about 65? It is obvious that 5 divides 65 also, since $5 \cdot 13 = 65$. This leads us to the following useful statement about divisibility.

Statement 1 *Suppose x divides y and x also divides z. Then x divides y + z.*

(This just says that if x divides each of two numbers, then it also divides their sum.)

†*Proof of Statement 1* If x divides y, this means that there is a whole number, m, such that

$$y = x \cdot m. \tag{1}$$

Similarly, if x divides z, this means that there is a whole number, n, such that

$$z = x \cdot n. \tag{2}$$

If we now add (1) and (2) together, we see that

$$y + z = x \cdot m + x \cdot n$$
$$= x(m + n) \quad \text{(by the distributive law).}$$

Now $(m + n)$ is the sum of two whole numbers, and is therefore itself a whole number. Why? Let us call it M. Thus, $M = m + n$. Then

$$y + z = x(m + n)$$

or

$$y + z = xM. \tag{3}$$

This last line (3) says that there is a number M such that $x \cdot M = y + z$. By Definition 3.1, we therefore have that x divides $y + z$.

Example 4 4 divides 8 and 4 divides 20. Statement 1 tells us that 4 divides $8 + 20$, or that 4 divides 28.

Example 5 6 divides 30 and 6 divides 12. Statement 1 tells us that 6 divides $30 + 12$, or that 6 divides 42.

Example 6 3 divides 12. 3 does not divide 5. Does 3 divide $12 + 5$, or 17? Obviously not!

The last example leads us to another useful statement.

Statement 2 *If x divides y, and x does not divide z, then x does not divide y + z.*

(This just says that if x divides one part of a sum, and x doesn't divide the second part, then it can't divide the entire sum.)

†*Proof of Statement 2*   This proof is a proof by contradiction, so you should review this technique before reading further (see p. 148).

There are two possibilities:

a) x does divide $y + z$, or

b) x does not divide $y + z$.

Either (a) or (b) *must* be true.

Let us try possibility (a) and suppose it is true. Then x divides $y + z$. By Definition 3.1 this means that there is a number, m, such that $x \cdot m = y + z$. Since we are given that x divides y, then this means that there is a number, n, such that $x \cdot n = y$. Now

$x \cdot m = y + z$

$x \cdot m = (x \cdot n) + z$.   (We just substitute for y.)

Let us subtract $x \cdot n$ from both sides of this equation. We get

$(x \cdot m) - (x \cdot n) = (x \cdot n) + z - (x \cdot n)$

$(x \cdot m) - (x \cdot n) = z$ (The $x \cdot n$'s on the right side cancel out.)

$\qquad x(m - n) = z$.   (Here we used the distributive law on the left side of the equation.)

Now $m - n$ is the difference of two whole numbers, and m is larger than n. (Why?) Thus, $m - n$ is also a whole number. Let us call it M. We then have

$x(m - n) = z$

$\qquad x \cdot M = z$.   (We substituted M for $m - n$.)

But this last line says that there is a number M which, when multiplied by x, will give z. By Definition 3.1 this means that x *divides* z.

However, since we were told that x *does not divide* z, we therefore have a *contradiction*. This means that possibility (a) is wrong. Since possibility (a) cannot be correct, then we conclude that possibility (b), which says that x does *not* divide $y + z$, is correct.

Example 7   6 divides 12. However, 6 does not divide 13. What about the sum of 12 and 13? Clearly, 6 does not divide 12 + 13, or 25.

An interesting application of Statement 2 will be given in the next section.

Tests for Divisibility   There are many situations in which we want to know whether one number divides another evenly. Take the following example. In a southern town, the sanitation department employs 291 men. The department works in 3-man crews. Can these 291 men be evenly divided into 3-man crews? You can answer this question by actually dividing 3 into 291. If you do, you find that 3 goes into 291 exactly 97 times. However, sometimes we do not want to spend the time required to do this. It turns out that we can often answer such questions without actually doing the division, but by using the following **divisibility tests**.

Test for divisibility by 2   *A number is divisible by 2 when the ones digit is 0, 2, 4, 6, or 8.*

Example 1   The number 4308 is divisible by 2 since the ones digit is 8.

The number 23456 is divisible by 2 since the ones digit is 6.

The number 51694 is divisible by 2 since the ones digit is 4.

Test for divisibility by 3   *A number is divisible by 3 when the sum of its digits is divisible by 3.*

Example 2   The number 52341 is divisible by 3 since the sum of its digits is 5 + 2 + 3 + 4 + 1, or 15, and 15 is divisible by 3.

The number 291 is divisible by 3 since the sum of its digits is 2 + 9 + 1, or 12, and 12 is divisible by 3.

Test for divisibility by 4   *A number is divisible by 4 when the number formed by the last two digits is divisible by 4.*

Example 3   The number 5344 is divisible by 4 since the last two digits form the number 44, and 44 is divisible by 4.

The number 6213 is *not* divisible by 4 since the last two digits form the number 13, and 13 is not divisible by 4.

Test for divisibility by 5   *A number is divisible by 5 when the ones digit is 0 or 5.*

Example 4   The number 42805 is divisible by 5 since the ones digit is 5.

The number 28130 is divisible by 5 since the ones digit is 0.

Test for divisibility by 9   *A number is divisible by 9 if the sum of its digits is divisible by 9.*

Example 5   The number 5346 is divisible by 9 since the sum of its digits is 5 + 3 + 4 + 6, or 18, and 18 is divisible by 9.

The number 3289 is *not* divisible by 9 since the sum of its digits is 3 + 2 + 8 + 9, or 22, and 22 is not divisible by 9.

A very useful application of these divisibility tests is in reducing fractions to lowest terms. For example, the fraction $\frac{1470}{21657}$ can be reduced. The divisibility test for 3 shows us that both the numerator and denominator can be divided by 3. Can the fraction be reduced by any other number?

These tests for divisibility can be proved by using divisibility theorems. Consult the suggested readings for these proofs.

EXERCISES   1. Find all the divisors of 36.

2. Find all the divisors of 180.

3. Find all the divisors of 129.

*4. Find all the divisors of 0.

*5. If the sum of the divisors of a number, other than itself, is equal to another number, and *vice versa*, then we say that the numbers are **amicable**, or **friendly**. Show that 220 and 284 are friendly numbers.

*6. If 1184 is one of two numbers of a pair of friendly numbers, find the other number.

7. a) Multiply the numbers 3, 4, and 5 together. Divide the result by 6.

   b) Multiply the numbers 9, 10, and 11 together. Divide the result by 6.

   c) Multiply the numbers 20, 21, and 22 together. Divide the result by 6.

   d) Using the results of parts (a), (b), and (c), can you make a general statement about the product of three consecutive numbers?

8. Test each of the following numbers to see whether it is divisible by 2, 3, 4, 5, or 9.

   a) 586   b) 3413   c) 863   d) 341

   e) 793   f) 4124   g) 987   h) 5430

9. a) Make up a test to determine whether a number is divisible by 6.

   b) Use this test to determine which of the numbers of Exercise 8 are divisible by 6.

10. a) Make up a test to determine whether a number is divisible by 8.

    b) Use this test to determine which of the numbers of Exercise 8 are divisible by 8.

11. Make up a five-digit number that is divisible by 4, 5, and 9.

12. Make up a six-digit number that is divisible by 5, 6, and 9.

13. Make up a six-digit number that is divisible by 5 but not by 4, 6, or 9.

14. Make up a rule for determining whether a number is divisible by 50.

*15. A leap year is a year whose date is divisible by 4. However, century years (that is, years that end in two zeros) are leap years only when their dates are divisible by 400. Which of the following are leap years?

    a) 1986   b) 1622   c) 1404

    d) 2000   e) 1400   f) 1776

*16. Make up a three-digit number that is divisible by the product of its digits.

---

5. PRIME NUMBERS

You will recall from previous chapters that a prime number is any number larger than 1 which is divisible by only itself and 1, assuming we divide only by positive numbers. The first few prime numbers are 2, 3, 5, 7, 11, 13, ... A number that is *not* prime is called **composite**.

Example 1   4 is not a prime number since it can be divided by 2 as well as by itself and 1. It is a composite number. We know 17 is a prime number since the only numbers that divide it are 17 and 1. But 12 is not a prime number. It can be divided by 2, 3, 4, and 6. It is a composite number.

How can we determine whether a number is prime or not? One method by which this can be done was developed by the Greek mathematician Eratosthenes (approximately 276–194 B.C.). His procedure is called the **sieve of Eratosthenes**, and it enables us to find all primes less than a given number.

We will illustrate the technique by finding all the primes up to 50. We write down all the numbers from 1 to 50.

$$
\begin{array}{cccccccccc}
\cancel{1} & \textcircled{2} & \textcircled{3} & \cancel{4} & \textcircled{5} & \cancel{6} & \textcircled{7} & \cancel{8} & \cancel{9} & \cancel{10} \\
\textcircled{11} & \cancel{12} & \textcircled{13} & \cancel{14} & \cancel{15} & \cancel{16} & \textcircled{17} & \cancel{18} & \textcircled{19} & \cancel{20} \\
\cancel{21} & \cancel{22} & \textcircled{23} & \cancel{24} & \cancel{25} & \cancel{26} & \cancel{27} & \cancel{28} & \textcircled{29} & \cancel{30} \\
\textcircled{31} & \cancel{32} & \cancel{33} & \cancel{34} & \cancel{35} & \cancel{36} & \textcircled{37} & \cancel{38} & \cancel{39} & \cancel{40} \\
\textcircled{41} & \cancel{42} & \textcircled{43} & \cancel{44} & \cancel{45} & \cancel{46} & \textcircled{47} & \cancel{48} & \cancel{49} & \cancel{50}
\end{array}
$$

First cross out 1, which is not a prime. Circle 2, which is a prime, and cross out every *second* number after it. Now circle the next uncrossed number, 3, which is a prime, and cross out every *third* number after it. (Some of these will already have been crossed out.)

Circle the next uncrossed number, 5, which is a prime, and cross off every fifth number after it.

Continuing in this manner, we find all the prime numbers up to 50. We find that all the prime numbers have been circled and all the nonprime numbers have been crossed off. Using the same technique, we can find all the prime numbers less than any given number. Of course, if the given number is fairly large, this process may become very long and tiresome.

Now that we know how to find prime numbers, let us look at a nonprime number such as 12. It can be written as 6×2. Notice that 2 is a prime, but 6 isn't. However, 6 can be written as 2×3, and 2 and 3 are both primes. Thus, we can write 12 as

$$12 = 2 \times 3 \times 2,$$

which is a product of prime numbers.

Can 12 be written as a product of primes in a different way from $2 \times 3 \times 2$? The answer is obviously yes. It can also be written as

$$12 = 2 \times 2 \times 3$$

or

$$12 = 3 \times 2 \times 2.$$

However, these are really the same as the original one, $2 \times 3 \times 2$, except for the order. We see, then, that 12 can be written as a product of primes in exactly *one* way, if order is not considered.

Similarly, 20 can be written as a product of primes in only one way (except for order) as $2 \times 2 \times 5$. And 15 can be written as a product of primes in only one way, 3×5.

Actually, every integer greater than 1 (that is, 2, 3, 4, etc.) can be written as a product of primes in exactly one way, if order doesn't count. This fact is called the **Fundamental Theorem of Arithmetic**, and we restate it formally in the following way.

Fundamental Theorem of Arithmetic

Any integer greater than 1 can be written as a product of primes in only one way, except for order.

Example 2

The number 21 can be written as 3×7, which is a product of primes.

The number 18 can be written as $2 \times 3 \times 3$, which is also a product of primes.

The number 100 can be written as $2 \times 2 \times 5 \times 5$, which is also a product of primes.

The number 385 can be written as $5 \times 7 \times 11$, which is also a product of primes.

Many interesting questions can be asked about prime numbers. For example, we may ask: How many prime numbers are there? Is there a formula that tells us which numbers are primes and which are not? In the remainder of this section, we will attempt to answer some of these questions.

Let us begin with the first question: How many prime numbers are there? This was answered over 2000 years ago by Euclid, who *proved* that there are infinitely many prime numbers. He used the method of proof called **proof by contradiction** (see p. 148).

*Euclid's Proof That the Number of Primes is Infinite

There are two possibilities:

Possibility A. There are only a finite (specific) number of primes.

Possibility B. There are infinitely many primes.

Suppose Possibility A is true. Then (if we have enough time), we can list *all* the primes. We let P_1 stand for 2, which is the first prime; P_2 will stand for 3, which is the second prime; P_3 will stand for 5, which is the third prime. Similarly, P_4 will denote the next prime, P_5 the prime after that, and so on. Eventually we will come to the last prime which we call P_n. (We must come to a last prime, since there are only a finite number of them.)

Let us now multiply all these primes together. We get

$$P_1 \cdot P_2 \cdot P_3 \cdot P_4 \cdots P_n.$$

(The dots stand for the primes between P_4 and P_n.)

Let Q be the number we get by adding 1 to this product. Thus

$$Q = P_1 \cdot P_2 \cdot P_3 \cdot P_4 \cdots P_n + 1.$$

What type of number is Q, prime or composite? If Q is *not* a prime number, then it must have divisors, and some of these are primes. (Why?) Let P denote one of these divisors which is a prime. Since P is prime, it is one of the primes P_1, P_2,

$P_3 \ldots P_n$ (since these are all the primes that there are). Therefore, P *must* divide the product $P_1 \cdot P_2 \cdots P_n$. However, we know that P cannot divide 1. (Why?)

Thus we have that P *divides* $P_1 \cdot P_2 \cdots P_n$ and P does not divide 1. By Statement 2 of the last section, P does *not* divide the sum

$$P_1 \cdot P_2 \cdot P_3 \cdots P_n + 1.$$

Since this sum is just Q, we conclude that P *does not divide Q*. But this contradicts our earlier statement that Q was divisible by the prime P.

The contradiction means that Possibility A is wrong. This leaves us with Possibility B, which must be correct. Thus we see that the number of primes is infinite.

Euclid's proof also answers the next question: Is there a *largest* prime number? Clearly, the answer must be no. Why?

The largest number that is definitely known to be prime (at least, up to the time that this book was written) is the number $2^{19937} - 1$. This means it is the number we get when multiplying 2 by itself 19937 times and then subtracting 1 from the result. If you don't believe that this is a prime number, why don't you try to multiply 2 by itself 19937 times and subtract 1 from the result. If you survive this, you may want to convince yourself that this number is prime by trying to find a number that divides it other than itself and 1. (Don't spend too much time on this!) This was actually shown to be a prime number by using a computer.

Obviously, Euclid's proof shows us that this is definitely *not* the largest prime that exists. There is no such number. The number $2^{19937} - 1$ is just the largest number that is known for sure to be a prime.

There is another question that has puzzled mathematicians for a long time, and that has not yet been answered: Is there a formula that tells us which numbers are prime and which are not, a formula that, when numbers are substituted into it, will produce only prime numbers? Many attempts have been made to find such a formula, but none have been successful.

The mathematician Pierre Fermat (1601–1665) believed that he had succeeded in finding such a formula. He thought that the formula $2^{2^n} + 1$ would give only prime numbers, no matter what was substituted for n. Let us try it and see what happens!

If $n = 0$, the formula becomes

$$
\begin{aligned}
2^{2^0} + 1 = 2^1 + 1 \quad &\text{(remember } 2^0 = 1\text{)} \\
= 2 + 1 \quad &\text{(since } 2^1 = 2\text{)} \\
= 3, \quad &\text{which is prime.}
\end{aligned}
$$

If $n = 1$, the formula is

$$
\begin{aligned}
2^{2^1} + 1 = 2^2 + 1 \quad &\text{(since } 2^1 = 2\text{)} \\
= 4 + 1 \quad &\text{(since } 2^2 = 4\text{)} \\
= 5, \quad &\text{which is prime.}
\end{aligned}
$$

Fermat

If $n = 2$, the formula gives

$$2^{2^2} + 1 = 2^4 + 1 \quad \text{(since } 2^2 = 4\text{)}$$
$$= 16 + 1 \quad \text{(since } 2^4 = 16\text{)}$$
$$= 17, \quad \text{which is prime.}$$

If $n = 3$, we get from the formula

$$2^{2^3} + 1 = 2^8 + 1 \quad \text{(since } 2^3 = 8\text{)}$$
$$= 256 + 1 \quad \text{(since } 2^8 = 256\text{)}$$
$$= 257, \quad \text{and this is prime.}$$

If $n = 4$, the formula yields

$$2^{2^4} + 1 = 2^{16} + 1 \quad \text{(since } 2^4 = 16\text{)}$$
$$= 65536 + 1 \quad \text{(since } 2^{16} = 65{,}536\text{)}$$
$$= 65537, \quad \text{and this too can be shown to be prime.}$$

So far, things look good. However, the next "Fermat number" is

$$2^{2^5} + 1 = 2^{32} + 1 \quad \text{(since } 2^5 = 32\text{)}$$
$$= 4{,}294{,}967{,}297,$$

and it has been shown that this number is *not* prime. In fact, it is divisible by 641. Try it!

It has also been shown that Fermat was wrong when $n = 6$. If $n = 6$, $2^{2^6} + 1$ is a composite number.

Another attempt to produce a "prime number generating" formula was made by the Swiss mathematician Leonhard Euler (1707–1783). In addition to being a great mathematician, Euler had an amazing ability to perform calculations in his head. For the last 17 years of his life he was blind. Yet he was able to do accurate calculations to 50 decimal places. His formula was

$$n^2 - n + 41.$$

If we let $n = 1$ in this formula, we get

$1^2 - 1 + 41 = 1 - 1 + 41 = 41$, which is prime.

If $n = 2$, we get $2^2 - 2 + 41 = 4 - 2 + 41 = 43$, which is prime.

If $n = 3$, we get $3^2 - 3 + 41 = 9 - 3 + 41 = 47$, which is prime.

This formula also gives primes if $n = 3, 4, 5$, and up to $n = 40$. However, if $n = 41$, we get

$$41^2 - 41 + 41 = 41^2,$$

which is *not* prime, since it equals $41 \cdot 41$, and thus has 41 as a divisor, in addition to itself and 1.

Euler (Although unable to produce a prime generating formula, Euler did produce a prime number of children, thirteen.)

Euler knew that his formula would not give primes *all* the time. However, he found that it seemed to give primes for about half of the possible values of *n*. No one has been able to prove for certain whether this percentage of primes remains the same as more and more values of *n* are tried.

Up to the present, no formula has been found that will give *only* prime numbers, and no one has come up with a formula that will give *all* the prime numbers.

EXERCISES

1. Write down the first 15 prime numbers.

2. Find two consecutive numbers, both of which are prime.

3. Find three consecutive numbers, none of which is prime.

4. Find four consecutive numbers, none of which is prime.

5. Using the sieve of Eratosthenes, find all the prime numbers up to 100.

6. Write the following numbers as a product of primes:
   a) 48   b) 81   c) 2000   d) 528   e) 305
   f) 360   g) 300   h) 870   i) 630   j) 1200

7. Can a prime number ever end in the digit 6? In 5?

8. The prime number 13 can be written as $4 \cdot 3 + 1$. Can you find three other primes that can be written in the form $4n + 1$, where *n* is any whole number?

9. The prime number 7 can be written as $4 \cdot 1 + 3$. Can you find three other primes that can be written as $4n + 3$, where *n* is any whole number?

10. Find three prime numbers that can be written in the form $2^P - 1$, where *P* is itself a prime number.

11. a) Find two values of *n* for which the formula $n! + 1$ (see p. 249 for the meaning of *n*!) gives primes.

    b) Find two values of *n* for which the formula $n! + 1$ gives composite numbers.

12. Two prime numbers that differ by 2 are called **twin primes**. For example, 3 and 5 are twin primes. Find three other pairs of twin primes.

13. Find a set of three prime numbers that differ from each other by 2. Can you find another such set?

14. Can a prime number ever be perfect? (See Section 2.)

15. a) List all the prime numbers that are less than 50. How many are there?

    b) List all the prime numbers that are less than 100. How many are there?

    There are 168 prime numbers that are less than 1000. There are 303 prime numbers that are less than 2000. There are 430 prime numbers that are less than 3000. How many prime numbers are there that are less than 1,000,000? As of now, the only way to answer this question is to write down the primes and count them. (Good luck!) No one has yet discovered a formula that will tell you how many primes there are that are less than a given number.

16. Pick a number, say, 100. Double it to get 200. There is at least one prime number between 100 and 200. One such prime number is 101. It has been proved that given any number n and its double $2n$, there is always at least one prime number between these two numbers.

  a) Find a prime number between 16 and 32.

  b) Find a prime number between 200 and 400.

  c) Find a prime number between 1000 and 2000.

17. a) Complete the following chart.

Prime number	Prime number + 1	Prime number − 1
5	6	4
7	8	6
11		
13		
17		
19		

  b) Divide each of the numbers in the second and third columns by 6 and consider their remainders (if any).

  *c) Examine the results of part (b). Make a general statement about prime numbers larger than 3. (*Hint:* It involves division by 6.)

STUDY GUIDE   In this chapter, the following ideas about numbers have been introduced.

Number patterns (p. 184)
Arithmetic progression (p. 186)
Geometric progression (p. 186)
Fibonacci numbers and their
  applications (p. 188)
Perfect numbers (p. 192)
Divisibility (p. 192)
Divisor (p. 192)
Divisibility tests (p. 194)
Amicable numbers (p. 196)
Friendly numbers (p. 196)
Prime numbers (p. 196)
Composite numbers (p. 196)
Sieve of Eratosthenes (p. 197)
Fundamental theorem of
  arithmetic (p. 198)
Euclid's proof that the number
  of primes is infinite (p. 198)
Largest prime number (p. 199)

Fermat's formula for primes (p. 199)
Euler's formula for primes (p. 200)
Twin primes (p. 201)

SUGGESTED FURTHER READINGS

Bell, E. T., *Men of Mathematics.* New York: Simon and Schuster, 1937. Chapter 4 contains a biography of Fermat and Chapter 9 contains a biography of Euler.

Bell, E. T., "The Queen of Mathematics." In James R. Newman, ed., *The World of Mathematics*, Vol. I, Part III. New York: Simon and Schuster, 1956. Chapter 6 discusses arithmetic.

Boyer, C., *A History of Mathematics.* New York: Wiley, 1968.

Gardner, M. See the following articles in *Scientific American*: "Calculating Prodigies" (April 1967); "Divisibility Rules" (September 1962); "Fibonacci Numbers" (August 1959, December 1966, March 1969); "Perfect Numbers" (March 1968); "Prime Numbers" (March 1964, March 1968, August 1970); "Unsolved Problems in Elementary Number Theory" (December 1973).

Hogatt, V., *Fibonacci and Lucas Numbers.* Boston: Houghton-Mifflin, 1969. Interesting properties of Fibonacci and other numbers are discussed.

Mathematics in the Modern World (Readings from *Scientific American*). San Francisco: W. H. Freeman, 1968. Article 11 is a biographical sketch of the mathematician Ramanujan and Article 14 is a general survey of the theory of numbers.

Newman, J., "Sirinvase Ramanujan." In James R. Newman, ed., *The World of Mathematics*, Vol. I, Part II. New York: Simon and Schuster, 1956. Chapter 13 contains a biography of the great mathematician Ramanujan.

Ore, O., *Number Theory and its History.* New York: McGraw-Hill, 1948.

Polya, G., "Heuristic Reasoning in the Theory of Numbers." In the *American Mathematical Monthly*, Vol. 66, 1955. Pages 375–384 discuss differences between primes.

Vorob'ev, N. N., *Fibonacci Numbers.* Waltham, Mass.: Blaisdell, 1961.

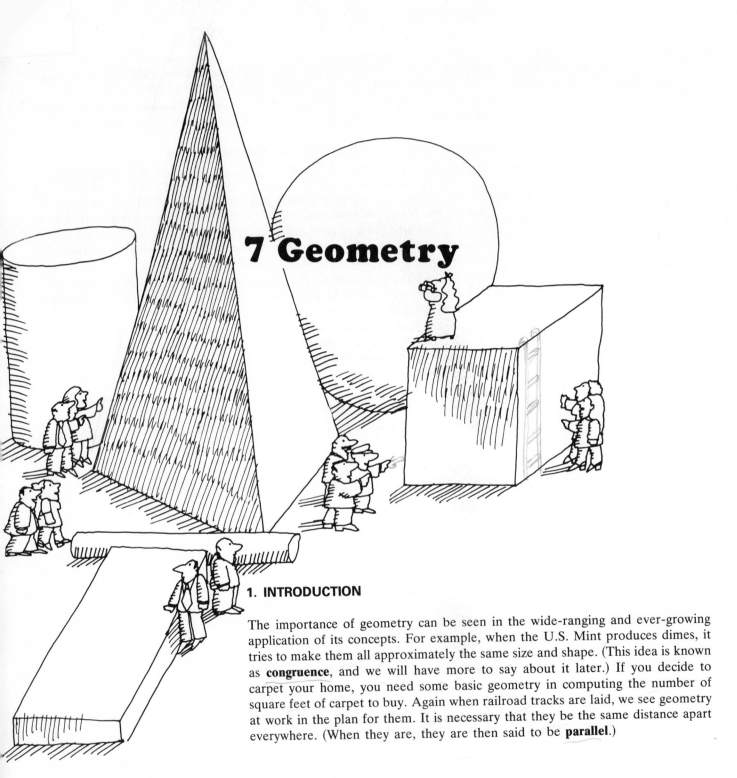

7 Geometry

1. INTRODUCTION

The importance of geometry can be seen in the wide-ranging and ever-growing application of its concepts. For example, when the U.S. Mint produces dimes, it tries to make them all approximately the same size and shape. (This idea is known as **congruence**, and we will have more to say about it later.) If you decide to carpet your home, you need some basic geometry in computing the number of square feet of carpet to buy. Again when railroad tracks are laid, we see geometry at work in the plan for them. It is necessary that they be the same distance apart everywhere. (When they are, they are then said to be **parallel**.)

Because geometry is so useful and necessary, it was one of the first branches of mathematics to be seriously studied. The science of geometry dates back at least to the ancient Babylonians and Egyptians who needed it for the measurement of land for taxation purposes and for building. The precision with which the pyramids were built clearly indicates the Egyptians' skill with geometry. Geometry was also used in studying astronomy, which was, in turn, important in constructing calendars.

The Babylonians and Egyptians developed considerable skill in practical geometry, that is, in calculating areas and volumes, etc. They also knew such facts as certain special cases of the Pythagorean Theorem. However, the geometric knowledge of the Babylonians and Egyptians was just a collection of facts accumulated through practical experience. The Greeks were the first people to undertake a formal study of geometry.

While there were many great Greek mathematicians, one of the best known was Euclid (about 300 B.C.) Although Euclid was a Greek, he was a mathematics professor at the University of Alexandria in Egypt. The story is told of a student who asked Euclid of what use it was to study geometry (a question quite familiar to all mathematics teachers). In reply, Euclid gave the student three pennies, since "he must make gain of what he learns."

Euclid's greatness lies not so much in his discovering new truths of geometry, but rather in his showing that all the known facts could be obtained from a few simple assumptions, using deductive logic. The modern high school geometry course is based upon Euclid's *Elements*.

In this chapter we will discuss some of the basic concepts of geometry, the work of Euclid, and also two other interesting geometries that differ from Euclid's in some of the basic assumptions.

2. POINTS, LINES, AND PLANES

The word **geometry** comes from the Greek *ge* meaning "earth" and *metria* meaning "measurement." Early humans observed that certain shapes, such as triangles, rectangles, circles, surfaces, etc., occurred frequently in nature. The study of geometry began with people's need to measure and understand the properties of these shapes.

Suppose you were asked, "What is a triangle?" You would probably say that a triangle is a figure bounded by three straight lines. This definition depends upon your knowing what a line is. What exactly is a line? Everyone, of course, has an idea of what a line is. You could draw one if asked to do so, but could you describe a line in words?

Line   A line is often said to be a collection of points. However, understanding such a definition depends upon your knowing what a point is. Moreover, the definition does not distinguish between lines and other geometric figures that can be thought of as collections of points. For example, a plane (flat surface) is also a collection of points.

Another definition of a line is that it is the shortest distance between two points. Again this depends upon your knowing what a point is. It also depends upon what is meant by "distance" and how it is measured.

No matter how we try to define a line, we encounter similar problems. Therefore, although we all know what a line is, we do not attempt to define it formally. We accept it as an *undefined term*.

Point   Now let us consider **point**. What is a point? The following definitions have been suggested.

1. A point is a dot.
2. A point is a location in space.
3. A point is something that has no length, breadth, or thickness.

If we analyze these definitions, we see that none of them is really satisfactory. Dots have varying thicknesses and can be measured with a precise instrument (and a magnifying glass). Thus when speaking of a point, we would have to specify what size dot we mean. This would involve measurement and length. Obviously this is not what we want.

We can make similar objections to the other definitions. Therefore, we accept point as another *undefined term.*

In mathematics, as in language in general, when we try to define a term, we find that the definition depends upon other terms, which in turn also have to be defined. Ultimately, we see that we must start with some basic terms which we do not define. These are called **undefined terms**. All other definitions are based upon these undefined terms. In geometry, point and line are undefined terms, as we have stated.

Mathematicians did not always recognize the need for undefined terms. In fact the most famous geometry book of the ancient world, Euclid's *Elements*, opens with twenty-three definitions. Many of these definitions involve other terms that have not been defined. As a result, they do not really define anything. For example, Euclid defines "point" as "that which has no part." However, no definition of what is meant by "part" is given. The modern mathematician recognizes the fact that we cannot define everything, and so we must start with some undefined terms, such as "point" or "line."

Because pictures are so useful in mathematics, a *point* is represented by a dot. The smaller we make the dot, the better it will represent the mathematical idea of a point.

Although *line* (by line, we will always mean a straight line) is undefined, it has certain important properties which we now state.

1. *A line is a set of points in space.* The points of the line are said to lie on the line and the line is said to contain them or pass through them. Points that lie on the same line are said to be **collinear**. In Fig. 1, the line l is drawn with

A computer-drawn geometric figure.

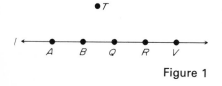

Figure 1

arrows pointing in both directions. This indicates that the line can be extended as far as we like in either direction.

The points A, B, Q, R, and V all lie on the line l. They are collinear. Point T does not lie on the line l.

Notation   Lines will be denoted by small letters such as l, m, n, ... Points are denoted by capital letters. A line containing two points, say A and B, can be denoted as $\overleftrightarrow{AB}$. The line of Fig. 1 can be denoted as line l, $\overleftrightarrow{AB}$, $\overleftrightarrow{AQ}$, $\overleftrightarrow{QV}$, $\overleftrightarrow{AV}$, etc.

2. *Given any two different points in space, there is exactly one line containing these two points.*

3. *Any point on a line divides a line into two parts.* Each of these parts is called a **half-line**. The dividing point is known as the **endpoint** of the half-line. The dividing point is not included in either of the half-lines.

Left half-line P Right half-line

Figure 2

Definition 2.1   *A half-line, together with its dividing point, is known as a **ray**.*

There will be many times when we will be interested in only certain finite parts of a line rather than the whole line or ray. For this reason, we introduce the following definition.

Definition 2.2   *Any two points A and B on a line, together with all the points on the line that lie between them, are called a **line segment**. We denote it by $\overline{AB}$.*

Example 1   The different terms and notations are illustrated below.

Description	Picture	Symbol
Line AB		$\overleftrightarrow{AB}$
Ray AB		$\overrightarrow{AB}$
Ray BA		$\overleftarrow{BA}$
Line segment AB		$\overline{AB}$

A solid dot, ·, means that the point is to be included. An open dot, ∘, means that the point is not to be included.

Notice the difference between ray $\overrightarrow{AB}$ and ray $\overleftarrow{BA}$. Ray $\overrightarrow{AB}$ starts at point A and extends in the direction of point B. Ray $\overleftarrow{BA}$ starts at point B and extends in the direction of point A.

Two different lines that contain the same point are said to **intersect** at the point. Lines l and m of Fig. 3 intersect at point P. Both lines contain this point.

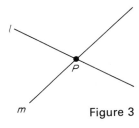

Figure 3

Plane   A **plane** can be thought of as a flat surface, such as the page of this book or the floor in your room. Mathematically, the word *plane* is an undefined term. Planes also have certain important properties.

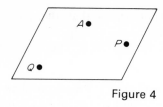

Figure 4

1. *A plane is a set of points.* We say that the points are on the plane and that the plane contains them. Points that are on the same plane are said to be **coplanar**.

2. *Any three noncollinear points determine one and only one plane.* The plane of Fig. 4 contains the three points A, P, and Q. This is the only plane that can contain these three points.

    We have all experienced four-legged tables that wobble because one of the feet is longer or shorter than the other three. This happens because the table can only be steady if all four feet are on the plane of the floor. When one leg is longer or shorter than the others, then the four feet do not lie in the same plane (the floor) and the table wobbles. A three-legged table is much more likely to be steady because it has only three feet and these three feet will always lie on some plane.

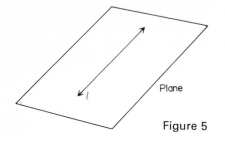

Figure 5

3. *If two points of a line are on a plane, then the whole line is on the plane.*

4. *A line on a plane divides the plane into two parts called half-planes.* The line does not belong to either of the half-planes. In Fig. 5, line l divides the plane into two half-planes.

5. *If we are given two planes, then they either meet in a line or do not meet at all.* Planes that do not meet at all are called parallel planes.

    Figure 6(a) shows two planes intersecting in line l. Figure 6(b) shows two parallel (nonintersecting) planes. The ceiling and floor of your room are examples of parallel planes. The ceiling and walls of your rooms are examples of intersecting planes.

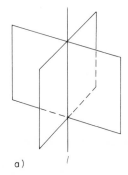

a) l

b)   Figure 6

Parallel Lines   If we are given two lines in the same plane that *never* meet, then such lines are called **parallel lines**. Railroad tracks are an example of parallel lines. The edges of the pages of this book are another example of parallel lines. If line l and line m are parallel, we write this as $l \parallel m$, that is, the symbol "$\parallel$" stands for "is parallel to."

    Parallel lines have played an important role in the development of geometry, as we shall see later.

EXERCISES

1. How many different lines can be drawn connecting three different noncollinear points?

2. How many different lines can be drawn connecting four different noncollinear points that lie on the same plane?

*3. How many different lines can be drawn connecting four different points that do not lie on the same plane, if no three of the points are collinear?

4. Consider the line shown on the left with the indicated points. Find the following.

a) $\overline{AB} \cap \overline{BC}$ b) $\overline{AB} \cup \overline{BC}$ c) $\overleftrightarrow{AB} \cap \overline{CD}$ d) $\overrightarrow{DC} \cup \overrightarrow{DE}$

e) $\overrightarrow{DC} \cap \overrightarrow{DE}$ f) $\overleftrightarrow{AB} \cap \overline{CD}$ g) $\overleftrightarrow{FE} \cap \overleftrightarrow{EA}$ h) $\overline{EF} \cap \overline{AB}$

i) $\overline{BC} \cup \overline{DE}$ j) $\overrightarrow{BC} \cup \overline{EF}$

5. Using the figure shown at the left, find each of the following.

a) $\overline{BC} \cup \overline{CD}$ b) $\overline{BC} \cup (\overline{CD} \cup \overline{BD})$ c) $\overline{BC} \cap \overline{CD}$

d) $\overrightarrow{BC} \cap \overleftrightarrow{AD}$ e) $(\overline{AB}) \cup \overline{BC}) \cup (\overline{CD}) \cup \overline{DA}$ f) $\overrightarrow{BC} \cap \overleftrightarrow{AC}$

g) $(\overline{AB} \cup \overline{BC}) \cap \overline{BD}$ h) $\overline{AD} \cup \overline{DC}$

*6. Is it possible to have two lines on the same plane that are not parallel and that do not intersect? Explain your answer. What if the lines are on different planes?

*7. If l is a line in plane P, and m is a line in plane Q, and plane P is parallel to plane Q, then is line l parallel to line m?

*8. Suppose we are given four different points, A, B, C, and D, which do not all lie on the same plane. How many different planes can be drawn connecting any three of the points if no three of the points are collinear?

3. ANGLES

What is an angle? We can all draw one as shown below.

We can define an angle formally in the following definition.

Definition 3.1 *An angle is the union of two rays that have a common endpoint. The rays are called the **sides** of the angle, and the endpoint is called the **vertex** of the angle.*

In Fig. 7, the angle is formed by the rays $\overrightarrow{AB}$ and $\overrightarrow{AC}$ and the vertex is A. We identify this angle in one of the following ways:

1. By writing $\angle A$. In this notation, $\angle$ stands for angle and A represents the vertex.

2. By writing $\angle BAC$ or $\angle CAB$. In this notation, the vertex is in the middle and the other two points are points on either side of the angle.

3. By putting a "1," a "2," etc., and calling it $\angle 1$ or $\angle 2$, as we have done in Fig. 7.

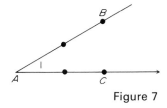

Figure 7

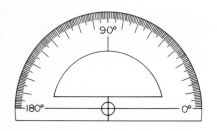

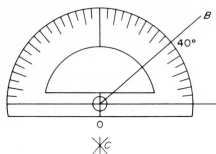

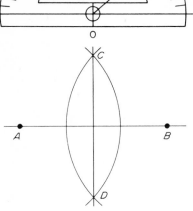

Angles are measured by an instrument called a **protractor**, such as the one pictured on the left. Angles are usually measured in units called **degrees** and these are denoted by the symbol "°." The scale on the protractor is marked from 0 to 180 degrees as shown. To measure an angle, the point marked O on the protractor is placed at the vertex and the 0° line is placed along one of the rays. The size of the angle is determined by the position of the second ray on the protractor. Thus, $\angle AOB$ below measures 40°. We denote this as $m(\angle AOB) = 40°$ where m stands for "the measure of."

Each degree is divided into 60 smaller units known as **minutes**. The symbol for minute is ′. Thus, 1° = 60′. Each minute is further divided in 60 smaller units known as **seconds**, which are denoted as ″. Therefore, we have

$$1° = 60', \text{ and}$$

$$1' = 60''.$$

This system of angular measurement can be traced back to the Babylonians, who used a base-60 number system (see Chapter 3).

Consider the following construction. Draw a straight line $\overleftrightarrow{AB}$ as shown. Now take a compass, place the point on A and draw an arc. Then put the point on B and, keeping the compass open the same amount, draw another arc that intersects the first arc in two places, C and D, as shown. (If they do not intersect, repeat the above procedure using larger compass settings.) Now join the two intersecting points.

Lines $\overleftrightarrow{CD}$ and $\overleftrightarrow{AB}$ form an angle. Measure it! It measures 90°. Such an angle is called a **right angle** and the lines $\overleftrightarrow{AB}$ and $\overleftrightarrow{CD}$ which form the right angle are said to be **perpendicular**.

Definition 3.2   *Two lines are said to be **perpendicular** if the angle at which they intersect is 90°, or a right angle.*

We denote this by using the symbol "⊥." Thus, if line $\overleftrightarrow{AB}$ is perpendicular to line $\overleftrightarrow{CD}$, we write $\overleftrightarrow{AB} \perp \overleftrightarrow{CD}$.

It is convenient to distinguish among the different kinds of angles, depending on their measure. We do this in the following definition.

Definition 3.3   *A **right angle** is an angle whose measure is 90°.*
*A **straight angle** is an angle whose measure is 180°.*
*An **acute angle** is an angle whose measure is between 0° and 90°.*
*An **obtuse angle** is an angle whose measure is between 90° and 180°.*

Example 1   An angle of 35° is an acute angle.
An angle of 138° 12′ 16″ is an obtuse angle.

Comment   According to our definition, an angle is merely a set of points. It consists of the points on the two rays (and the vertex, of course).

Comment   The measurement of an angle as described above requires that the measure of any angle be between 0° and 180°. Why?

It is not always convenient to restrict the measure of an angle to between 0° and 180°.

To extend angular measure beyond the 0°–180° restriction, we can think of an angle as being formed in the following way. Draw any ray and call this the **initial side**. Rotate this side a certain amount and stop. The place where we stop is another ray, which we call the **terminal side**. Some angles formed in this manner are shown here. The arrow indicates the direction and amount of the rotation. In each case, the vertex is at point *A*.

Figure 8

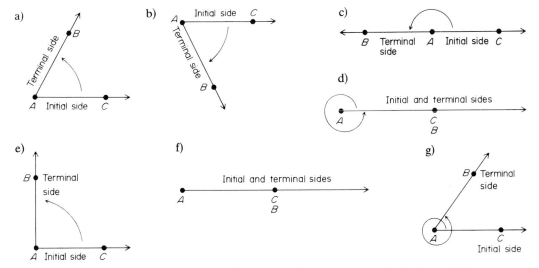

In Fig. 8(a), the rotation is counterclockwise and its measure is considered to be positive.

In Fig. 8(b), the rotation is clockwise and its measure is thought of as being negative.

Notice that the angle of Fig. 8(d) represents one complete revolution and therefore the terminal side is the same as the initial side. It is agreed that in one complete revolution, there are 360 degrees or 360°. This standard of measurement also comes from the Babylonians and was related to their studies in astronomy.

Notice that the angle of Fig. 8(c) contains one-half of a complete revolution, or measures 180°. This conforms to the method first discussed for measuring angles using protractors.

In Fig. 8(g), we have an angle whose measure is larger than 360°. Using rotations, we can have angles of any size, both positive and negative, as shown.

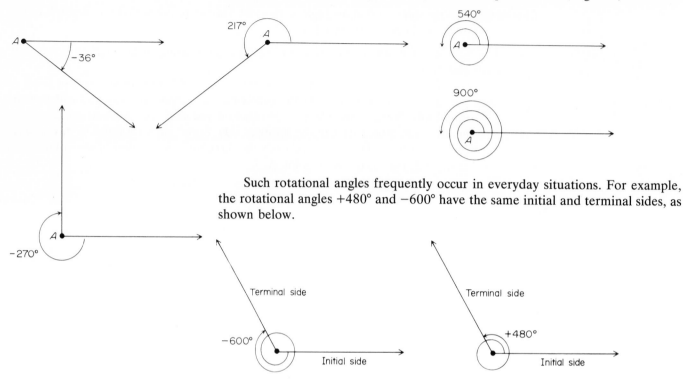

Such rotational angles frequently occur in everyday situations. For example, the rotational angles +480° and −600° have the same initial and terminal sides, as shown below.

If we consider these angles in terms of the initial and terminal sides, Definition 3.1 tells us that the two angles are the same since they are determined by the same rays and the same vertex. However, from the rotational point of view, they are obviously different. Suppose your car is stuck near the edge of a cliff and someone is giving you directions on how to proceed safely. It makes a big difference whether he tells you to rotate the wheels of the car +480° or −600°.

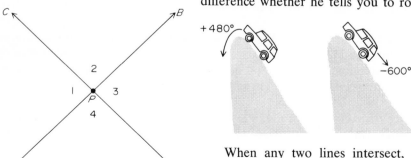

When any two lines intersect, four angles are formed, as shown in the diagram on the left.

Angles 1 and 2 have a common ray, $\overset{\bullet}{\overrightarrow{PC}}$, and we call them **adjacent angles**. Similarly, angles 2 and 3 have a common ray, $\overset{\bullet}{\overrightarrow{PB}}$, and they are also adjacent angles. There are two other pairs of adjacent angles in this diagram. Name them.

In the same illustration, angles 2 and 4 are nonadjacent. We call them **vertical angles**. Similarly, angles 1 and 3 are called vertical angles. This leads us to the following definitions.

Definition 3.4   *Two angles are said to be **adjacent** if they have a common ray.*

Definition 3.5   *When two lines intersect, the nonadjacent angles formed are said to be **vertical** angles.*

We know that we can measure something—for instance, the length of this piece of paper—with standard rulers, of different units. The length can be given in either inches or centimeters. Similarly, weight can be measured in different units—in either pounds or grams. So far in this chapter, we have been using degrees as the unit of measurement for angles, but there is another unit that is often used in mathematics, science, and engineering. This is called the **radian**.

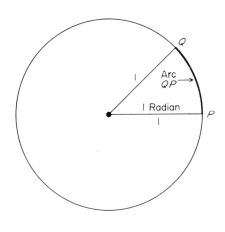

Consider a circle whose radius is 1 in. (a radius is any line drawn from the center of the circle to the circle). Draw a radius in this circle. Now take a piece of string 1 in. long and place it *along the circle* with one end at the point where the radius drawn meets the circle (point P). Put the other end of the string at point Q. Draw another radius from the center to point Q. The arc QP (portion of the circle between points P and Q) obviously has length equal to 1 in. The angle between the two radii (plural of radius) is assigned a measure of one **radian**.

We know that in one complete rotation there are 360°. How many radians are there in one complete rotation? Let us see. It can be shown that if a string were to be placed around the entire circle, it would measure 2π in. As you know, π is an irrational number that cannot be written exactly as a decimal. It is approximately equal to 3.1416.... An arc of 1 in. gives an angle of 1 radian. Therefore an arc of 2π in. (the whole circle) gives an angle of 2π radians.

Thus in one complete rotation there are 2π radians or 360°, depending upon which unit you are using.

$$\boxed{2\pi \text{ radians} = 360°} \tag{1}$$

Since 2π radians is 360°, we can find how many degrees there are in 1 radian by dividing by 2π. We get

$$\frac{\cancel{2\pi}}{\cancel{2\pi}} \text{ radians} = \frac{360°}{2\pi}. \qquad \left(\frac{2\pi}{2\pi} \text{ equals } 1 \right)$$

$$\boxed{1 \text{ radian} = \frac{180°}{\pi}.} \tag{2}$$

Similarly, if we divide (1) by 360, we get

$$\frac{2\pi}{360} \text{ radians} = \frac{360°}{360}.$$

$$\boxed{\frac{\pi}{180} \text{ radians} = 1 \text{ degree.}} \qquad (3)$$

If we use 3.14 as an approximation for π, we get

$$1 \text{ radian} = \frac{180°}{\pi}$$

$$= \frac{180°}{3.14} \qquad \text{which is approximately } 57°.$$

Also,

$$1 \text{ degree} = \frac{\pi}{180} \text{ radians}$$

$$= \frac{3.14}{180} \text{ radians,} \qquad \text{which is approximately } 0.0174 \text{ radians.}$$

Thus 1 radian is approximately 57 degrees, and 1 degree is approximately 0.0174 radians.

Formulas (2) and (3) enable us to convert from degrees to radians and from radians to degrees, as shown in the following examples.

Example 2   Convert $\frac{\pi}{4}$ radians to degree measure.

Solution   We use Formula (2) which says that

$$1 \text{ radian} = \frac{180}{\pi} \text{ degrees.}$$

We multiply both sides by $\frac{\pi}{4}$ and get

$$\frac{\pi}{4}(1 \text{ radian}) = \frac{\pi}{4}\left(\frac{180}{\pi} \text{ degrees}\right)$$

$$\frac{\pi}{4} \text{ radians} = \frac{180}{4} \text{ degrees}$$

$$\frac{\pi}{4} \text{ radians} = 45°.$$

Example 3   Convert 60° to radian measure.

Solution   We use Formula (3) which says that

$$1 \text{ degree} = \frac{\pi}{180} \text{ radians.}$$

We multiply both sides by 60 and get

$$60 \,(1 \text{ degree}) = 60 \left(\frac{\pi}{180} \text{ radians} \right)$$

$$60 \text{ degrees} = \frac{60\pi}{180} \text{ radians}$$

$$60 \text{ degrees} = \frac{\pi}{3} \text{ radians.}$$

EXERCISES   1. Using a protractor, construct angles having the following measures.

    a) 30°   b) 46°

    c) 160°   d) 90°

2. Using a protractor, construct the following rotational angles.

    a) 220°   b) 380°

    c) 450°   d) −200°

    e) −360°   f) −800°

3. In the following diagrams, find all the adjacent angles and all the vertical angles.

    a)   b)

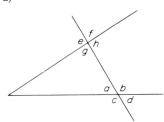   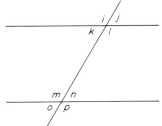

    c)   d)

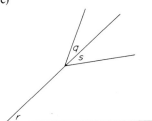   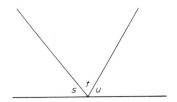

4. Convert each of the following angles to degree measure.

a) $\dfrac{3\pi}{4}$ b) $\dfrac{-7\pi}{8}$ c) 10π d) $\dfrac{11\pi}{6}$

e) 0 f) $-\dfrac{8\pi}{5}$ g) $\dfrac{-\pi}{2}$ h) $\dfrac{4\pi}{3}$

i) $-\dfrac{5\pi}{3}$ j) $\dfrac{\pi}{3}$ k) -6π l) 3

5. Convert each of the following angles to radian measure.

a) $5°$ b) $120°$ c) $-135°$ d) $1040°$

e) $70°$ f) $-125°$ g) $215°$ h) $-75°$

i) $1°$ j) $300°$ k) $-45°$ l) $\pi°$

---

4. CURVES, TRIANGLES, AND POLYGONS

Using straight lines, we can construct figures in a plane which are known as **polygons**. Several examples of polygons are shown below.

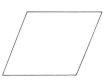

In order to define a polygon formally, we first introduce the idea of a simple closed curve.

Definition 4.1 *A **simple closed curve** is any curve that can be drawn without lifting the pencil* (or *other writing instrument*) *and that has the following properties:*

1. *The drawing starts and stops at the same point.*

2. *No point is touched twice* (*with the exception of the starting point*).

The following examples will illustrate this definition.

a)

b)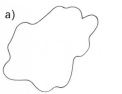

c)

d)

e)

f)

g)

In the figure shown on the previous page, (a), (b), and (c) are simple closed curves. Figure (d) is not a simple closed curve because it does not start and stop at the same point. (It is not closed.) Figure (e) is not a simple closed curve because point P is touched twice in drawing the curve. (To see this, try to draw it.) Figures (f) and (g) are not simple closed curves. Why not?

We can now define what we mean by a polygon.

Definition 4.2   *A **polygon** is a simple closed curve consisting of straight line segments.*

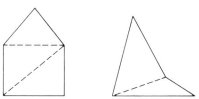

The simplest kind of polygon is a **triangle**, which consists of *three* lines called **sides**. Triangles are very important in the study of geometry since any polygon can be broken up into triangles, such as the figures shown at the left. The Babylonians and Egyptians often used this idea in measuring land that was in the shape of a polygon. They divided it into triangles and measured each part individually.

There are many different kinds of triangles. Three types that are important are equilateral, isosceles, and right triangles. These are defined as follows.

Definition 4.3   *An **equilateral triangle** is a triangle with three sides whose measures are equal.*
*An **isosceles triangle** is a triangle with two sides whose measures are equal.*
*A **right triangle** is a triangle that contains a 90° angle.*

It can be shown that if a triangle is equilateral, then all the angles are equal and will measure exactly 60°.

It can also be shown that if a triangle is isosceles, then the two angles opposite the equal sides are also equal.

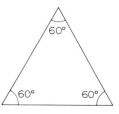

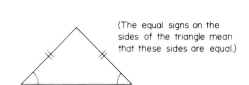

(The equal signs on the sides of the triangle mean that these sides are equal.)

The Pythagorean theorem (see p. 7) states that in a right triangle, such as the one below, $a^2 + b^2 = c^2$.

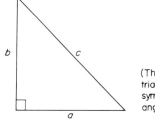

(The right angle of this triangle is indicated by the symbol "⌐" in it. All right angles are similarly marked.)

a)

b)

c)

Dennis Stock, Magnum Photos

d)

Dennis Stock, Magnum Photos

Scala

Rene Burri, Magnum Photos

f)

Geometric shapes in architecture.

a) Transamerica building. Courtesy of Transamerica
   Corporation.
b) Giotto's Campanile, Florence.
c) Solomon Guggenheim Museum.
d) A commune building.
e) Habitat.
f) Pentagon. Courtesy of U.S. Navy.

Consider this triangle.

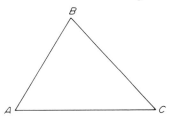

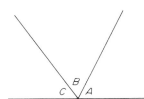

Trace each of the angles on separate pieces of paper and cut them out. Now place the cut-out angles with the vertices (plural of vertex) together, and the sides adjacent to each other, as shown on the left.

Notice that the three angles together add up to a straight angle (180°). This will be true for any triangle. What this means is that the sum of the angles of any triangle is 180°. This is a basic idea in the geometry of Euclid. We will see later that it is not true in non-Euclidean geometries.

EXERCISES

1. Which of the following are simple closed curves?

a)   b)   c)

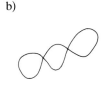

d)   e)   f)

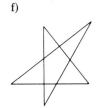

2. Which of the following are polygons?

a)   b)   c)   d)

3. Draw two triangles that intersect in exactly one point.

4. Draw two triangles that intersect in exactly two points.

5. Draw two triangles that intersect in exactly three points.

6. Draw any four-sided polygon (which is known as a **quadrilateral**). Using tracing paper, copy each of the angles and cut them out. Now place the cut-out angles with the vertices together and sides adjacent.

   a) What is the sum of these four angles?

   b) What can you say about the sum of the angles of any quadrilateral?

7. Draw a quadrilateral and a triangle such that the intersection is exactly

   a) one point.   b) two points.   c) three points.

8. State the precise meaning of each of the following terms. (If you do not know, look them up in a dictionary or geometry book.)

   a) parallelogram   b) rhombus   c) rectangle   d) square

   e) trapezoid   f) pentagon   g) hexagon   h) octagon

   i) duodecagon   j) scalene triangle

9. Is an angle a simple closed curve? Explain.

---

5. SIMILAR AND CONGRUENT TRIANGLES

Consider these three triangles.

a) 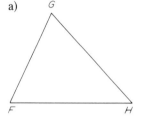 b)   c)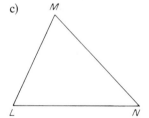

Figure 9

The triangles of Figs. 9(a) and 9(c) have exactly the same size and shape. If you cut out △LMN[1] and place it on top of △FGH (with ∠L falling on top of ∠F, and ∠M falling on top of ∠G), then △LMN will fit *exactly* over △FGH. Each can be considered a carbon copy of the other. We say that these two triangles are **congruent**. The symbol "≅" will stand for the words "is congruent to." Thus we write △LMN ≅ △FGH, and we read this as "triangle LMN is congruent to triangle FGH." Although we have described what we mean by congruent triangles, in modern textbooks on geometry, "congruent" is considered an undefined term. (See p. 206.)

Now consider △FGH and △IJK. Clearly they are not congruent since they are of different size. However, △IJK has exactly the same shape as △FGH. Triangle FGH is an enlarged version of △IJK. Triangles that have the same shape (but not necessarily the same size) are called **similar triangles**.

---

1. The symbol "△LMN" stands for "triangle LMN."

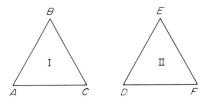

Figure 10

Comment Congruent triangles are also similar triangles because they have the same shape. Similar triangles are not necessarily congruent because, although they have the same shape, their sizes may differ.

Congruent and similar triangles have many important properties which we now state.

In Fig. 10, triangles I and II are congruent. If we compare them, we find that side $\overline{AB}$ is equal in length to the **corresponding** side $\overline{DE}$. The length of a side of a triangle is called its **measure**. Thus the length of side $\overline{AB}$ is called the measure of side $\overline{AB}$ and is denoted by the symbol $m(\overline{AB})$. Similarly, side $\overline{(BC)}$ is equal to corresponding side $\overline{EF}$. Also, side $\overline{AC}$ corresponds to side $\overline{DF}$, and they are equal.

Property 1. When two triangles are congruent, their corresponding sides are equal.

Now measure angle A and the corresponding angle D. They are equal. Similarly, $m(\angle C) = m(\angle F)$ and $m(\angle B) = m(\angle E)$.

Property 2. When two triangles are congruent, their corresponding angles are equal.

Figure 11

In Fig. 11, side $\overline{DE} = 3$ units, side $\overline{EF} = 4$ units, and side $\overline{DF} = 2$ units. Now measure side $\overline{AB}$, $\overline{BC}$, and $\overline{AC}$. You should get 9, 12, and 6 units, respectively. Note that side $\overline{AB}$ is 3 times side $\overline{DE}$. Similarly, side $\overline{BC}$ is 3 times side $\overline{EF}$ and side $\overline{AC}$ is 3 times side $\overline{DF}$. Each side of $\triangle ABC$ is 3 times its corresponding side in $\triangle DEF$. We express this fact by saying that all three sides are **proportional**. In this example, we have

$$\frac{m(\overline{AB})}{m(\overline{DE})} = \frac{m(\overline{BC})}{m(\overline{EF})} = \frac{m(\overline{AC})}{m(\overline{DF})}$$

$$\frac{9}{3} = \frac{12}{4} = \frac{6}{2} = 3.$$

This leads us to the following property of similar triangles.

Property 3. If two triangles are similar, then the corresponding sides are proportional.

Measure angles A and D with a protractor. You should find $m(\angle A) = m(\angle D)$. Similarly, if you measure the other angles, you will find $m(\angle B) = m(\angle E)$ and $m(\angle C) = m(\angle F)$. This leads us to the following property.

Property 4. If two triangles are similar, then the corresponding angles are equal in measure. The converse (see p. 28) of this statement is also true.

Example 1   In the similar triangles at the left, find the measure of the lengths of the unmarked sides.

Solution   Since the triangles are similar, we know by Property 3 that the sides are proportional. Thus,

$$\frac{m(\overline{AB})}{m(\overline{DE})} = \frac{m(\overline{BC})}{m(\overline{EF})}$$

$$\frac{m(\overline{AB})}{3} = \frac{8}{4}.$$

Multiplying both sides by 3, we get

$$3\,\frac{m(\overline{AB})}{3} = 3 \cdot \frac{8}{4}$$

$$m(\overline{AB}) = 3 \cdot \frac{8}{4}$$

$$= 3 \cdot 2 = 6.$$

Similarly, we have

$$\frac{m(\overline{AC})}{m(\overline{DF})} = \frac{m(\overline{CB})}{m(\overline{FE})}$$

$$\frac{m(\overline{AC})}{5} = \frac{8}{4}.$$

Multiplying both sides by 5, we get

$$5\,\frac{m(\overline{AC})}{5} = 5 \cdot \frac{8}{4}$$

$$m(\overline{AC}) = 5 \cdot \frac{8}{4}$$

$$= 5 \cdot 2 = 10.$$

Example 2   In the following similar triangles, find the measure of the unmarked angles.

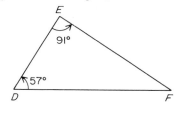

∠ A corresponds to ∠ D
∠ B corresponds to ∠ E

Solution   Since the sum of the angles of a triangle is 180° (see p. 219), we know that

$$m(\angle D) + m(\angle E) + m(\angle F) = 180°$$
$$57° + 91° + m(\angle F) = 180°$$
$$148° + m(\angle F) = 180°.$$

Thus, $$m(\angle F) = 32°.$$

By Property 4, the angles of $\triangle ABC$ must be equal in measure to those of $\triangle DEF$. Thus,

$$m(\angle A) = m(\angle D) = 57°, \; m(\angle B) = m(\angle E) = 91°, \text{ and } m(\angle C) = m(\angle F) = 32°.$$

Example 3   If a lamppost 30 ft high casts a shadow of 40 ft, how high is a pole next to it if its shadow is 10 ft long?

Solution   First we draw a diagram. We will assume that the lamppost and the pole stand vertically and that the sun strikes each at the same angle. These angles are marked with an "*x*" in the diagram. In the diagram, each triangle has two equal angles, the right angle and the angle marked *x*. Thus the third angles are equal. (Why?) Since the triangles have their corresponding angles equal, the triangles are similar. It follows that their sides are proportional. Thus, we have

Shadow of lamppost

$$\frac{m(\overline{DE})}{m(\overline{AB})} = \frac{m(\overline{DE})}{m(\overline{AC})}$$

$$\frac{m(\overline{DE})}{30} = \frac{10}{40}.$$

Multiplying both sides by 30, we get

$$m(\overline{DE}) = 30 \cdot \frac{10}{40} = 7\tfrac{1}{2} \text{ ft.}$$

There are many other interesting and useful properties of congruent and similar triangles, some of which will be discussed in Section 7.

Shadow of pole

EXERCISES   1. In each of the following pairs of similar triangles, find the sides or angles labeled with a question mark.

a)

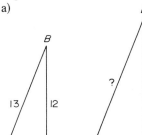

b)

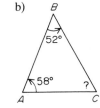   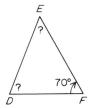

c)   d)

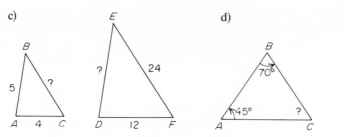

2. a) Are all equilateral triangles congruent?

   b) Are all equilateral triangles similar?

3. a) Are all isosceles triangles congruent?

   b) Are all isosceles triangles similar?

4. The idea of congruent triangles can be extended to any geometric figures. With this in mind, state whether each of the following pairs of figures appears to be congruent.

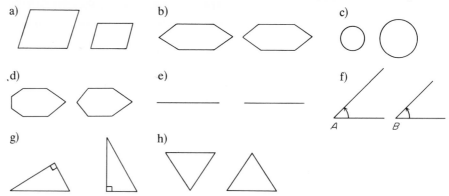

5. Perry Walker has a wallet photo of his girl friend that is 2 in. wide and 3 in. long. He wants to have it enlarged so that it is 18 in. wide. How long will it be?

6. A father and son are walking. If the father's shadow is 4 ft and his 5-ft son casts a shadow of $3\frac{1}{2}$ ft, how tall is the father?

---

6. MEASURE AND AREA   Measure the line segment $\overline{AB}$ with a ruler.

A   B

You should find that it is three in. long. In measuring the line segment, you compared it to the markings on the ruler. Now how did the company that made the ruler know how to mark it? Obviously they used some standard measure, but where did that come from? Units of measurement, whether they be of length, weight, volume, or other measure, have been determined in the following way.

Scientists agree on a basic unit for measuring a particular quantity. For example, in the United States at present, the standard unit of measurement of length is the yard. The "standard yardstick" is kept at the National Bureau of Standards in Washington, D.C. It is made of metal and is maintained at a constant temperature to prevent the metal from expanding. All other yardsticks are copies of this standard yardstick.

Most of the world today uses the metric system of measurement. In this system, the meter is the accepted unit of length. Originally, the meter was defined as one ten-millionth of the distance measured along a meridian through Paris from the North Pole to the equator. A standard meter bar was constructed but in fact it differed slightly (about 0.023%) from its intended length. Copies of the standard meter are kept in the major cities of the world. In 1961, the General Conference of Weights and Measures defined the meter in terms of the wavelength of a particular isotope of krypton. A meter measures approximately 39.37 inches. Similar standard measures have been agreed on for other quantities such as weight, liquid measure, etc. In Appendix A, we show how to convert measurements from the metric system to the "foot-pound" system and vice versa.

In dealing with figures in a plane, we often want to measure quantities other than just length of line segments. For example, we may want to measure the area of a polygon. We all have some idea of what is meant by area. If someone were to ask you to find the area of this page, you could do so with little difficulty. What do we mean by area? How do we measure it?

To answer these questions, imagine that we have a room that measures 9 ft by 6 ft as shown on the left.

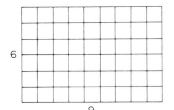

We wish to cover the floor with square tiles that measure 1 ft by 1 ft. It seems reasonable to say that each tile has unit area. If you are measuring length in feet, then the unit used for measuring area is square feet. Thus each of these tiles has area equal to 1 sq ft.

How many of the tiles will we need to cover the floor, assuming that no tiles will overlap? To answer this, we place 9 of these along the longer side of the floor and 6 along the shorter side of the floor, as shown. We continue to place tiles next to each other until the entire floor is covered. We see that it takes 54 tiles to completely cover the floor. Since one tile has an area of 1 sq ft, it makes sense to say that the entire floor will have an area of 54 sq ft. We note that $9 \times 6 = 54$, so we know that the area of the floor is its length times its width.

If we wanted to measure the area of any other rectangle (for example, a polygon with four sides and four right angles), then we could follow the same procedure. In each case, it would lead us to the conclusion that the area is length times width. Thus for rectangles, we agree that

Area of rectangle = length × width.

In symbols, we have

$A = l \cdot w.$

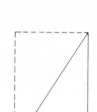

Height

Base

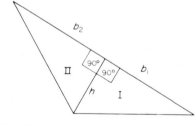

In our previous example, we do not actually have to place tiles on the floor to determine its area. We simply multiply the length by the width, getting 9×6, or 54, sq ft.

How do we measure the area of a right triangle such as the one pictured on the left?

In this case we cannot place square tiles over the triangle and expect to cover it exactly. The unit squares will cover either too much or too little. Instead, we can use the following procedure. Take tracing paper, copy the triangle, and then cut it out. Place the cut-out triangle (the dotted one) alongside the original triangle as shown. We see that the two triangles together make a rectangle whose area we already know how to calculate by using the formula Area = length × width. Since our rectangle is two triangles, we see that the area of the original triangle is half the area of the rectangle. Thus,

$$\text{Area of triangle} = \frac{1}{2}\text{ length} \times \text{width of rectangle.}$$

Note that the length of the rectangle equals the base of the triangle, and that the width of the rectangle equals the height of the triangle. Therefore the area of the triangle can be written as

$$\text{Area of triangle} = \frac{1}{2}\text{ base} \times \text{height.}$$

In symbols,

$$A = \frac{1}{2}\,b \cdot h.$$

Now suppose we are given any triangle (not necessarily a right triangle) and we wish to measure its area. From one of the vertices, we can draw a perpendicular to the opposite side, as shown in Fig. 12. The resulting figure will consist of two right triangles whose area we already know how to calculate. In each case, the area of triangle I is $\frac{1}{2}\,b_1 h$, and the area of triangle II is $\frac{1}{2}\,b_2 h$. Thus, the total area is

$$\frac{1}{2}\,b_1 h + \frac{1}{2}\,b_2 h$$
$$= \frac{1}{2}(b_1 + b_2)h \qquad \text{(by the distributive property)}$$
$$= \frac{1}{2}(\text{entire base})(\text{height})$$
$$= \frac{1}{2}\,b \cdot h.$$

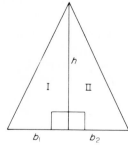

Figure 12

Therefore, the area of *any* triangle is given by $A = \frac{1}{2}bh$, where h is the perpendicular drawn from a vertex to the opposite side, b.

Comment   The area of any polygon can be calculated by dividing it into a collection of triangles, finding the area of each triangle individually, and then adding the areas together.

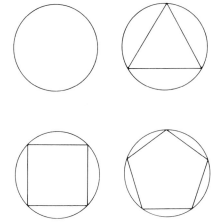

What about figures such as circles that cannot be broken up into triangles? It still seems reasonable to talk about the area of such figures, but what exactly do we mean by it? Let us consider a circle.

We can draw a triangle inside it as shown, which of course does not cover the circle completely.

We can now replace the triangle with a square. This will cover still more of the circle.

Now replace the square with a pentagon (a five-sided polygon). This will cover still more of the circle.

If we continue in this way, drawing polygons with more and more sides in the circle, we find that, each time, we cover more and more of the circle. (We can, of course, find the area of each polygon using the techniques discussed earlier.)

It should be clear that as the number of sides of the polygon increases, the area of the polygon gets closer to the area of the circle. By using this procedure, it can be shown that the area of the circle is equal to πr^2, where r is the radius of the circle.

Comment   Area is one measure of a closed plane figure. Another measure of such a figure is its **perimeter**, which for a polygon is just the sum of the length of its sides. The perimeter of a circle is called the **circumference** and in more advanced mathematics it is shown that the measure of the circumference of a circle is equal to $2\pi r$.

EXERCISES   1. Find the area of each of the following closed plane figures.

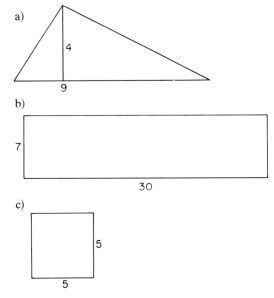

a)

4

9

b)

7

30

c)

5

5

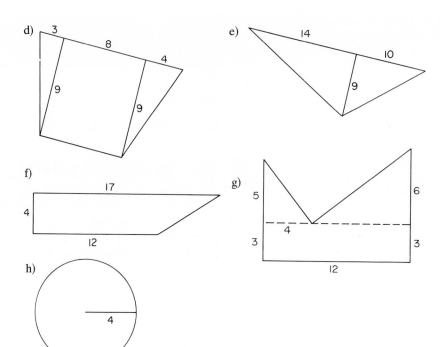

d)

e)

f)

g)

h)

2. Two rooms both have area equal to 150 sq ft. One room measures 15 ft in length and the other measures 12 ft in length. Which room is wider and why?

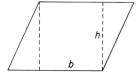

*3. By breaking up a parallelogram, as shown in the figure on the left, try to derive a formula for its area.

*4. A trapezoid is a four-sided figure, two of whose sides are parallel as shown below. By breaking up a trapezoid as indicated, try to derive a formula for its area.

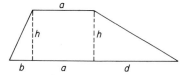

7. EUCLIDEAN GEOMETRY

Euclid's great work was his *Elements*, which was written in approximately 300 B.C. It was a collection of all elementary mathematics, including geometry, and was actually an introductory textbook. When Euclid wrote the *Elements*, all of the geometry in it was already known. His contribution consisted of organizing the available material into a mathematical system.

The *Elements* had an enormous influence on western civilization. For 2000 years, it was considered the best and most thorough example of mathematical reasoning. Early editions of the *Elements* were hand-copied in Greek, Latin, and Arabic, and the first printed version appeared in 1482. Since then, about 1000 different editions have been published. Most of our modern editions of Euclid are based on an edition of Euclid's *Elements* revised by Theon, a fourth-century mathematician. Theon is also known as the father of the first important woman mathematician, Hypatia.

Euclid started (and so we have too) with certain basic terms such as *point* and *line*. However, unlike modern mathematicians, he did not see the need for undefined terms. He believed that it was possible to define all terms, and the *Elements* begins with twenty-three definitions. Unfortunately, these definitions are not adequate because they depend on other terms that have not been previously defined. For example, *point* is defined as "that which has no part." In order to understand this definition, you first have to know what "part" is. Again, *line* is defined as "breadthless length." To understand this, you have to know what "breadth" and "length" mean. These were never defined by Euclid. His definitions are "circular" in the sense discussed in Section 2 of this chapter. Apparently Euclid did not see this, and he considered his definitions satisfactory.

While Euclid did not see that there were certain terms that had to be undefined, he did recognize that there were certain statements that had to be accepted without proof. Euclid intended to give deductive proofs of known mathematical facts such as the Pythagorean theorem. These proofs depend on other facts, which in turn depend on still others, and so on. This could go on forever unless you agree to stop somewhere. Therefore, it is necessary to start with certain statements that must be accepted without proof. These are called **postulates** or **axioms** and Euclid assumed ten. They are sometimes thought of as statements that are self-evident, statements that must be true because they are obvious. There is some indication that Euclid may have regarded at least some of his postulates in this way.

However, it often turns out that "obvious" truths are false. A good example of this is that it was once obvious to almost everyone that the sun revolved around the earth; one could watch it moving through the sky from east to west as the day passed. Today we all know that this obvious truth is false. Furthermore, in recent years scientists and philosophers have shown that it is very difficult to be certain about anything at all. Thus, present-day mathematicians do not claim that the postulates they use are true. They merely say that these postulates are what they are assuming. Postulates are just a starting point. If they turn out to be true statements about the physical world, so much the better. If not, it is still interesting to see what follows logically from them.

Euclid's ten postulates[2] or axioms were the following.

1. A straight line may be drawn connecting any two points.
2. A line segment can be extended indefinitely to form a line.

---

2. See Carl Boyer, *A History of Mathematics.* New York: John Wiley, 1968, pp. 116–117.

3. A circle may be drawn with any center and any radius. (The radius of the circle is a line segment drawn from the center to the circle.)

4. All right angles are equal.

5. Given a line and a point not on the line, then only one line can be drawn parallel to the first line. (Remember, parallel lines are lines which do not intersect.) This version of Euclid's fifth postulate, which was popularized by John Playfair, is the one that appears in many high school geometry texts. There are other versions of this postulate.

6. Things equal to the same things are equal to each other.

7. If equals are added to equals, then the sums are equal.

8. If equals are subtracted from equals, then the differences are equal.

9. Things which coincide with one another are equal to one another.

10. The whole is greater than any of its parts.

Starting from these postulates, Euclid was able to prove deductively (see pp. 7–10) many important and interesting statements called **theorems**. Some of the important theorems are the following.

1. Vertical angles are equal.

2. The base angles of an isosceles triangle are equal.

3. If the base angles of a triangle are equal, then the triangle is isosceles. (This is the converse of Theorem 2.)

4. If three sides of a triangle are equal to three sides of another triangle, then the triangles are congruent (SSS).

5. If two sides and the angle between them of one triangle are equal to two sides and the angle between them of another triangle, then the triangles are congruent (SAS).

6. If two angles and an included side of one triangle are equal to two angles and an included side of another triangle, then the triangles are congruent (ASA).

7. Given a line and a point not on the line, then a perpendicular can be drawn from the point to the line.

8. The sum of the angles of any triangle equals 180°.

9. The Pythagorean theorem.

10. Two different lines intersect in, at most, one point.

Although, as we pointed out earlier, Euclid's work was long considered a perfect example of deductive reasoning, within the past century it has been discovered that his reasoning is often incomplete. These logical gaps occur because he makes certain unstated assumptions based on diagrams. These assumptions cannot always be justified logically. In order to fill in the logical gaps, additional postulates must be introduced. This can and has been done, notably by the German mathematician David Hilbert (1862–1943).

David Hilbert (1862–1943)

One example of the kind of gap that occurs in Euclid's reasoning is the following "proof" that there exists a triangle with two right angles. (Of course this is ridiculous in Euclidean geometry, since the sum of the three angles of a triangle must be 180°. If two angles of a triangle are each 90°, then their sum alone is 180°, and when we add the third angle, the total is more than 180°.)

Proof that there exists a triangle with two right angles

Figure 13

Take two circles that meet in points *A* and *B* as in Fig. 13. Let $\overline{AC}$ and $\overline{AD}$ be their diameters drawn from *A*. (A diameter is a line segment through the center of the circle that bisects the circle). Draw line segment $\overline{CD}$ meeting the circles at points *E* and *F*, as given. It can be shown in Euclidean geometry that ∠*AFC* is a right angle. (Any angle inscribed in a semicircle is a right angle.)[3] Similarly, ∠*AED* is a right angle (because it also is inscribed in a semicircle).

We now have triangle *AEF* with two right angles. What is wrong with this proof?

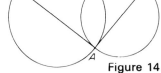

Figure 14

If you draw your own diagram very carefully, you will discover that $\overline{CD}$ passes through point *B* as shown in Fig. 14. Thus we see that points *E* and *F* are exactly the same as point *B*. So triangle *AEF* does not even exist. The carelessly drawn diagram of Fig. 13 was misleading.

You may think that the above proof was rigged and that Euclid himself would never have made such an error. However, many of the proofs in the *Elements* and in high school geometry texts (which are based on Euclid) contain similar faults. The following proof is taken from a geometry text.

Theorem   *The base angles of an isosceles triangle (Fig. 15) are equal.*

Given: *ABC* with $\overline{AC} = \overline{BC}$.
To prove: ∠*A* = ∠*B*.

Proof

Statements	Reason
1. Draw the bisector of ∠*C* this is the line that divides ∠*C* into two equal angles).	1. Every angle has a bisector.
2. Extend the bisector of ∠*C* to meet line segment $\overline{AB}$ at point *D*.	2. A line may be extended.
3. In △*ACD* and △*BCD*, *AC* = *BC*.	3. Given.
4. ∠1 = ∠2.	4. An angle bisector divides the angle into two equal angles.
5. *CD* = *CD*.	5. Anything is equal to itself.
6. △*ACD* is congruent to △*BCD*.	6. Theorem 5 (p. 230) SAS.
7. ∠*A* = ∠*B*.	7. If two triangles are congruent, then the corresponding parts are equal.

Figure 15

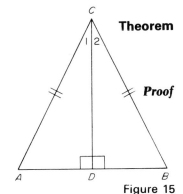

Diameter

3. "An angle inscribed in a semicircle" is an angle such as ∠*A* shown on the left. Its vertex is on the circle and its rays (sides) pass through the endpoints of a diameter.

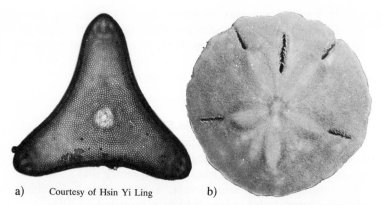

a) Courtesy of Hsin Yi Ling

b)

Richard A. Davis, Jr.

c)

William H. Amos

d)

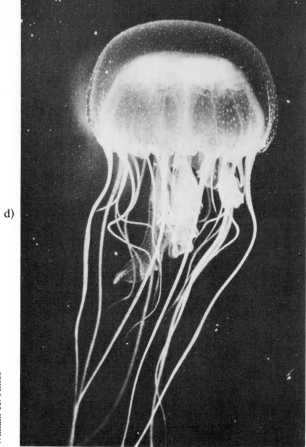

e)

Lee H. Somers

Geometric shapes in nature.

a) Photomicrograph of a diatom.
b) Sand dollar.
c) Sea urchin.
d) Jellyfish.
e) Starfish.

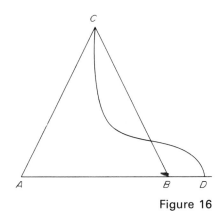

Figure 16

This proof contains an error similar to the error in the previous proof. What is wrong? The problem is in the second step. Our reason for this step is that a line may be extended. This is Postulate 2 (p. 229). However, this postulate does not tell us that when this line is actually extended, it will meet line segment $\overline{AB}$. The diagram certainly suggests that it does. But there is nothing that logically forces us to conclude that the bisector will meet $\overline{AB}$ at point D as shown. In Fig. 16, the bisector does *not* meet line segment $\overline{AB}$.

You may say that straight lines simply do not behave like line $\overline{CD}$ of Fig. 16. But what do you mean by a straight line? If you mean a straight pencil mark on paper or a straight chalk mark on the blackboard, then you are right. However, we are not discussing chalk or pencil marks. We are discussing lines, which have only the properties assumed in the postulates and no others. It does not follow *logically* from the postulates that a line cannot behave as $\overline{CD}$ does in Fig. 16. Mathematicians are concerned with what follows *logically* from the postulates and not with what *appears* to be true in a picture.

Some of Euclid's other proofs contain similar faults. However, by adding suitable postulates to Euclid's original ten, modern mathematicians have been able to correct the faults. Let us not underestimate the work of Euclid because of these gaps. It took 2000 years for critics to discover and correct them.

Euclid's work is an example of what is known today as a **mathematical system**, which we will discuss in detail in Section 9.

8. NON-EUCLIDEAN GEOMETRY

Euclid may have considered some of his postulates as obviously true. There is evidence, however, that Euclid was not entirely convinced of the truth of at least one postulate, namely, the fifth. This is called the parallel postulate and it states:[4]

Euclid's Parallel Postulate

Given a line and a point not on the line, then only one line can be drawn parallel to the first line.

Euclid did not use the parallel postulate in proving theorems until he could not continue any further without it, and it is this apparent hesitation to use Postulate 5 that suggests that he was not completely satisfied with it. Why not? Well, let us look at the postulate closely.

The postulate states that it is possible to construct a line parallel to a given line. This means that the two lines will *never* meet, no matter how far they are extended. Now, it is not humanly possible to go on extending lines forever. So how can we really know that these two lines will *never* meet at some point? Perhaps they would meet billions and billions of miles away from the starting point.

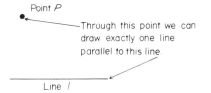

---

4. Euclid stated this postulate differently. The version given here is that of the English mathematician Playfair.

Other mathematicians were also doubtful about the truth of the parallel postulate and for 2000 years after the *Elements* first appeared, there were several attempts made to *prove* this postulate rather than accept it as an unproved statement. However, these attempts all involved other unproved assumptions that were actually the same as the parallel postulate, but were just stated differently.

In the seventeenth century, an Italian monk named Girolamo Saccheri (1667–1733) approached the problem in a new way. He used the method called *proof by contradiction* (see p. 148). Saccheri wanted to prove that Euclid's parallel postulate was true. Obviously there are only two possibilities.

Possibility 1. Euclid's parallel postulate is false.

Possibility 2. Euclid's parallel postulate is true.

He assumed that Euclid's parallel postulate was false (Possibility 1) and tried to arrive at a contradiction. Believing that he had actually reached the contradiction, he concluded that Possibility 2 is correct, that is, that the parallel postulate is true. He published his results in *Euclides Vindicatus*, which means "Euclid vindicated," or "Euclid proved true." Apparently, Saccheri had great faith in Euclid and was very anxious to show that the parallel postulate was true. In fact, Saccheri made an error and did not really obtain a contradiction at all. Thus, he did not prove the parallel postulate as he thought he had.

Saccheri's work is of great interest because he actually proved many theorems in the non-Euclidean geometries developed in the next century by Bolyai and Lobachevsky. So, while he failed to prove Euclid's geometry true, he paved the way for a new and important approach to geometry.

Saccheri's approach, as we have said, was to assume that the parallel postulate was false. If, indeed, it is false, then one of the following two situations is possible.

Possibility 1. Given a line *l* and a point *P* not on *l*, *at least two lines* can be drawn parallel to it.

Possibility 2. Given a line *l* and a point *P* not on *l*, *no* lines can be drawn parallel to *l*.

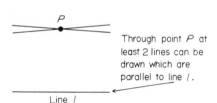

Through point *P* at least 2 lines can be drawn which are parallel to line *l*.

Line *l*

J. Bolyai (1802–1860), a Hungarian army officer, and N. J. Lobachevsky (1793–1856), a Russian mathematician, each independently developed a geometry based on Possibility 1. They used all the postulates of Euclid except the fifth, and in its place they substituted Possibility 1. This is known as the *Lobachevskian parallel postulate.* Then from this new set of postulates, they proceeded to prove theorems. The geometry that they developed in this way is called *Lobachevskian geometry*, and it is different from Euclid's geometry in many startling ways. Some of the theorems of this geometry are as follows.

1. Given a line *l* and a point *P* not on *l*, then through point *P* *infinitely* many lines can be drawn parallel to it (compare this with the Lobachevskian parallel postulate).

2. The sum of the angles of any triangle is *less than 180°*.

3. Different triangles have different angle sums.

4. The sum of the angles of any quadrilateral is *less than 360°* [compare this with Exercise 6(b), p. 220].

5. There are no rectangles. (This follows from the last theorem. Why?)

6. If two triangles are similar, then they must also be congruent. (This means that if two figures have the same shape, then they must also have the same size. Thus, in a Lobachevskian world, miniature or enlarged copies of objects would be impossible to produce without distortion. In order to be accurate, all photographs would have to be lifesize!)

7. Parallel lines are not spaced an equal distance apart.

It is interesting to note that the great German mathematician K. F. Gauss (1777–1855) also developed a geometry that is based on the Lobachevskian parallel postulate, but he did not want to publish it at the time because the ideas of Euclid were so widely accepted.

In 1854, B. Riemann (1826–1866), a German mathematician, introduced a different non-Euclidean geometry. He replaced Euclid's parallel postulate by Possibility 2, which is now called the *Riemann parallel postulate*. When the parallel postulate is replaced by Possibility 2, it is also necessary to give up some of Euclid's other assumptions. There is actually a choice as to which assumptions can be abandoned. You can give up Postulate 1 (p. 229), or you can give up the principle, discussed in Section 2, that a line separates a plane into two half-planes.

The theorems of Riemann's geometry are also surprising and interesting. Some of them are as follows.

1. Parallel lines do not exist.

2. The sum of the angles of any triangle is greater than 180°.

3. There are no rectangles.

4. If two triangles are similar, then they must also be congruent. (This means that if the two figures have the same shape, then they also have the same size.)

5. A line is not separated by a point into two half-lines.

6. Two different lines intersect in *two* points. (This theorem is true only if you make the choice to abandon Postulate 1.)

B. Riemann (1826–1866)

Geometries make statements about physical objects—figures, shapes, areas, distances, etc. Thus they can be used to explain the physical world in which we live. The theorems of Riemannian and Lobachevskian geometry seem very strange to us. On first seeing these theorems, it is natural to think that they cannot possibly be true in the real world. In fact, some of them actually seem to contradict our own experiences. (For example, haven't we all seen rectangles with our own eyes? Yet these don't exist in either Riemannian or Lobachevskian geometry.) It was partly this feeling that convinced Saccheri that he had proved Euclid's geometry to be true.

However, it has been shown that if Euclid's geometry is *logically* correct, then so are Riemann's and Lobachevsky's. Thus none of these three has any "logical superiority." Nevertheless, for 2000 years, Euclid's geometry had been accepted as an absolutely accurate description of the physical world. So it was difficult for mathematicians and scientists to accept the possibility that it might not be correct, and that one of the non-Euclidean geometries might describe the world more accurately. Thus, for a long time, most mathematicians believed that the non-Euclidean geometries were interesting logical works, but could not have any application to the real world.

In this connection, it is interesting to note that the great German mathematician Karl Friedrich Gauss (1777–1855) was actually the first to realize that Euclid's parallel postulate was not necessarily true. He created a non-Euclidean geometry. However, he did not publish his results, partly because he was afraid of being ridiculed.

Gauss tried to test the "truth" of his geometry in the following way. In Euclidean geometry, the sum of the angles of a triangle is *exactly* 180°. In non-Euclidean geometry it is either less than or greater than 180°. (In Gauss's version it was less than 180°.) So he tried to measure the angle sums of triangles to see whether they would turn out to be exactly 180° or less than 180°. Now if you draw a triangle on a piece of paper, measure the angles and add them up, you will find that the angle sum seems to be 180° (if you do it carefully). But measurements are only approximate, no matter how carefully they are made. Even worse, in Gauss's geometry, the smaller the triangle is, the closer the angle sum is to 180°. For example in Gauss's geometry, a small triangle such as the one on the left might have angle sum equal to 179.99999999999999°. This is so close to 180° that it would be impossible to measure any difference between this triangle and one that was exactly 180°. Thus Gauss needed very large triangles. He got them by putting three people on three different mountains. Each one measured the angle between the lines of sight from himself to the other two observers, as shown in Fig. 17.

In this figure, observer 1 measured angle *A*, observer 2 measured angle *B*, and observer 3 measured angle *C*. The sum of these three angles turned out to be 179°59′58″. Did this mean that Gauss's geometry was correct because the result was less than 180°? No, the difference between his result and 180° was too small to be conclusive and might have been due to a measuring error. What was needed was an enormously large triangle, such as is found in astronomy.

In fact, the discoveries of twentieth-century physics gave this kind of support to non-Euclidean geometry. In his work on relativity, Einstein used non-Euclidean geometry and obtained far better results than with Euclidean geometry. That is, predictions made based on Einstein's theory agreed more closely with observed facts if non-Euclidean geometry was used. Then even the strongest

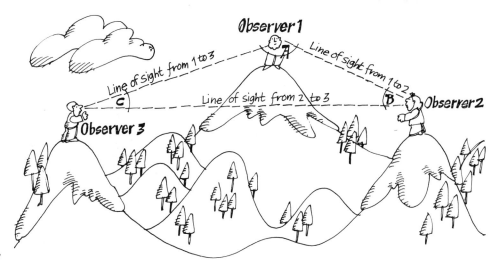

Figure 17

supporters of Euclid had to acknowledge the applicability of non-Euclidean geometry to the real world.

Now the greater usefulness of non-Euclidean geometry in relativity theory is certainly a strong argument for its truth. But it does not *prove* that it *is* true.

Perhaps you are wondering why Euclidean geometry is still taught in all schools and used in engineering and practical applications if it is probably not true. The reason is simple—it works! Why does it work? As we have said with regard to Gauss's triangle experiment, in small areas, there is no significant measureable difference between the results of Euclidean and non-Euclidean geometries. The earth is a very small region when compared to the whole universe. So on earth, Euclidean geometry is as "true" as any of the others for practical use. Since it is familiar and the easiest to use, we do so with perfectly good results.

The discovery and acceptance of non-Euclidean geometry has had an impact even outside mathematics. Most of science uses Euclidean geometry and the doubt about its truth necessarily led to doubt about the truth of the scientific conclusions based on it. In fact, scientists began to doubt whether it was possible to even find absolute scientific truths. The modern view is that such truths are not possible, and that scientific "laws" are just approximate descriptions of the way we see the physical world. When the "absolutely true" geometry of Euclid turned out to be not so true as people had thought, scholars in other areas began to question their "truths." In fact, philosophers began to ask whether we can ever discover truths in general. This has led to a reexamination of what we can "know" in all areas of human knowledge—history, economics, law, ethics, etc. The debate about what we can know is still in progress. It will probably continue for many years to come.

*9. ABSTRACT MATHEMATICAL SYSTEMS

The geometries that we discussed in the last section are all examples of **abstract mathematical systems**.

Definition 9.1

*An **abstract mathematical system** begins with a set of **undefined terms**. Other terms are then defined using these terms. There is also a set of **axioms** (sometimes called **postulates** or **assumptions**) which state certain properties of, and relationships between, the terms. **Theorems** are logically derived from the axioms and definitions. Thus an abstract mathematical system consists of undefined terms, axioms, and theorems.*

Abstract mathematical systems arise in different ways. Sometimes our knowledge about the real world is organized by mathematicians into abstract mathematical systems in order to make it easier to study and use. In this way, Euclid organized the known geometry of his time into such a system. His work was, as we have already pointed out, by no means perfect, and has been revised and improved through the years by others, notably Hilbert.

Many scientific theories, such as Newton's, can be considered as abstract mathematical systems that arise from our study of the physical world.

Non-Euclidean geometries are abstract mathematical systems that developed, not from known facts about the physical world, but rather as a mathematical questioning of Euclidean geometry. It was only later that their applicability to the study of science became apparent.

In Chapter 5, when we studied groups, we were really considering an abstract mathematical system. The undefined terms were "set," "element," etc. The axioms were the four group properties.[5] We did not prove any theorems, but in a course on group theory, as it is called, this would be done. Group theory arose from mathematical studies, rather than those of physical science. But it has been shown to have important applications to many areas of science.

As we indicated in Definition 9.1, an abstract mathematical system has undefined terms whose meaning we do not give. It also has assumed statements (axioms) whose truth we do not prove or even know. Finally, because an abstract mathematical system consists of these undefined terms, axioms, and the theorems based on them, we cannot say that such a system is true or false. For this reason, the English mathematician and philosopher Bertrand Russell (1872–19) once said that mathematics is "the subject in which we never know what we are talking about nor whether what we are saying is true."

Why study these systems at all? If we replace the undefined terms with *concrete interpretations*, then the axioms and theorems become statements about the real world. Thus they may then be considered true or false. Such interpretations of abstract mathematical systems are called **models**. Any one system may have many models. This increases their usefulness. A detailed discussion of systems and models can be found in the suggested further readings for this chapter.

---

5. Actually, in the advanced study of group theory, other axioms may also be introduced.

STUDY GUIDE   In this chapter, the following ideas about geometry were discussed.

Point (p. 206)	Simple closed curve (p. 216)
Undefined term (p. 206)	Triangle (p. 217)
Line (p. 205)	Equilateral, isosceles, and right
Half-line (p. 207)	triangles (p. 217)
Collinear points (p. 206)	Similar and congruent triangles (p. 220)
Ray (p. 207)	Area (p. 225)
Line segment (p. 207)	Euclid's *Elements* (p. 228)
Intersecting lines (p. 207)	Postulates or axioms (p. 229)
Plane (p. 207)	Theorems (p. 230)
Coplanar points (p. 208)	Mathematical system (p. 233)
Half-plane (p. 208)	Non-Euclidean geometry (p. 233)
Parallel lines (planes) (p. 208)	Lobachevskian geometry (p. 234)
Angle (p. 209)	Riemannian geometry (p. 235)
Protractor (p. 210)	Abstract mathematical system (p. 238)
Degree measure (p: 210)	Models (p. 238)
Right angle (p. 210)	
Perpendicular lines (p. 210)	
Straight angles (p. 210)	
Acute angles (p. 210)	
Obtuse angles (p. 210)	
Rotational angle (p. 211)	
Adjacent and vertical angles (p. 213)	
Radian measure (p. 213)	
Polygon (p. 216)	

SUGGESTED FURTHER READINGS

Aaboe, A., *Episodes from the Early History of Mathematics.* New York: Random House, 1964. Chapters 2 and 3 deal with early Greek geometry.

Adler, C., *Modern Geometry*, 2nd ed. New York: McGraw-Hill, 1967.

Adler, I., *A New Look at Geometry.* New York: American Library, 1966.

Bell, E. T., *Men of Mathematics.* New York: Simon and Schuster, 1961. Chapter 14 contains a biography of Gauss; chapter 16 contains a biography of Lobachevsky.

Blumenthal, L. M., *A Modern View of Geometry.* San Francisco: W. H. Freeman, 1961. Chapter 1 discusses the development of non-Euclidean geometry.

Choquet, G., *Geometry in a Modern Setting.* Paris: Hermann, 1961. Section 57 deals with the definition of angles.

Jacobs, H. R., *Geometry.* San Francisco: W. H. Freeman, 1974.

Kline, M., *Mathematics: A Cultural Approach.* Reading, Mass.: Addison-Wesley, 1962. Chapter 6 discusses the nature and uses of Euclidean geometry and chapter 26 discusses non-Euclidean geometries and their significance.

Mathematics in the Modern World (Readings from *Scientific American*). San Francisco: W. H. Freeman, 1968. Article 16 discusses geometry, article 17 discusses projective geometry, and article 26 discusses geometry and intuition.

Rouse-Ball, W. W. *A Short Account of the History of Mathematics.* New York: Dover, 1960.

Sawyer, W. W., *Prelude to Mathematics.* Baltimore, Md.: Penguin Books, 1959. Chapter 6 discusses geometries other than Euclid's and chapter 13 discusses finite geometries.

Schaaf, W., *Our Mathematical Heritage.* New York: Collier, 1966. Pages 115–129 discuss geometry and empirical science.

Synge, J. L., *Science; Sense and Non-sense.* New York: W. W. Norton, 1950. Pages 26–30 contain an imaginary discussion between Euclid and a 12-year-old boy.

Toth, I., "Non-Euclidean Geometry Before Euclid." In *Scientific American* (November 1969). This article suggests that the ancient Greeks knew about non-Euclidean geometry.

Van der Waerden, B. L., *Science Awakening.* New York: Oxford University Press, 1961. Contains a discussion of Babylonian and Egyptian geometry.

Wolfe, H. E., *Introduction to Non-Euclidean Geometry.* New York: Dryden Press, 1945.

Weyl, Hermann, "Symmetry." In James R. Newman, ed., *The World of Mathematics*, Vol. I., Part IV. New York: Simon and Schuster, 1956.

8 Probability

1. INTRODUCTION

Most people start the day by listening to the weather forecast. The announcer may say: "There is a 90% *probability* of rain today." What is meant by this statement? Either it will rain or it won't rain.

We also frequently hear expressions such as "I'll *probably* get an A in this course," or "I'll *probably* call her for a date," or "In all *probability*, you are right." In this chapter, we will be discussing the meaning of probability and how it is used.

The mathematical study of probability can be traced back to the mathematician Jerome Cardan (1501–1576). The illegitimate child of a distinguished lawyer,

Cardan

Cardan became a famous doctor, who treated many prominent people throughout Europe. On various occasions he was also a professor of medicine at several Italian universities. While practicing as a doctor, he also studied, taught, and wrote mathematics.

Although he was extremely talented, Cardan's personality and personal life appear to have been less than perfect. He was very hot-tempered. In fact, he is said to have cut off one of his son's ears in a fit of rage. (His sons seem to have followed their father's example. One of them poisoned his own wife.)

Cardan was also an astrologer. There is a legend that claims that he predicted the date of his death astrologically and to guarantee its accuracy, he drank poison on that day. That's one way of being right!

Cardan suffered from many illnesses that prevented him from enjoying life. To forget his troubles, he gambled daily for many years. His intense interest in gambling led him to write a book on the subject. This work, called *The Book on Games of Chance*, is really a textbook for gamblers, complete with tips on how to succeed in cheating. In this book we find the beginnings of the study of probability.

The development of mathematical probability was further helped along its way by the Frenchman, the Chevalier de Méré. Like Cardan, he was a gambler. He was also an amateur mathematician and was interested in the following problem: Suppose a gambling game must be interrupted before it is finished. How should the players divide up the money that is on the table? He sent the problem to his friend, the mathematical genius Blaise Pascal (1623–1662).

When Pascal received the Chevalier de Méré's gambling problem, he sent it to his friend, the great amateur mathematician Pierre Fermat (1602–1665). The two men wrote to each other on this subject. It is this correspondence that is the starting point for the modern theory of probability. Many other gifted mathematicians were attracted by the work that Pascal and Fermat had begun.

Later, as the subject of statistics developed, it was discovered that a knowledge of probability is essential to the statistician. Today, the use of probability in gambling is just one of its minor applications. The importance of probability lies in its wide range of application to such nonmathematical fields as medicine, psychology, economics, and business, to name a few.

2. COUNTING PROBLEMS

Suppose we toss 2 coins. What are the possible outcomes? There are four possibilities.

Coin 1	Coin 2
Head	Head
Head	Tail
Tail	Head
Tail	Tail

If we let H stand for head and T stand for tail, then the set of these outcomes can be written as {HH, HT, TH, and TT}.

If we were to flip a coin three times, then we would have 8 possible outcomes. These form the set {HHH, HHT, HTH, HTT, THH, THT, TTH, and TTT}.

In these two examples, and in other similar problems, it is rather simple to list and count all the possible outcomes. In other situations there may be so many possible outcomes that it may be impractical or impossible to list all the possibilities. For problems like that, we will introduce an easy rule that can be used. The following examples will illustrate the above ideas.

Example 1   A die (the plural is dice) is tossed once. The possible outcomes are 1, 2, 3, 4, 5, and 6.

Example 2   If two dice are tossed, then there are 36 possible outcomes. These are

1, 1	1, 2	1, 3	1, 4	1, 5	1, 6
2, 1	2, 2	2, 3	2, 4	2, 5	2, 6
3, 1	3, 2	3, 3	3, 4	3, 5	3, 6
4, 1	4, 2	4, 3	4, 4	4, 5	4, 6
5, 1	5, 2	5, 3	5, 4	5, 5	5, 6
6, 1	6, 2	6, 3	6, 4	6, 5	6, 6

Example 3   Four men and five women have signed up for mixed doubles at the No Strings Tennis Club. (In mixed doubles, there are two teams competing against each other, and each team consists of one man and one woman). The men are Stu, Drew, Lou, and Hugh. The women are Nell, Adele, Anabel, Clarabel, and Maybelle. How many different teams can be arranged?

Solution   There are 20 possibilities. These are

Men	Women	Men	Women
Stu	Nell	Lou	Nell
Stu	Adele	Lou	Adele
Stu	Anabel	Lou	Anabel
Stu	Clarabel	Lou	Clarabel
Stu	Maybelle	Lou	Maybelle
Drew	Nell	Hugh	Nell
Drew	Adele	Hugh	Adele
Drew	Anabel	Hugh	Anabel
Drew	Clarabel	Hugh	Clarabel
Drew	Maybelle	Hugh	Maybelle

There are four men and five women. No Strings can select either Stu, Drew, Lou, or Hugh as the man for any team. If they select Stu, then they can select any one of the five women to be his partner. Thus, there are five possible teams on which Stu can be the male partner. Similarly, there are five teams on which Drew can be the male partner, five teams for Lou, and five for Hugh. This makes a total of 4 × 5, or 20, teams. If there were 5 men and 6 women, then each of the 5 men would have 6 possible partners, and there would be 5 × 6, or 30, possible teams.

This leads us to the following useful rule.

Rule   *If one thing can be done in m ways, and if, after this is done, something else can be done in n ways, then there are a total of m · n possible ways of doing both things (in the stated order).*

Blouses	Skirts
Beige	Beige
White	Blue denim
Black	Gray
	Red

Example 4   Sally Swinger is planning to go away for the weekend and is taking with her three blouses and four skirts. Sally will wear any of the blouses with any of the skirts. How many different outfits will Sally have if the colors of the clothes are as shown on the left.

Solution   For each outfit, Sally can select any one of three blouses and any one of four skirts. This gives her a total of 3 × 4, or 12, possible outfits.

Example 5   Assume that we have a deck of cards that consists of only four aces, four kings, four queens, and four jacks. We first select one card from this deck and then, without replacing it, select another. How many different outcomes are there?

Solution   On the first draw, any one of 16 cards may be selected. There are now only 15 cards left for the second draw. This gives us a total of

$16 \times 15 = 240$ possible outcomes.

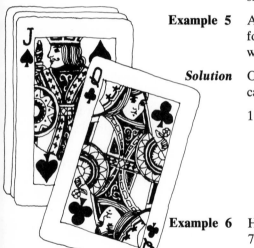

Example 6   How many different three-digit numbers can be formed using only the numbers 5, 7, 8, or 9, if

a) repetitions are allowed?

b) repetitions are not allowed?

c) you can't start with 5, but repetitions are allowed?

Solution   a)   The first digit can be 5, 7, 8, or 9; that is, it can be chosen in four different ways. Similarly (since repetition is permitted), the second digit can be chosen in four different ways. The same is true for the third digit. This gives us

$4 \times 4 \times 4 = 64$

possible three-digit numbers. Notice that we are using the same rule as given above, but we have extended it to three possible things. The same rule can obviously be extended to any number of possible things.

b)   Again there are four different ways of selecting the first digit. Once we select a digit (whatever it is), it can no longer be used. Thus for the second digit, we have only *three* choices. For the last digit there are only *two* choices. Why? Therefore, we have a total of

$4 \times 3 \times 2 = 24$ possible three-digit numbers.

c)   Since 5 cannot be used as the first digit, there are *three* choices for the first digit. However, *any* number, including 5, may be used for both the second and third digits. Thus for each of these, there are four possible choices. This gives us a total of

$3 \times 4 \times 4 = 48$ possible three-digit numbers.

Example 7   In Example 4, Sally Swinger could choose any one of three blouses and any one of four skirts. The solution to this problem can be pictured in a diagram as follows.

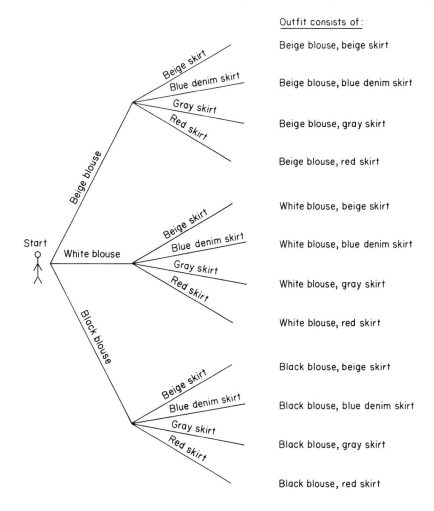

Outfit consists of:

Beige blouse, beige skirt

Beige blouse, blue denim skirt

Beige blouse, gray skirt

Beige blouse, red skirt

White blouse, beige skirt

White blouse, blue denim skirt

White blouse, gray skirt

White blouse, red skirt

Black blouse, beige skirt

Black blouse, blue denim skirt

Black blouse, gray skirt

Black blouse, red skirt

This diagram shows each blouse paired with all possible skirts. Such a diagram is called a **tree diagram**. We construct it as follows. We draw a branch for each blouse. Each branch then breaks up into four smaller branches corresponding to the four skirts. The number of possible outcomes is obtained by counting the total number of smaller branches on the right.

Example 8 A coin is tossed four times. Using a tree diagram, find the total number of possible outcomes.

Solution There are 16 little branches on the right of the diagram as shown below. So there is a total of 16 possible diagrams.

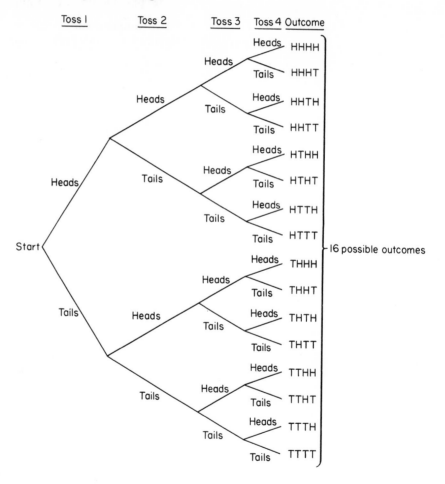

EXERCISES 1. If a coin is flipped 5 times, how many possible outcomes are there? List them.

2. If a die is tossed 3 times, how many possible outcomes are there?

3. Isabel Ringing is getting dressed. She can select any one of 7 dresses, 4 pairs of shoes, and 3 handbags (any of the dresses can be worn with any of the shoes or handbags). In how many different ways can Isabel select an outfit?

4. In how many different ways can the letters of each of the following words be arranged if (a) repetition is allowed; (b) repetition is not allowed? (*Note:* Each arrangement does not necessarily have to form a word.)

   (i) pot   (ii) bust   (iii) shlep

5. The cafeteria at Contaminated College offers the following menu.

Main course	Dessert
Steak	Ice cream
Hamburger	Jello
Hot dogs	Pie
	Chocolate pudding
	Melon

   How many different meals can be chosen from this menu, assuming that exactly one main course and one dessert will be ordered at any given time?

6. Five students are standing in line waiting to register for a math course. In how many different ways can they stand in line?

7. License plates in a midwestern state are of the following type: The first place must be a 9; the second place must be a letter; the third and fourth places can be any numbers except 9 with no repetitions allowed. How many different license plates can be made?

8. A thief is attempting to break into a bookstore to steal copies of this math book. There is a simple combination lock on the door. The lock has four open spaces and for each you rotate a dial so that any letter from A to Z inclusive shows in each space. The lock can be opened only if the correct letter shows in each space. How many combinations does the thief have to try before the lock opens?

9. How many numbers greater than 5000 can be formed from the digits 2, 4, 7, and 8 if no repetitions are allowed?

10. Professor Smith has told his class that they must read a play, a biography, a novel, and a poem. In how many ways can a student do the assignment if he or she must choose from the following list?

   Plays
   Romeo and Juliet
   Our Town
   Major Barbara
   The Crucible
   The Cherry Orchard

   Biographies
   The Confessions of Rousseau
   Soul on Ice
   Lawrence of Arabia
   The Lusts of Casanova

   Novels
   War and Peace
   Slaughterhouse 5
   Steppenwolf
   One Flew over the Cuckoo's Nest

   Poems
   The Ancient Mariner
   Paradise Lost
   The Waste Land

11. A deck of cards contains only the 2, 3, 4, and 5 of spades, clubs, and hearts. In how many ways can two cards be selected without replacement if

    a) any two cards can be chosen?

    b) both cards must be spades?

    c) the first card must be a spade, and the second must be either a club or a heart?

    In this problem, order does not count in parts (a) and (b).

12. Three boys and four girls at Love University have registered with the General Mating Dating Service. The boys are Stan, Dan, and Van. The girls are Mindy, Cindy, Windy, and Lindy. How many different dates can GM arrange?

13. Bruce is at the laundromat doing the family wash. There are vending machines that sell detergents and fabric softeners as listed below.

Detergents	Fabric softeners
Hide	Brownie
Fib	Stay Ruff
Axiom	Old Sof
Hot Power	
Gall	
Gash	

    Bruce is about to purchase a box of detergent and a bottle of fabric softener. In how many different ways can this be done?

---

3. PERMUTATIONS

In some of the exercises of the previous section, order was important, while in others it was not. For example, in Exercise 6 order *was* important. On the other hand, in Exercise 3 order was not important. This leads us to the useful idea of **permutations**. We state this as a definition.

Definition 3.1   *A **permutation** is any arrangement of objects **in a certain order.***

Example 1   How many permutations are there of the letters in the word "sex?"

Solution   There are 6. They are {sex, sxe, xse, xes, exs, esx}.

Example 2   How many different 3-letter permutations can be formed using the letters of the word "drug?"

Solution   There are 24. They are

drg	dgu	grd	gud	rug	rdg	urg	udr
dgr	dru	gdr	gru	rgu	rdu	ugr	ugd
dug	dur	gdu	gur	rgd	rud	urd	udg

In the previous example we were interested in the number of possible permutations of 3 things that can be formed out of a possible 4 things. The symbol we use for this is $_4P_3$. We read this as "the number of permutations of 4 things taken 3 at a time."

If we were interested in the number of possible permutations of 2 things that can be formed out of a possible 4 things, then we would write this as $_4P_2$. More generally, we have the following.

Notation   The symbol $_nP_r$ means the number of permutations of n things taken r at a time. The symbol $_nP_n$ means the number of permutations of n things taken n at a time. This, of course, simply represents the number of different ways of arranging these n things.

There is a simple formula that allows us to calculate $_nP_r$ for any values of n and r. Before giving this formula, we introduce the symbolism $n!$, read as "n factorial."

For example, $4!$, read as "4 factorial," means $4 \cdot 3 \cdot 2 \cdot 1$. Thus

$$4! = 4 \cdot 3 \cdot 2 \cdot 1 = 24.$$

Also,

$$5! = 5 \cdot 4 \cdot 3 \cdot 2 \cdot 1 = 120,$$

and

$$7! = 7 \cdot 6 \cdot 5 \cdot 4 \cdot 3 \cdot 2 \cdot 1 = 5040,$$

$$1! = 1.$$

The symbol $0!$ is taken to be equal to 1.

Now we are ready for the formula for the number of permutations of n things taken r at a time.

Formula 1

$$_nP_r = \frac{n!}{(n - r)!}$$

Example 3   Find $_5P_3$.

Solution   The symbol $_5P_3$ means the number of permutations of 5 things taken 3 at a time. Using the above formula, we have $n = 5$ and $r = 3$, so that we get

$$_5P_3 = \frac{5!}{(5 - 3)!}$$

$$= \frac{5!}{2!}$$

$$= \frac{5 \cdot 4 \cdot 3 \cdot 2 \cdot 1}{2 \cdot 1}$$

$$= \frac{5 \cdot 4 \cdot 3 \cdot \cancel{2} \cdot \cancel{1}}{\cancel{2} \cdot \cancel{1}}$$

$$= 5 \cdot 4 \cdot 3 = 60.$$

Thus, $_5P_3 = 60$.

Example 4   Find $_6P_2$.

Solution   The symbol $_6P_2$ means the number of permutations of 6 things taken 2 at a time. Using Formula 1, we see that $n = 6$ and $r = 2$. Thus,

$$_6P_2 = \frac{6!}{(6-2)!}$$

$$= \frac{6!}{4!}$$

$$= \frac{6 \cdot 5 \cdot 4 \cdot 3 \cdot 2 \cdot 1}{4 \cdot 3 \cdot 2 \cdot 1}$$

$$= \frac{6 \cdot 5 \cdot \cancel{4} \cdot \cancel{3} \cdot \cancel{2} \cdot \cancel{1}}{\cancel{4} \cdot \cancel{3} \cdot \cancel{2} \cdot \cancel{1}}$$

$$= 6 \cdot 5 = 30.$$

Therefore $_6P_2 = 30$.

Example 5   In Example 2, we found all the 3-letter permutations of the 4-letter word "drug." There were 24 of them. We could have used Formula 1 to obtain this answer. Since we have 4 letters to start with, $n = 4$. We are selecting 3-letter permutations. So $r = 3$. Thus we want $_4P_3$ which equals

$$\frac{4!}{(4-3)!} = \frac{4!}{1!}$$

$$= \frac{4 \cdot 3 \cdot 2 \cdot \cancel{1}}{\cancel{1}}$$

$$= 24.$$

This confirms our previous answer, which we got by just listing the permutations.

Example 6   Betty Butterfingers has just typed 5 letters and 5 envelopes. Before she can insert the letters into the envelopes, she drops them on the floor and they get all mixed up. When Betty picks them up, she inserts the letters into the envelopes without looking at them. In how many different ways can this be done?

Solution   When Betty picks up a letter, she has to select 1 out of 5 envelopes into which to put it. Thus $n = 5$ and $r = 5$. The total number of different ways that she can do this is

$$_5P_5 = \frac{5!}{0!}$$

$$= \frac{5 \cdot 4 \cdot 3 \cdot 2 \cdot 1}{1} \quad \text{(Remember } 0! = 1.\text{)}$$

$$= 120.$$

Betty can then insert the letters into the envelopes in 120 different ways.

GURU   UURG
GRUU   UUGR
GUUR   URGU
UGRU   RGUU
URUG   RUGU
UGUR   RUUG

Next we consider a slightly different permutation problem. How many different 4-letter words can be formed from the word "GURU?" Since there are two U's and we cannot tell them apart, Formula 1 has to be changed somewhat. Let us first list all the possible permutations. There are twelve of them, as shown on the left.

Had we used Formula 1, we would have obtained

$$_4P_4 = \frac{4!}{(4-4)!} = \frac{4!}{0!}$$

$$= \frac{4 \cdot 3 \cdot 2 \cdot 1}{1} \qquad \text{(Remember } 0! = 1.\text{)}$$

$$= 24.$$

Why did we get only 12 when Formula 1 gives 24? A little thought shows us that since we cannot tell the two U's apart, half of the 24 permutations of the formula will be repetitions. We therefore do not count them. So we end up with half of 24, or 12, different permutations. For example, if we label the two U's as U_1 and U_2, then two possible permutations given by Formula 1 are U_1U_2GR and U_2U_1GR. However, we cannot tell these apart (since when writing these, we do not really label the U's with 1 and 2). Thus we count these two possibilities as just one permutation.

This example leads us to the following formula for the number of permutations of n things, when some of them are alike.

Formula 2

> Suppose we have n things of which p are alike, q are alike, r are alike, etc. Then the number of different permutations is
>
> $$\frac{n!}{p!\,q!\,r!\ldots}.$$
>
> (It is understood that $p + q + r + \cdots = n$.)

Example 7   How many different permutations are there of the word (a) Shnook? (b) Tennessee?

Solution   a) Since "shnook" has 6 letters, then $n = 6$. The "o" is repeated twice, so $p = 2$. Formula 2 then tells us that the number of permutations is

$$\frac{6!}{2!} = \frac{6 \cdot 5 \cdot 4 \cdot 3 \cdot 2 \cdot 1}{2 \cdot 1}$$

$$= \frac{6 \cdot 5 \cdot 4 \cdot 3 \cdot \cancel{2} \cdot \cancel{1}}{\cancel{2} \cdot \cancel{1}} = 360.$$

There are 360 permutations.

b) "Tennessee" has nine letters, so n is 9. There are 4 e's, 2 n's, and 2 s's, so p is 4, q is 2, and r is 2. Formula 2 tells us that the number of permutations is

$$\frac{9!}{4!\,2!\,2!} = \frac{9 \cdot 8 \cdot 7 \cdot 6 \cdot 5 \cdot 4 \cdot 3 \cdot 2 \cdot 1}{4 \cdot 3 \cdot 2 \cdot 1 \cdot 2 \cdot 1 \cdot 2 \cdot 1}$$

$$= \frac{9 \cdot \overset{2}{\cancel{8}} \cdot 7 \cdot 6 \cdot 5 \cdot \cancel{4} \cdot \cancel{3} \cdot 2 \cdot \cancel{1}}{\cancel{4} \cdot \cancel{3} \cdot 2 \cdot \cancel{1} \cdot 2 \cdot 1 \cdot 2 \cdot \cancel{1}}$$

$$= 3780.$$

So there are 3780 permutations.

EXERCISES

1. Evaluate each of the following symbols.

a) $3!$ b) $\dfrac{7!}{5!}$ c) $\dfrac{23!}{23!}$ d) $\dfrac{0!}{1}$

e) $\dfrac{1}{0!}$ f) $\dfrac{8!}{5!\,3!}$ g) $\dfrac{8!}{3!\,5!}$ h) $\dfrac{8!}{4!\,4!}$

i) ${}_6P_4$ j) ${}_5P_3$ k) ${}_2P_2$ l) ${}_4P_0$

m) ${}_6P_5$ n) ${}_9P_9$ o) ${}_6P_1$ p) ${}_0P_0$

2. How many different permutations are there of the letters in each of the following words?

a) grass   b) gonorrhea   c) bikini

d) birdbrain   e) bubblehead   f) differentiate

3. In how many different ways can the police department of Metropolis arrange suspects in its lineup if each lineup consists of six people?

4. Sergeant Splendid enters the barracks where there are ten soldiers resting. He needs four "volunteers," one to mop the floors, one to peel potatoes, one to scrub the walls, and one to wash dishes. In how many ways can he get his group of volunteers?

5. In how many different ways can the manager of a baseball team arrange his batting order of nine players if

a) any player can bat in any position?

b) the pitcher must bat in the ninth position?

6. Smokey is lost in the woods and has 6 flags with which to signal for help. If each message consists of 4 flags hung in a row, how many different messages can he send?

7. Many people are standing in a cashier's line. The manager decides to start a new line at another cash register. Seven people rush over to the new register. In how many different ways can these 7 people line up?

8. Hy Pocondriac is arranging the bottles in his medicine cabinet. On one shelf he is going to put three different bottles of pain relievers and four different bottles of cold remedies. In how many different ways can he do this if

a) the bottles can be arranged in any order?

b) the pain relievers are to be placed together and the cold remedies are to be placed together?

9. At a certain school, students are given five qualities a teacher should have and are asked to rank them in order of importance. In how many different ways can this be done?

10. At a wedding, a photographer wishes to take a picture of the bride's and groom's parents together. In how many different ways can they line up for the picture?

11. A daily feature in a leading newspaper gives the reader a scrambled 6-letter word that the reader must unscramble to make a meaningful word. How many different permutations are there of the six letters?

12. In the Miss America contest, there are five finalists who are rated by the judges as winner, first runner-up, second runner-up, third runner-up, and fourth runner-up. If 5 girls are picked as the finalists, in how many different ways can they finish?

13. How many different 7-digit telephone numbers are possible if

a) the first digit cannot be a 1, but there are no other restrictions?

b) no repetitions are allowed and there are no other restrictions?

c) there are no restrictions at all?

d) the first digit must be a 9 with repetitions allowed?

14. At a straight banquet table, a guest speaker and 7 other banquet guests of honor are to be seated. In how many different ways can this be done if

a) anyone can sit anywhere?

b) the guest speaker must sit in the middle?

c) the guest speaker must sit in the middle and another guest who has to leave early must sit at the extreme left?

15. In a certain country, license plates start either with the digit 5 or 6. This is followed by two letters and then two numbers. How many different license plates can be issued if

a) repetitions are allowed?

b) no repetitions are allowed in the last 2 digits?

c) no repetitions are allowed at all?

16. The Monotone Symphony Orchestra has rehearsed nine different musical selections. They are giving several concerts, each consisting of four different pieces. How many different programs are possible? (Assume that order counts.)

17. I. M. Nasty, a leading movie critic, is asked to list, in order of preference, the ten best movies that he has seen during 1976. If he has seen 50 movies during the year, in how many different ways can he select the 10 best movies?

4. COMBINATIONS   Suppose Mike is in a record shop. He has enough money to buy only 3 records by the latest popular singing group, the Rockheads. The store has 5 different records by this group. In how many ways can Mike make his selection?

In this situation, we are again interested in selecting 3 out of 5 things. However, this time we are *not* interested in the order in which the selection is made. We call a selection of this kind a **combination**.

Definition 4.1   *A **combination** is any selection of things where the order is not important.*

Notation   The number of combinations of n things taken r at a time will be denoted as $_nC_r$.

Let us go back to Mike in the record shop. He must select 3 out of 5 records. This can be done in $_5C_3$ ways. Let us try to calculate $_5C_3$. If the records are labeled as A, B, C, D, and E, and if order counts, then there are $_5P_3$ possible ways of selecting 3 records out of a total of 5. This gives

$$_5P_3 = \frac{5!}{(5-3)!}$$

$$= \frac{5!}{2!}$$

$$= 60.$$

This figure takes order into account, since it is the number of permutations. In our case, we do not care about the order. Thus if he selects records A, B, and C, then all of the following permutations represent the same purchase: *ABC, CAB, ACB, BAC, BCA, CBA.* These six permutations are thus considered *one* combination. The same is true for any other combination of 3 records. Therefore, to get the correct number of combinations, we divide the 60 by 6 and obtain 10. Notice that 6 is 3! Thus $_5C_3$ is

$$\frac{_5P_3}{3!} = \frac{5!}{(5-3)!\,3!}$$

$$= 10.$$

Thus, we have the following formula.

Formula 3

$$_nC_r = \frac{n!}{(n-r)!\,r!}$$

The following examples will illustrate how the formula is used.

Example 1   Eight workers at the Broken Record Music Corporation are unhappy about their working conditions. They wish to complain to the management. If management will listen to a committee of only 3 people, in how many ways can such a committee be formed?

Solution   Since the order of selecting people for the committee is not important, the answer is the number of combinations of 8 things taken 3 at a time. We thus want $_8C_3$ which is

$$_8C_3 = \frac{8!}{(8-3)!\,3!} = \frac{8!}{5!\,3!}$$
$$= \frac{8\cdot7\cdot6\cdot5\cdot4\cdot3\cdot2\cdot1}{5\cdot4\cdot3\cdot2\cdot1\cdot3\cdot2\cdot1}$$
$$= 56.$$

Thus, 56 different committees can be formed.

Example 2   Tom Cartel is going to Raskin-Bobbins to buy 4 pints of ice cream. Raskin-Bobbins sells 28 different flavors of ice cream, and the smallest amount they will sell of any one flavor is one pint. Tom wants to try as many different flavors as he can. In how many different ways can he buy 4 different flavors?

Solution   Since order is not important, we want $_{28}C_4$. Formula 3 tells us that this is

$$\frac{28!}{(28-4)!\cdot4!} = \frac{28!}{24!\cdot4!} = 20{,}475.$$

Thus, Tom can select the 4 different flavors in 20,475 ways.

Example 3   In how many different ways can a committee of 3 men and 4 women be formed from a group of 8 men and 6 women?

Solution   We must select any 3 men from a possible 8, and order does not matter. This can be done in $_8C_3$ ways.

Then we must select any 4 women from a possible 6, again where order does not count. This is $_6C_4$.

Since any group of men can be combined with any group of women to form the entire committee, we have a total of

$$_8C_3\cdot{_6C_4} = \frac{8!}{(8-3)!\,3!}\cdot\frac{6!}{(6-4)!\,4!}$$
$$= \frac{8!}{5!\,3!}\cdot\frac{6!}{2!\,4!}$$
$$= 56\cdot15 = 840.$$

Thus, 840 committees can be formed.

Another useful technique for computing the number of possible combinations is **Pascal's triangle**. The triangle is shown on the left. It is not hard to see how this triangle is constructed. Each row has a 1 on either end. All the other entries

```
              1
           1     1
        1     2     1
     1     3     3     1
   1    4     6     4     1
 1    5    10    10    5     1
```

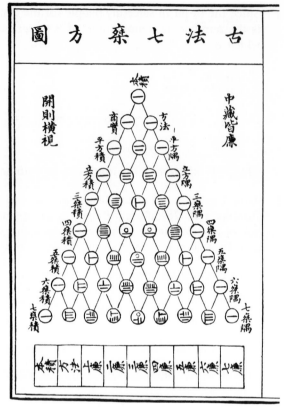

The Pascal triangle as depicted in 1303 at the front of Chu Shih-Chieh's *Ssu Yuan Yii Chien.* It is entitled "The Old Method Chart of the Seven Multiplying Squares" and tabulates the binomial coefficients up to the eighth power.

are obtained by adding the numbers immediately above it directly to the right and left as shown below by the arrows in the diagram. Thus to get the entries for the sixth row we add 1 and 5 to get 6.

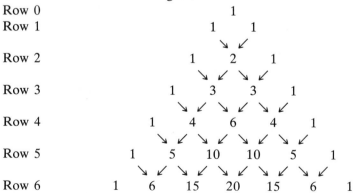

Then we add 5 and 10 to get 15. We next add 10 and 10 to get 20, and so on. To complete the row, we add 1's on each end. The numbers must be lined up exactly as shown in the diagram.

A triangle of numbers such as this is called Pascal's triangle. It can have as many rows as you want. This triangle was known to the Chinese for several centuries before Pascal's time. However, it is named for Pascal because of the many interesting applications he found for it.

Blaise Pascal had demonstrated his mathematical talent at an early age. He proved a very important theorem in geometry when he was only 16 years old. His interest in mathematics was not only in geometry. When he was about 18 years old, he built the first successful computing machine. He also did valuable work in physics, notably confirming the fact that air has weight.

Throughout his life, Pascal suffered from severe illness and hardly a day passed without pain. He was deeply religious and when he was almost killed by a runaway horse in 1654, he regarded his narrow escape as a sign from God. As a result, he devoted himself even more than ever to religious meditation and writing. His great work is *Pensées*, which deals largely with philosophy and religion. Pascal died in 1662 at the age of only 39. He is famous both as a mathematician and philosopher.

Let us now see how Pascal's triangle can be used to solve problems in combinations.

Example 4   In a recent mining accident, a volunteer rescue squad consisting of 3 people was needed. Seven people volunteered. In how many different ways could a rescue squad be formed?

Solution   Since order is not important, we want $_7C_3$. First let us evaluate $_7C_3$ by using Formula 3. We have

$$_7C_3 = \frac{7!}{(7-3)!\,3!}$$
$$= 35.$$

Thus there are 35 possible rescue squads that can be formed. Now let us evaluate $_7C_3$ using Pascal's triangle. We have 7 people to select from so we write the first 7 rows of Pascal's triangle as shown below.

Row 0								1							
Row 1							1		1						
Row 2						1		2		1					
Row 3					1		3		3		1				
Row 4				1		4		6		4		1			
Row 5			1		5		10		10		5		1		
Row 6		1		6		15		20		15		6		1	
Row 7	1		7		21		35		35		21		7		1

Figure 1

Look at Row 7. Since we must select 3 people, we go to the third entry (from the left) *after* the end 1. This entry is 35 and this is our answer. Thus again we see that $_7C_3$ is 35. We used the third entry (after the end 1) in row 7 because we wanted

$_7C_3$. If we had wanted $_7C_5$, we would have used the fifth entry (after the end 1) in row 7. This entry is 21. Thus $_7C_5$ is 21.

In general, to find the value of $_nC_r$, we go to row n. Then we select the rth row (after the end 1) from the left. This entry is $_nC_r$. In this procedure, *we always label the first row as row* 0.

Example 5   Using Pascal's triangle, find (a) $_6C_4$; (b) $_6C_0$; and (c) $_5C_5$.

Solution   We will use the Pascal triangle shown in Fig. 1.

a)   To find $_6C_4$, go to row 6. Then go across to the fourth entry from the left (after the end 1). This entry is 15. Thus, $_6C_4 = 15$.

b)   To find $_6C_0$ go to row 6. Then go across to the "0'th entry" from the left (after the end 1). This means that we must remain at the 1. Thus, $_6C_0 = 1$.

c)   To find $_5C_5$ we go to row 5. Then we go across to the fifth entry from the left (after the end 1). This entry is 1. Thus $_5C_5 = 1$.

Although we have used Pascal's triangle to evaluate $_nC_r$, there are many other interesting and important applications of this triangle. Consult the suggested further readings given at the end of the chapter for such applications.

EXERCISES

1. Using Formula 3, evaluate each of the following symbols.

   a) $_7C_5$ b) $_8C_6$ c) $_4C_3$ d) $_{10}C_9$ e) $_4C_4$

   f) $_7C_1$ g) $_6C_2$ h) $_8C_4$ i) $_3C_4$

2. Check each of the answers obtained in Exercise 1 by using Pascal's triangle.

3. Professor U. R. Small drives a car that holds only four passengers (excluding driver). Eight of his students want to hitch a ride with him. In how many different ways can the car be filled?

4. In how many different ways can a jury of 12 people be selected from a panel of 16 prospective jurors?

5. Nine astronauts are on a planet and find that their spaceship has enough fuel to take only two of them safely back to earth. In how many ways can they choose the two lucky astronauts?

6. How many committees can be formed consisting of 4 freshmen and 5 sophomores out of a total of 7 freshmen and 9 sophomores?

7. How many different poker hands consisting of 5 cards can be formed from a deck of 52 cards?

8. John has a nickel, a dime, a quarter, a half-dollar, and a dollar piece in his pocket. He decides to give the cab driver a tip consisting of 3 coins. How many different sums of money can the cab driver get as a tip?

9. Connie Sumer is buying a new car and has decided that the color should be blue. The car she wants is available in 4 different shades of blue. Each car color can be matched with any one of 6 upholstery colors. In how many ways can Connie select her car?

10. A jewelry designer is making a necklace out of precious stones. She has 10 pearls, 7 rubies, and 4 emeralds. The necklace is to contain 3 pearls, 6 rubies, and 2 emeralds. In how many different ways can she select the jewels for the necklace? (Assume that the order in which the jewels appear on the necklace is not important.)

11. Hedda Lettuce has brought her 15 children to the Horror House at a local amusement park. Only 4 children can go on the ride at one time. In how many different ways can Hedda select 4 children for the first ride?

*12. How many different committees of 7 people can be formed from 7 faculty members and 4 students if each committee must have at least 3 faculty members?

13. Nora is taking an exam on which there are 8 essays. She must answer any 5 out of the 8. In how many different ways can she do this?

14. a) Evaluate $_{10}C_7$.

b) Evaluate $_{10}C_3$.

c) How do the answers in parts (a) and (b) compare?

15. A parking lot has room for only 15 regular-size cars and 10 spaces for compact cars. If 30 regular-size cars and 20 compact cars want to park, in how many different ways can the attendant fill up the lot?

16. Look at the diagonals in the Pascal's triangle shown in Fig. 2.
a) Add the number in diagonal 1.

b) Add the numbers in diagonal 2.

c) Add the numbers in diagonal 3.

d) Add the numbers in diagonal 4, and so on throughout.

e) What do you notice about the results?

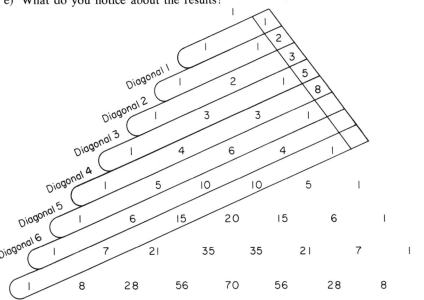

Figure 2

5. DEFINITION OF PROBABILITY

We are now ready to define what is meant by the concept of probability.

Suppose we toss an honest coin many times and observe the number of heads that appear. The results are summarized in the chart below.

Number of heads appearing	4	26	49	250	498	5,001
Number of tosses	10	50	100	500	1,000	10,000

We see that in each case the number of heads appearing is approximately $\frac{1}{2}$ the number of tosses. If we flip the coin a million times, we would expect to get approximately 500,000 heads. If we now toss a coin once, we would say that *the probability of getting a head is* $\frac{1}{2}$.

Note that when we flip the coin, there are two possible outcomes, heads and tails, both of which are equally likely. We are interested in only one of these outcomes, namely, heads. If heads occurs, we will call this a *favorable* outcome. The probability in this case is the number of favorable outcomes divided by the total number of possible outcomes, that is, it is 1 divided by 2, or $\frac{1}{2}$.

Suppose we were now to roll a die once. What is the probability of getting a 4? There are six possible outcomes, all of which are equally likely. These are 1, 2, 3, 4, 5, and 6. If the die is fair, then we would expect to get a 4 approximately $\frac{1}{6}$ of

In this Japanese wine-drinking game, players use a marked die and boxes with matching marks. When the die is rolled, would you say the winner or the loser drinks the wine?

Peabody Museum of Salem, M. W. Sexton.

the time, since there are 6 possible outcomes and only 1 of these, namely the 4, is favorable. We would then say that *the probability of getting a 4 is* $\frac{1}{6}$.

We can now make this concept of probability more specific. Before doing so, however, we will give a definition that makes it easier to talk about probability.

Definition 5.1   *The set of all possible outcomes of an experiment is called the **sample space**. We will usually not be concerned with the entire sample space but rather with only some of these outcomes. Such a collection will be referred to as an **event**. (In other words, an event is a subset of the sample space.)*

Example 1   If we toss a coin once, then the possible outcomes are H and T. Thus the sample space is {H, T}.

If we were to toss the coin twice, then the sample space would be {HT, TH, HH, TT}. The event "no heads" would be {TT}.

Example 2   If a die is tossed once, then the sample space is {1, 2, 3, 4, 5, 6}.

The event "even number" is {2, 4, 6}.
The event "odd number" is {1, 3, 5}.
The event "number greater than 4" is {5, 6}.
The event "number divisible by 3" is {3, 6}.

We are now ready to define probability.

Definition 5.2   *If an event can occur in any one of n equally likely ways and if f of these are considered as favorable outcomes, then the **probability** of getting a favorable outcome is*

$$\frac{\text{number of favorable outcomes}}{\text{total number of outcomes}} = \frac{f}{n}.$$

$$\text{prob (favorable event)} = \frac{f}{n}.$$

Example 3   A card is drawn from a 52-card deck. What is the probability of getting (a) a heart? (b) a black card? (c) an ace? (d) the ace of spades?

Solution   Since there are 52 cards in the deck, the total number of outcomes is 52.

a)   There are 13 hearts in a deck, so there are 13 favorable outcomes. Using Definition 5.2, we get

$$\text{prob (hearts)} = \frac{13}{52} = \frac{1}{4}.$$

b)   Half of the deck consists of black cards, so there are 26 favorable outcomes. Hence,

$$\text{prob (black card)} = \frac{26}{52} = \frac{1}{2}.$$

c)   There are four aces in a deck, so there are 4 favorable outcomes. We then have

$$\text{prob (ace)} = \frac{4}{52} = \frac{1}{13}.$$

d)   There is only one ace of spades in a deck. We therefore have only one favorable outcome. Thus

$$\text{prob (ace of spades)} = \frac{1}{52}.$$

Example 4   A fair die is tossed once. What is the probability of getting

a)   an odd number larger than 1?

b)   a number larger than 3?

c)   a prime number? (Remember, a prime number is any number larger than 1 that is exactly divisible only by itself and 1.)

d)   a number larger than 6?

Solution   There are six possible outcomes. These are 1, 2, 3, 4, 5, and 6.

a)   We know that 1, 3, and 5 are the possible odd numbers; 3 and 5 are larger than 1. So there are two possible favorable outcomes. Hence,

$$\text{prob (odd number larger than 1)} = \frac{2}{6} = \frac{1}{3}.$$

b)   There are three favorable outcomes 4, 5, and 6, so that

$$\text{prob (number larger than 3)} = \frac{3}{6} = \frac{1}{2}.$$

c)   The prime numbers between 1 and 6 are 2, 3, and 5. There are three of them. Thus,

$$\text{prob (prime number)} = \frac{3}{6} = \frac{1}{2}.$$

d)   Since there are no numbers larger than 6 on a die, the number of favorable outcomes is 0. So

$$\text{prob (number larger than 6)} = \frac{0}{6} = 0.$$

Example 5   In a state lottery, each ticket has 5 numbers. If you get all 5 numbers right, you win $50,000. If you get the last 4 numbers in the correct order, you win $10,000. If you get the last 3 numbers in the correct order, you win $5000. If you get the last 2 numbers in the correct order, you win $1000. What is the probability of winning (a) $50,000? (b) $10,000? (c) $5,000?

Solution   Since there are 5 numbers on each ticket ranging from 00001 to 99,999, there is a total of 99,999 possible outcomes.

a)   Only 1 ticket has the winning number, so there is just 1 favorable outcome. Thus,

$$\text{prob (winning \$50,000)} = \frac{1}{99,999}.$$

b)   In order to win $10,000, we must calculate how many tickets have the last 4 numbers in the correct order. We do not care about the first number. So the first number could be any one of 10 possible numbers. Hence there are 10 tickets that have the last 4 digits in the correct order. However, one of these (the one with the correct first number) is the big winner. So we don't count it. Thus there are only 9 favorable outcomes. Since there are 99,999 total possible outcomes we get

$$\text{prob (winning \$10,000)} = \frac{9}{99,999} = \frac{1}{11,111}.$$

c)   To win $5000 we must calculate how many tickets have the last 3 numbers in the correct order. This time we don't care about the first 2 numbers. The first place can be filled in 10 ways, and so can the second. So there is a total of 10×10, or 100, possible tickets with the last 3 digits correct. We must disregard 10 of these since 1 is the big winner and 9 are $10,000 winners. This then leaves us with 90 favorable outcomes out of a total of 99,999 possible outcomes. Therefore

$$\text{prob (winning \$5,000)} = \frac{90}{99,999} = \frac{10}{11,111}.$$

Example 6   In Example 5 on p. 4, we stated that doctors in the United States have been treating patients who are suffering from mental depression with the drug lithium. They have found that approximately 80% of all such patients treated with lithium reported feeling better. What is the probability that a person who is suffering from mental depression and is treated with lithium will feel better?

Solution   Since the doctors claim an 80% improvement rate when treated with lithium, this means that out of every 100 people treated, 80 will improve. Thus,

$$\text{prob (improvement)} = \frac{80}{100} = \frac{4}{5}.$$

Example 7   Greg, Rita, William, Frank, Yolanda, and Dawn are six students who are enrolled in a math honors course at State University. The departmental policy is to award a $100 prize to each of the top two students. What is the probability that Dawn and William will receive the prize?

Solution   We first find the total number of different ways in which the two winners can be selected. This is $_6C_2$ (the number of ways of selecting two out of six people where order does not count). Using Formula 3 of the previous section, we get

$$_6C_2 = \frac{6}{(6-2)!\,2!} = \frac{6}{4!\,2!} = 15.$$

Of these 15 ways of selecting the two winners, only one consists of Dawn and William. Thus,

$$\text{prob (Dawn and William win prize)} = \frac{1}{15}.$$

Example 8   What is the probability that your math teacher will be fired on April 31?

Solution   Since April has exactly 30 days, your math teacher cannot be fired on April 31. There are no favorable outcomes. Thus,

prob (your math teacher gets fired on April 31) = 0.

*Something that can never happen is called the **null** event. Its probability is 0.*

Example 9   Ma Bell has just been admitted to the maternity ward at a hospital to have a baby. What is the probability that it is a boy or a girl?

Solution   There are only two possible outcomes (boy or girl), so that $n = 2$. Both of these are favorable, so $f = 2$. Thus,

$$\text{prob (boy or girl)} = \frac{2}{2} = 1.$$

It is obvious that a favorable outcome *must* occur in this case.

*Something that is certain to occur is called the **definite** event. Its probability is 1.*

Comment   An event may never occur in which case its probability is 0. An event may occur for certain, in which case its probability is 1. There are events that may or may not occur, and these will have probability between 0 and 1. Thus *the probability of any event is always somewhere between 0 and 1 and possibly including 0 or 1.*

Now that we have computed the probability of several different events, we can go back to the question we raised earlier: "What do we mean by probability?"

Let us analyze the weatherforecaster's prediction that the probability of rain is 90%. We first point out that 90% can also be written as $\frac{90}{100}$. The weatherforecaster means that, in the past, when the clouds and winds have been as they are today, then it has rained 90 times out of 100. In other words, on 100 days, when conditions have been as they are today, the event of rain has occurred on 90 of these days. Thus, based on past experience, he or she predicts rain for today with a probability of $\frac{90}{100}$, or 90%.

In a similar manner, when the doctor tells you that you have a 50–50 chance of surviving the operation, he means that based on past experience, out of every 100 patients that he operated on, 50 pulled through and 50 didn't. Thus the probability of surviving is 50 out of 100, or $\frac{1}{2}$.

In general, if the probability of any event is $\frac{f}{n}$, this means that, in the long run, out of every n trials there will be f favorable outcomes. Thus probability represents the percentage of the time that the event will happen in the long run. This is sometimes called the **relative frequency** of the event.

This definition of probability is based on the number of favorable events occurring in many repeated trials. Obviously, this involves collecting statistical data about the number of events. For this reason, this approach is often called **statistical probability**. Some books use another approach and define probability using axioms. This approach is often called **axiomatic probability**. The interested reader can consult any standard text on probability for a more detailed discussion of axiomatic probability.

EXERCISES

1. Tickets numbered 1–12 are placed in a hat. One ticket is drawn from the hat. What is the probability that the number that appears on the ticket is

    a) an even number?

    b) the number 3?

    c) a number larger than 6?

    d) the number 13?

    e) a number less than 20?

2. A card is selected from an ordinary deck of cards. Find the probability that it is

    a) a heart?

    b) an ace?

    c) a black card?

    d) a card higher than 8? (aces are considered as ones)

3. A die is tossed twice. What is the probability of getting

    a) a sum of 2?

    b) a sum between 3 and 10, including these two numbers?

    c) a sum of 14?

    d) a sum larger than 6?

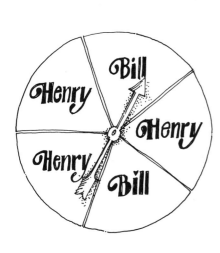

4. Katherine cannot decide which boy to date, Henry or Bill. She decides to use the spinner shown here. What is the probability that she

    a) dates Henry?   b) dates Bill?   c) dates José?

5. What is the probability that your teacher's birthday is

   a) February 30?

   b) April Fool's Day? (Assume that he or she was not born in a leap year.)

6. A coin is tossed 4 times. What is the probability that it will come up heads on exactly 2 of the tosses?

7. Terrorists are holding six hostages. The hostages are David Minter, Louis Fazio, Miguel Ramos, Michael Jones, Linda Butler, and Mohammed Jordan. Two hostages are to be released to negotiate with authorities. What is the probability that

   a) these two people will be Miguel and Linda?

   b) Mohammed will not be one of the two people selected?

   c) one of the two people will be Michael?

8. The Staten Island County Jail has four cells in a row. Four prisoners, Bill, Arthur, Martin, and Jim, are arrested and each is put in a separate cell. What is the probability that they will appear in the cells in the order shown?

9. What is the probability that Arthur will be in cell 1 in Exercise 8?

10. What is the probability that Arthur will not be in either cell 2 or 3 in Exercise 8?

11. Scientists have determined that the number of possible genetic makeups that a child of one couple can have is 2^{48}. Mr. and Mrs. Pascal already have one child. Find the probability that their next child (assuming they have one) has the same genetic makeup as the first child.

12. Mr. A. Klapp was hit on the head by a falling brick. He reasons that since there are two possible outcomes, live or die, the probability that he will live is $\frac{1}{2}$. Do you agree with his reasoning? Explain your answer.

13. Which of the following numbers cannot represent the probability of an event? (a) 0.0013; (b) −3/5; (c) 5/3; (d) 0; (e) 0.13.

14. Mr. and Mrs. Smith have four children. What is the probability that three of them are girls?

15. A die is loaded so that the outcomes have the relative frequencies shown below.

Outcome	Relative frequency
1	1/3
2	1/24
3	1/8
4	1/12
5	1/6
6	1/4

What is the probability of getting

a) an odd number?

b) an odd number less than 4?

c) an even number?

16. A little child picks up a telephone and dials a seven-digit number. What is the probability that the baby dials his home number? (Assume that there are no restrictions at all and that repetition is allowed.)

17. If you go to a party attended by 63 people, what is the probability that you will win the door prize?

18. Statistics collected in Euphoria indicate that for every 100,000 ten-year-old children, 97,206 will live to the age of 25, and 73,429 will live to the age of 65. Find the probability that a person of

a) age 10 will live to be 25 years old?

b) age 25 will live to be 65 years old?

19. Many people believe that when the 13th of a month falls on a Friday, this represents an unlucky event. Furthermore, they believe that this does not occur too often. To check the truth of this belief, let us consider the following facts. The calendar changes every year. By this we mean that if your birthday falls on a Monday this year, then next year it will fall on a Tuesday or Wednesday (depending on whether it is a leap year or not). However, our calendar repeats itself every 400 years. There are 4800 months during this period. The 13th day of the month in each of these 4800 months occurs on the different days of the week according to the following chart.

Day of week	Sun.	Mon.	Tues.	Wed.	Thurs.	Fri.	Sat.
How often the 13th day of month occurs on this day	687	685	685	687	684	688	684

a) Using the above chart, find the probability that the 13th day of the month will occur on a Friday.

b) Is this probability greater than, less than, or equal to the probability of it falling on any other day of the week?

6. RULES OF PROBABILITY

Consider the following problem. A card is selected from an ordinary deck of 52 cards. What is the probability that it is either a heart *or* a black card? Obviously a card cannot be both a heart and a black card at the same time. We say that the events of drawing a heart and of drawing a black card are **mutually exclusive**. If A represents the event of drawing a heart and if B represents the event of drawing a black card, then $A \cap B = \varnothing$.

The probability of getting a heart is $\dfrac{13}{52}$, or $\dfrac{1}{4}$, since there are 13 hearts out of a possible 52 cards.

The probability of getting a black card is $\dfrac{26}{52}$, or $\dfrac{1}{2}$.

Since there are 13 hearts and 26 black cards, this gives us a total of 39 favorable outcomes. Since there are 52 cards in the deck, our answer is

$$\text{prob (heart or black card)} = \frac{39}{52} = \frac{3}{4}.$$

Note that if we add the probability of a heart and the probability of a black card, we get the following:

$$\text{prob (heart)} + \text{prob (black card)} = \frac{1}{4} + \frac{1}{2}$$

$$= \frac{3}{4}.$$

It would require the use of a computer to determine the probability of (a row on) any one of these cards being completed ahead of the others, even though we are given the fact that some numbers have already been called. The fascination of "Bingo" lies in its complete (to the average person) unpredictability.

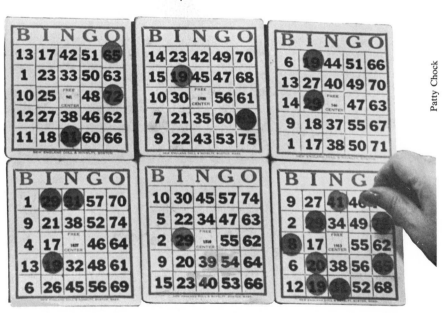

Patty Chock

Thus we see that

prob (heart or black card) = prob (heart) + prob (black card).

This leads us to the following.

Definition 6.1   *Two events, A and B, are said to be **mutually exclusive**, if both A and B cannot occur at the same time. In terms of sets this means $A \cap B = \varnothing$.*

Formula 4

> If A and B are mutually exclusive events, then
>
> prob $(A$ or $B)$ = prob (A) + prob (B).

Example 1   A card is drawn from a deck of 52 cards. What is the probability of getting a 7 or a picture card?

Solution   The events "getting a 7" and "getting a picture card" are mutually exclusive. We can therefore use Formula 4. We first calculate prob (getting a 7). Since there are four 7's out of a total of 52 cards,

$$\text{prob } (7) = \frac{4}{52} = \frac{1}{13}.$$

Also, we know that there are 12 picture cards in a 52-card deck. Thus,

$$\text{prob (picture card)} = \frac{12}{52} = \frac{3}{13}.$$

By Formula 4, we get

prob (7 or picture card) = prob (7) + prob (picture card)

$$= \frac{1}{13} + \frac{3}{13}$$

$$= \frac{4}{13}.$$

Example 2   At the Fresh Air Fund Charity Bazaar, there is a table at which 30 unmarked surprise packages are being sold. Six of the packages contain transistor radios, 3 of the packages contain perfume, 10 of the packages contain wallets, 5 of the packages contain ashtrays, and 6 of the packages contain shavers. No package contains more than one item and all the packages are wrapped identically. What is the probability that Ann, who buys the first package, gets either a radio or perfume?

Solution   Since Ann is buying only one package, the events "getting a radio" and "getting perfume" are mutually exclusive. Thus Formula 4 can be applied. Since there are 30 packages altogether, 6 of which are radios and 3 of which are perfume, then prob (gets a radio) = $\frac{6}{30}$, and

prob (gets perfume) = $\frac{3}{30}$.

Therefore,

prob (gets a radio or gets perfume) = prob (gets a radio) + prob (gets perfume)

$$= \frac{6}{30} + \frac{3}{30} = \frac{9}{30} = \frac{3}{10}$$

Therefore, the probability that Ann gets a radio or perfume is $\frac{3}{10}$.

Example 3   A mailman cannot read the address on a letter. He is not sure but thinks that the address is either 390 Main Street or 890 Main Street. The probability that he will deliver it to 390 Main Street is $\frac{1}{3}$ and the probability that he will deliver it to 890 Main Street is $\frac{2}{5}$. What is the probability that he will deliver the letter to 890 Main Street or 390 Main Street?

Solution   Since the mailman cannot deliver the letter to both addresses (at the same time) we are dealing with mutually exclusive events. Formula 4 can be used. Therefore

prob (390 or 890 Main St.) = prob (390) + prob (890)

$$= \frac{1}{3} + \frac{2}{5} = \frac{5}{15} + \frac{6}{15} = \frac{11}{15}.$$

The probability that he delivers it to one of these addresses is $\frac{11}{15}$.

Example 4   Renée has been taking a birth control pill that claims to be 97% effective. This means that out of every 100 women who take this pill as prescribed, 97 will not become pregnant. What is the probability that Renée will become pregnant?

Solution   The events "becoming pregnant" and "not becoming pregnant" are mutually exclusive, so we can use Formula 4. Obviously one of these events must occur. Therefore the event "pregnant or not pregnant" is the certain event and has a probability of 1. By Formula 4, we have

prob (pregnant or not pregnant) = prob (pregnant) + prob (not pregnant)

$$1 = \text{prob (pregnant)} + \frac{97}{100}.$$

Subtracting $\dfrac{97}{100}$ from both sides, we get

$$1 - \frac{97}{100} = \text{prob (pregnant)}$$

$$\frac{3}{100} = \text{prob (pregnant)}.$$

Thus, the probability that Renée will become pregnant is $\dfrac{3}{100}$ or 0.03.

Now let us consider the following problem. What is the probability of drawing from a deck of cards an ace *or* a spade? We first notice that the events "drawing an ace" and "drawing a spade" are *not* mutually exclusive since the ace of spades is both an ace and a spade. If we let A stand for the event of drawing an ace and let B stand for the event of drawing a spade, then $A \cap B \neq \varnothing$. Thus Formula 4 cannot be used. For situations of this type, we introduce the following.

Formula 5

> If A and B are *any* events, then
> prob (A or B) = prob (A) + prob (B) − prob (A and B).

Let us apply this formula to the above problem. We know that

$$\text{prob } (A) = \frac{4}{52} \quad \text{and} \quad \text{prob } (B) = \frac{13}{52}.$$

We now calculate prob (A and B). This event occurs only when the card drawn is an ace of spades. This has already been calculated on p. 262. Thus prob (A and B) = $\dfrac{1}{52}$. Using Formula 5, our answer is

$$\text{prob } (A \text{ or } B) = \text{prob } (A) + \text{prob } (B) - \text{prob } (A \text{ and } B)$$

$$= \frac{4}{52} + \frac{13}{52} - \frac{1}{52} = \frac{16}{52} = \frac{4}{13}.$$

The probability of drawing an ace or a spade is $\dfrac{4}{13}$.

Example 5   Josiah S. Carberry is the world's most traveled man. He has booked a flight with Dodo Airlines. The probability that his luggage will be lost is $\dfrac{1}{6}$ and the probability that the plane will be delayed is $\dfrac{3}{8}$. If the probability of both of these happening is $\dfrac{1}{24}$, what is the probability that his luggage will be lost or that the plane will be delayed?

Solution   Since these events are not mutually exclusive (as both of these could easily happen with Dodo Airlines), we will use Formula 5.

prob (lost luggage or plane delayed) = prob (lost luggage) + prob (plane delayed)

$$- \text{prob (lost luggage } and \text{ plane delayed)}$$

$$= \frac{1}{6} + \frac{3}{8} - \frac{1}{24} = \frac{4}{24} + \frac{9}{24} - \frac{1}{24}$$

$$= \frac{12}{24} \text{ or } \frac{1}{2}.$$

Thus, the probability that Josiah will lose his luggage or that his plane will be delayed is $\frac{1}{2}$.

Example 6   Mr. and Mrs. Schlamazel are going on a skiing trip. The probability that Mr. Schlamazel will break a leg is $\frac{1}{5}$. The probability that Mrs. Schlamazel will break a leg is $\frac{2}{3}$. The probability that both Schlamazels will break a leg is $\frac{3}{5}$. What is the probability that either Schlamazel will break a leg?

Solution   prob (either Schlamazel breaks a leg) = prob (Mr. Schlamazel breaks a leg)

$$+ \text{prob (Mrs. Schlamazel breaks a leg)}$$

$$- \text{prob (both Schlamazels break a leg)}$$

$$= \frac{1}{5} + \frac{2}{3} - \frac{3}{5}$$

$$= \frac{4}{15}.$$

Thus, the chance of either Schlamazel breaking a leg is $\frac{4}{15}$.

Comment   Formula 4 is just a special case of Formula 5. Formula 5 applies to *any* events A and B. If these events happen to be mutually exclusive, then A and B cannot happen together. Thus prob $(A$ and $B)$ is 0. In this case, Formula 5 becomes

prob $(A$ or $B)$ = prob (A) + prob (B) − prob $(A$ and $B)$

prob $(A$ or $B)$ = prob (A) + prob (B) − 0

prob $(A$ or $B)$ = prob (A) + prob (B).

This is exactly the same as Formula 4.

EXERCISES

1. Determine which of the following events are mutually exclusive and which are not.

   a) becoming pregnant and getting a headache

   b) cheating on an exam and getting caught

   c) having blue eyes and having brown eyes

   d) being male and being female

   e) being crazy and being a math teacher

   f) having type A blood and having type O blood

   g) having a beard and having a mustache

2. Charlie Starkist is fishing. The probability that he catches a trout is $\frac{3}{8}$ and the probability that he catches a bluefish is $\frac{1}{4}$. What is the probability that he catches either, if the probability that he catches both is $\frac{1}{16}$?

3. In a certain community, the following data have been collected. The probability of having gonorrhea is 0.24. The probability of having syphilis is 0.18. The probability of having both syphilis and gonorrhea is 0.02. What is the probability that a person in this community has either syphilis or gonorrhea?

4. Bill has applied for two part-time jobs, a morning job and an evening job. The probability that he gets the morning job is $\frac{3}{10}$ and the probability that he gets the evening job is $\frac{1}{20}$. If the probability that he gets both jobs is $\frac{3}{40}$, what is the probability that he gets either job?

5. A visitor to the United Nations in New York City meets a guard and asks for directions. The probability that the guard speaks French is 0.2 and the probability that the guard speaks German is 0.3. What is the probability that the guard speaks either language, if the probability that he speaks both is 0.1?

6. According to statistics, approximately 21 out of every 1000 Federal Income Tax returns filed in 1974 by Chicago residents were audited by the Internal Revenue Service. What is the probability that a Chicago resident who filed a 1974 Federal tax return will *not* be audited?

7. An absentminded professor often forgets to put gas in his car, and to turn off the lights when he parks. One morning, his car won't start. The probability that the car has no gas is 0.6, and the probability that the battery is dead is 0.4. If the probability that he has either no gas *or* a dead battery is 0.9, what is the probability that he has both no gas *and* a dead battery?

8. The probability that Mark will drive his mother crazy is 0.73. The probability that he will drive his father crazy is 0.48. The probability that he will drive both of his parents crazy is 0.21. What is the probability that he will drive one of his parents crazy?

9. Bruce has bet a lot of money on two horses, whose names are Joe and Frances. The probability that Joe wins the race is 0.21 and the probability that Frances wins the race is 0.25. What is the probability that one of Bruce's two horses comes in first?

10. In a certain mining community, the following statistics have been accumulated. The probability that a miner has black lung disease is 0.41. The probability that a miner has arthritis is 0.27. The probability that a miner has either disease is 0.33. What is the probability that a miner has both?

11. Rochelle is on her way home from school and stops off at the local McDonald's. The probability that she orders a hamburger is 0.64. The probability that she orders a Coke is 0.8. The probability that she orders either a hamburger or a Coke is 0.93. What is the probability that she orders both a hamburger and a Coke?

12. Refer back to Example 2 on p. 269. Suppose Maurice buys the second box. If Ann gets a wallet, what is the probability that Maurice will get

    a) a wallet also?

    b) a bottle of perfume?

13. Uncle Sam has just received 100 $1 bills on which the last three digits of the serial numbers go from 201 to 300. He drops them on the floor and a passerby kindly volunteers to help him pick them up. If the passerby picks up one of those bills, what is the probability that it ends in an even number?

14. On a certain time machine, the probability that the schnazola will break down is $\frac{3}{7}$. The probability that both the schnazola and the bamboozal will break down is $\frac{1}{8}$. The probability that either the schnazola or the bamboozal will break down is $\frac{2}{3}$. What is the probability that the bamboozal will break down?

---

7. ODDS AND MATHEMATICAL EXPECTATION

Gamblers are always interested in the odds of a game or a race. They are also interested in the amount of money to be won. In this section we will investigate the meaning of these ideas and learn how to calculate them.

To best understand these ideas, let us consider a man at Aqueduct Raceway. Nine horses have been entered in the big race. Our man places $10 on the horse

Liverwurst. We will assume that each horse has an equal chance of winning. Therefore the probability that the man will win is $\frac{1}{9}$. Gamblers prefer to speak in terms of **odds**. They would say that the odds in favor of his winning are 1 to 8 and the odds against his winning are 8 to 1. The 8 represents the eight chances of losing. Thus we have the following definitions.

Definition 7.1   *The **odds in favor** of an event occurring are p to q where p is "the number of favorable outcomes" and q is "the number of unfavorable outcomes."*

Definition 7.2   *The **odds against** an event occurring are q to p where q and p are the same as in Definition 7.1.*

We illustrate these definitions with several examples.

Example 1   What are the odds in favor of drawing an ace from a full deck of 52 cards on one draw?

Solution   Since there are 4 aces and 48 non-aces, Definition 7.1 tells us that the odds in favor of drawing an ace are 4 to 48.

Example 2   What are the odds in favor of winning at the Hopeless Wheel of Fortune which is divided into 10 equal parts, each with a different color, if someone bets the colors red and blue and only one color wins?

Solution   Since there are 2 favorable and 8 unfavorable outcomes, Definition 7.1 tells us that the odds in favor of winning are 2 to 8.

Example 3   What are the odds *against* throwing a 2 or a 12 in throwing a pair of dice?

Solution   When a pair of dice is thrown, there are 36 possible outcomes (p. 243). Two are favorable and 34 are nonfavorable. Thus Definition 7.2 tells us that the odds against getting a 2 or a 12 are 34 to 2.

When gambling, the amount of money to be won in the long run is called the **mathematical expectation**. It is defined as follows.

Definition 7.3   *Suppose an event has several possible outcomes with probabilities p_1, p_2, p_3, and so on. Suppose on the first event the payoff is m_1, on the second event the payoff is m_2, on the third event the payoff is m_3, etc. Then the **mathematical expectation** of the event is*

$$m_1 p_1 + m_2 p_2 + m_3 p_3 + \ldots$$

The following examples will show how this definition is applied.

Example 4   A die is tossed once. If a 1 comes up, then Joe will win $10 and if a 4 comes up, he will win $7. What is his mathematical expectation?

Solution   When a die is tossed once, then the probability of getting a 1 is $\frac{1}{6}$. Similarly, the probability of getting a 4 is $\frac{1}{6}$. Using Definition 7.3, we find that the mathematical expectation is

$$10 \cdot \frac{1}{6} + 7 \cdot \frac{1}{6} = \frac{17}{6} = \$2.83 \text{ (when rounded)}.$$

Example 5   The local chapter of the American Cancer Society is planning to hold a bazaar to raise funds. If the bazaar is held outdoors, \$100,000 is expected to be raised. If the bazaar is held indoors, then \$75,000 is expected to be raised. The probability that it rains, forcing the bazaar to be held indoors, is $\frac{3}{5}$ and the probability that the weather is suitable for an outdoor bazaar is $\frac{2}{5}$. How much money can they expect to raise?

Solution   We will use Definition 7.3. We have

$$100,000 \cdot \frac{2}{5} + 75,000 \cdot \frac{3}{5} = 40,000 + 45,000$$
$$= 85,000$$

Thus they can expect to raise \$85,000.

Example 6   A wheel of fortune at an amusement park is divided into 4 colors—red, blue, yellow, and green. The probabilities of the spinner landing in any of these colors are $\frac{3}{10}, \frac{4}{10}, \frac{2}{10}$, and $\frac{1}{10}$. A player can win \$4 if it stops on red and \$2 if it stops on green, and lose \$2 if it stops on blue and \$3 if it stops on yellow. Trudy has decided to try her luck at the wheel. What is her mathematical expectation?

Solution   We indicate the possible outcomes and the corresponding probabilities by the following chart.

Outcome	Probability	Amount of money won or lost
Red	$\frac{3}{10}$	+4
Blue	$\frac{4}{10}$	−2
Yellow	$\frac{2}{10}$	−3
Green	$\frac{1}{10}$	+2

Bruce Anderson

What would you think is the probability of two people in a crowd having the same birthday? Mathematically the probability that *at least* two people, in a crowd of 25, have a common birthday is greater than 1/2. The probability increases to about 1 (almost a certainty) in a crowd of 60 people.

Thus, her mathematical expectation is

$$(4)\frac{3}{10} + (-2)\frac{4}{10} + (-3)\frac{2}{10} + (2)\frac{1}{10} = \frac{12}{10} - \frac{8}{10} - \frac{6}{10} + \frac{2}{10}$$
$$= 0.$$

Her mathematical expectation is 0. What does this 0 mean? We interpret this to mean that in the long run she will win 0 dollars or break even.

Comment   Some gamblers base their decision whether or not to play a particular game solely on the game's mathematical expectation. Obviously, if the game has a negative mathematical expectation, a gambler should not play since he or she will lose money in the long run. There would be little point in playing a game whose mathematical expectation is 0 since in the long run the amount of money that can be won is 0.

Mathematical expectation can also be applied to non-money situations, as the following example will illustrate.

Example 7   An observer for an energy conservation group has collected the following statistics on the number of occupants per car that pass through a certain tollgate.

Number of passengers in car (including driver)	1	2	3	4	5	6
Probability	0.37	0.29	0.18	0.09	0.05	0.02

What is the expected number of occupants per car?

Solution   We apply Definition 7.3 and multiply each of the possible outcomes by its probability. We get

$$1(0.37) + 2(0.29) + 3(0.18) + 4(0.09) + 5(0.05) + 6(0.02)$$

which equals 2.22. Thus the expected number of occupants per car is 2.22.

Pascal used mathematical expectation to make a "wager with God." As we have said, Pascal was extremely religious. He reasoned that leading a religious life will result in eternal happiness. The value of eternal happiness is infinite. Therefore, the expectation is

$$m \cdot p = \text{(value of eternal happiness)(probability of obtaining eternal happiness)}.$$

Since the value of eternal happiness is infinite, the product $m \cdot p$ is also infinite, even if the probability of obtaining eternal happiness is small. Thus, it pays to lead a religious life as the expectation is infinite.

EXERCISES

1. A publisher ships 100 copies of *Mathematics as a Second Language* to a bookstore. Due to a printer's error, five of these are defective. A student buys one of these books. What are the odds against getting a defective book?

2. It is known that 12 of the 1000 cars produced on April 1st by an auto manufacturer have defective engine mounts. If Ralph Raider buys one of these cars, what are the odds in favor of his getting a defective car?

3. A traveler is carrying 13 $10 bills, 3 of which are counterfeit. He does not know that any of his bills are counterfeit. If the traveler gives a cabdriver a $10 tip, what are the odds that the cabdriver gets a counterfeit bill?

4. Due to the budget crisis, the mayor of a large city has decided to fire 97 of the 1000-man sanitation force. All the workers have the same chance of being fired. Mr. Clean is one of the 1000 workers. What are the odds against his getting fired?

5. Mary needs an additional nickel to buy cigarettes from a vending machine. (No change is given by the machine.) There are 12 people standing nearby. She decides to ask *only one* person for the correct change. What are the odds in favor of her getting the change, if 5 of the people actually have the change?

6. A die is tossed twice. If "doubles" (the same number on both tosses) comes up, Harvey wins $50. Otherwise he loses $10. What is his mathematical expectation?

7. Eddie is a doorman in a large apartment house. Based on past experience, he has determined that the number of people who enter the building during the hours of 1:00 A.M. and 2:00 A.M. is as follows.

Number of people entering building	0	1	2	3	4	5	6
Probability	0.07	0.19	0.28	0.18	0.14	0.08	0.06

What is the expected number of people who enter the building during these hours?

8. The doorman in Exercise 7 has found from past experience that the tenants in the building will give different Christmas gifts with the probability shown below.

Tenant	Amount of gift	Probability
Mr. Kok	$10	0.83
Mr. Chandler	1	0.98
Ms. Szeto	20	0.93
Mrs. Lake	25	0.78
Mr. Newmark	15	0.86
Mr. Gordon	50	0.63
Mr. Ablon	5	0.74
Mr. Sormani	21	0.57

What is the amount of money that the doorman can expect to receive from these tenants for Christmas?

9. A magician claims that he has supernatural powers in being able to always guess the correct color of objects. To test his claim, he is blindfolded and a box containing 5 red marbles, 2 green marbles, 8 black marbles, and 4 white marbles is placed in front of him. He selects a marble and guesses its color. If he is right, then he wins different amounts of money depending upon the color. Red wins $2, green wins $8, black wins $5, and white wins $4. What is his mathematical expectation?

10. 700 lottery tickets are sold at a charity bazaar. The tickets are sold for $3 apiece. The grand prize is $1000. Alice has just bought 5 tickets. What is her mathematical expectation?

11. A large company that operates drive-in movies is planning to open a new theatre in one of two locations. The management of the company believes that if it opens the new theatre in location A, its annual profit will be $65,000 if successful, and its annual loss will be $8,000 if it is not. For location B, the annual profit will be $90,000 if successful, and the annual loss will be $25,000 if it is not. Where should the company open the new theatre and why? (Assume that the probability of success at either location is $\frac{1}{2}$.)

---

8. RANDOM NUMBERS

In a large city, the social services department has decided to make a detailed study of 100 welfare cases. These 100 cases are to be chosen completely *at random* from among the 20,000 families currently receiving aid. It is important to the department that each family have an equal chance of being selected for the study. How can the department select the families for the study? There are several ways in which this can be done.

1. Write each family name on a piece of paper and put the names in a box. After mixing the contents thoroughly, select 100 names from the box. While this approach might seem sensible, it is unlikely that we would obtain a truly random mix. For one thing, most people would not stick their hand in to the bottom of the box to make a selection. Thus the names on the bottom are less likely to be selected than the names on top. Moreover, if you do not

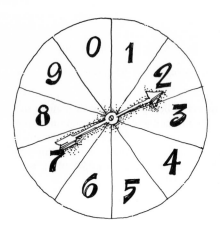

replace each name after a selection, then the 100th name drawn has a greater likelihood of being selected than the first name drawn. (Can you see why this is so?) Even if we correct these difficulties, which clearly can be done, this method is obviously inefficient.

2. Arrange all the names alphabetically and number them from 1 to 20,000. Then select every 200th name. Again this does not give a truly random selection. By selecting every 200th name, we are actually making sure that those families with numbers 1–199, 201–399, etc., never have a chance of being selected.

3. Assign a 5-digit case number to each family, starting with the number 00001. Then use a spinner like the one shown here, to generate *random numbers* as follows. Use the spinner 5 times to obtain 5 digits in order. The 5-digit number obtained by this process is the first case number to be selected. We continue in this way until we get the 100 case numbers that we want.

It is unlikely that this method will generate completely random numbers either. Can you find some reasons why it will not? Basically, this last method is good. That is, generating random numbers and using the families whose numbers have been selected is reasonable. The difficulties arise because we generate the random numbers using the spinner. We can overcome this difficulty by using a different technique to generate the random numbers. There are various ways of doing this, most of which use computers. Random numbers generated in this manner are often listed in *Tables of Random Digits*.

Therefore, the best way for the social services department to select the families for the study is to use such a table to generate random case numbers, and then use the families whose numbers come up in this way.

On p. 281, we give a table of random digits. To use the table, we can start at any column and on any line. Thus, if we use column 1 and read off the numbers, we get 10480, 22368, 24130, and so forth. Since the families have numbers only up to 20,000, we skip those numbers that are over 20,000. Therefore, we skip the number 22368. The same is true about the number 24130. The next number that we accept is on line 13; it is 09429, then 10365, and so forth. Proceeding in this manner, we can obtain a random sample by selecting those families whose numbers are 10480, 09429, 10365, 07119, 02368, 01011, 07056 . . . . When we get to the bottom of column 1, we go to column 2 and follow the same procedure. We stop when we get the 100 numbers that we need.

Example 1   During 1976, a large auto manufacturer received 350 complaints from customers about the quality of the service performed by one particular dealer. The company decides to investigate some of these complaints by selecting a random sample of 18 of these complaints and thoroughly investigating them. By using column 4 of Table 1, which customers will we select?

Solution   We first number the customers' complaint letters from 1 to 350. Then we use column 4 of the table of random digits. Although the table gives 5-digit numbers,

Table 1. *Table of random digits.*

Line	Col. (1)	(2)	(3)	(4)	(5)	(6)	(7)	(8)	(9)	(10)	(11)	(12)	(13)	(14)
1	10480	15011	01536	02011	81647	91646	69179	14194	62590	36207	20969	99570	91291	90700
2	22368	46573	25595	85393	30995	89198	27982	53402	93965	34095	52666	19174	39615	99505
3	24130	48360	22527	97265	76393	64809	15179	24830	49340	32081	30680	19655	63348	58629
4	42167	93093	06243	61680	07856	16376	39440	53537	71341	57004	00849	74917	97758	16379
5	37570	39975	81837	16656	06121	91782	60468	81305	49684	60672	14110	06927	01263	54613
6	77921	06907	11008	42751	27756	53498	18602	70659	90655	15053	21916	81825	44394	42880
7	99562	72905	56420	69994	98872	31016	71194	18738	44013	48840	63213	21069	10634	12952
8	96301	91977	05463	07972	18876	20922	94595	56869	69014	60045	18425	84903	42508	32307
9	89579	14342	63661	10281	17453	18103	57740	84378	25331	12566	58678	44947	05585	56941
10	85475	36857	53342	53988	53060	59533	38867	62300	08158	17983	16439	11458	18593	64952
11	28918	69578	88231	33276	70997	79936	56865	05859	90106	31595	01547	85590	91610	78188
12	63553	40961	48235	03427	49626	69445	18663	72695	52180	20847	12234	90511	33703	90322
13	09429	93969	52636	92737	88974	33488	36320	17617	30015	08272	84115	27156	30613	74952
14	10365	61129	87529	85869	48237	52267	67689	93394	01511	26358	85104	20285	29975	89868
15	07119	97336	71048	08178	77233	13916	47564	81056	97735	85977	29372	74461	28551	90707
16	51085	12765	51821	51259	77452	16308	60756	92144	49442	53900	70960	63990	75601	40719
17	02368	21382	52404	60268	89368	19885	55322	44819	01188	65255	64835	44919	05944	55157
18	01011	54092	33362	94904	31273	04146	18594	29852	71585	85030	51132	01915	92947	64951
19	52162	53916	46369	58586	23216	14513	83149	98736	23495	64350	94738	17752	35156	35749
20	07056	97628	33787	09998	42698	06691	76988	13602	51851	46104	88916	19509	25625	58104
21	48663	91245	85828	14346	09172	30168	90229	04734	59193	22178	30421	61666	99904	32812
22	54164	58492	22421	74103	47070	25306	76468	26384	58151	06646	21524	15227	96909	44592
23	32639	32363	05597	24200	13363	38005	94342	28728	35806	06912	17012	64161	18296	22851
24	29334	27001	87637	87308	58731	00256	45834	15398	46557	41135	10367	07684	36188	18510
25	02488	33062	28834	07351	19731	92420	60952	61280	50001	67658	32586	86679	50720	94953
26	81525	72295	04839	96423	24878	82651	66566	14778	76797	14780	13300	87074	79666	95725
27	29676	20591	68086	26432	46901	20849	89768	81536	86645	12659	92259	57102	80428	25280
28	00742	57392	39064	66432	84673	40027	32832	61362	98947	96067	64760	64584	96096	98253
29	05366	04213	25669	26422	44407	44048	37937	63904	45766	66134	75470	66520	34693	90449
30	91921	26418	64117	94305	26766	25940	39972	22209	71500	64568	91402	42416	07844	69618
31	00582	04711	87917	77341	42206	35126	74087	99547	81817	42607	43808	76655	62028	76630
32	00725	69884	62797	56170	86324	88072	76222	36086	84637	93161	76038	65855	77919	88006
33	69011	65795	95876	55293	18988	27354	26575	08625	40801	59920	29841	80150	12777	48501
34	25976	57948	29888	88604	67917	48708	18912	82271	65424	69774	33611	54262	85963	03547
35	09763	83473	73577	12908	30883	18317	28290	35797	05998	41688	34952	37888	38917	88050
36	91567	42595	27958	30134	04024	86385	29880	99730	55536	84855	29080	09250	79656	73211
37	17955	56439	90999	49127	20044	59931	06115	20542	18059	02008	73708	83517	36103	42791
38	46503	18584	18845	49618	02304	51038	20655	58727	28168	15475	56942	53389	20562	87338
39	92157	89634	94824	78171	84610	82834	09922	25417	44137	48413	25555	21246	35509	20468
40	14577	62765	35605	81263	39667	47358	56873	56307	61607	49518	89656	20103	77490	18062
41	98427	07523	33362	64270	01638	92477	66969	98420	04880	45585	46565	04102	46880	45709
42	34914	63976	88720	82765	34476	17032	87589	40836	32427	70002	70663	88863	77775	69348
43	70060	28277	39475	46373	23219	53416	94970	25832	69975	94884	19661	72828	00102	66794
44	53976	54914	06990	67245	68360	82948	11398	42878	80287	88267	47363	46634	06541	97809
45	76072	29515	40980	07391	58745	25774	22987	80059	39911	96189	41151	14222	60697	59583
46	90725	52210	83974	29992	65831	38857	50490	83765	55657	14361	31720	57375	56228	41546
47	64364	67412	33339	31926	14883	24413	59744	92351	97473	89286	35931	04110	23726	51900
48	08962	00358	31662	25388	61642	34072	81249	35648	56891	69352	48373	45578	78547	81788
49	95012	68379	93526	70765	10592	04542	76463	54328	02349	17247	28865	14777	62730	92277
50	15664	10493	20492	38391	91132	21999	59516	81652	27195	48223	46751	22923	32261	85653

Page 1 of *Table of 105,000 Random Digits*, Statement No. 4914, May, 1949, File No. 261-A-1, State Commerce Commission, Washington, D.C.

we simply use the first 3 digits of the column. Thus we select the numbers 20, 166, 79, 102, 332, 34, 81, 99, 143, 242, 73, 264, 129, 301, 73, 299, 319, and 253. Therefore, the customers with these numbers will have their complaints investigated.

Example 2   60 students have registered for a statistics course. The chairperson of the department wishes to start a new section. She asks for 15 volunteers, but no one volunteers to transfer to the new section. She decides to randomly select 15 students. By using column 5 of Table 1, how can this be done?

Solution   She should first assign each student a number from 1 to 60. Then she should use the first 2 digits of the numbers in column 5. Thus she should select those students whose numbers are 30, 7, 6, 27, 18, 17, 53, 49, 48, 31, 23, 42, 9, 47, and 13.

EXERCISES

1. In a certain state, license plates have 5 digits. The state motor vehicle department is interested in knowing how many of these cars are equipped with radial tires. It decides to send a letter to 100 randomly selected car owners. If columns 1 and 2 of Table 1 are used, which car owners will be selected to receive this letter?

2. A new drug is being tested on its ability to overcome drowsiness. Two hundred people have volunteered to take this drug to test its effectiveness. Each person is given a number from 1 to 200. For experimental purposes, only 15 people will receive the new drug. The remaining 185 will be given a sugar pill. If columns 2 and 3 are used which of the volunteers will be selected to receive the new drug?

3. During 1975, a manufacturer produced 9000 stereo sets, each with a serial number from 1 to 9000. The company decides to send a questionnaire to 30 owners of these sets. Using the seventh column of Table 1, to which owners should the company send the questionnaire?

4. A new movie has just been seen by 4000 people. The producer wants to know how the people reacted to a particular horror scene. Since each person already has a ticket stub with a number from 1 to 4000 on it, he decides to select a random sample of 25 people. If columns 9 and 10 are used, which ticket holders will be selected?

5. There are 328 licensed restaurants in Newville. The health department has decided to investigate the sanitary conditions of these restaurants by randomly selecting 20 of these restaurants. The licenses are numbered consecutively from 1 to 328. If columns 8 and 9 of Table 1 are used, which restaurants should be investigated?

---

9. APPLICATION TO GENETICS

One very interesting application of probability is in the science of *genetics*. This science is concerned with which traits can be inherited. The pioneer in this field was Gregor Mendel. As we mentioned in Chapter 1, Mendel performed many experiments with garden peas. As a result of his experiments, Mendel was able to state the basic laws of heredity.

Specifically, Mendel crossbred plants from wrinkled seeds with plants from smooth seeds. The resulting plants all had smooth seeds. However, when he

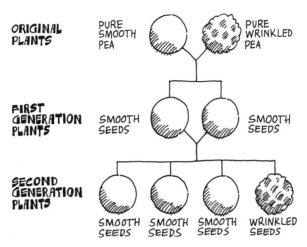

ORIGINAL PLANTS   PURE SMOOTH PEA   PURE WRINKLED PEA

FIRST GENERATION PLANTS   SMOOTH SEEDS   SMOOTH SEEDS

SECOND GENERATION PLANTS   SMOOTH SEEDS   SMOOTH SEEDS   SMOOTH SEEDS   WRINKLED SEEDS

Figure 3

crossbred these new plants with one another, a strange thing happened. Three-fourths of the seeds from the resulting plants were smooth and one-fourth were wrinkled (see Fig. 3).

He concluded that certain *genes* are *dominant*. When a gene is dominant, then the trait that it represents will always appear, no matter which gene it is paired with. Such was the case for smooth seeds. On the other hand, there are other genes that are *recessive*. If a trait is recessive, it must be matched with the same type of gene for that trait to appear. Wrinkled seeds are recessive. This explains why when wrinkled seeds were matched with smooth seeds, the resulting plants all had smooth seeds. The smooth-seed gene "dominated" the wrinkled-seed one. On the other hand, in the second generation "mixed" plants that now had both wrinkled and smooth genes were matched with each other. Those that contained any smooth-seed genes produced smooth-seed plants, and those that contained wrinkled-seed genes *only* produced wrinkled seeds.

From his observations on the peas, Mendel concluded that certain characteristics (or traits) were determined by a single pair of genes (one for each parent). The same is true in human beings. For example, it is known that a baby's eye color is determined by certain genes, one obtained from each parent. There are other inherited traits that are determined by several genes from each parent. In this section, we concern ourselves only with those characteristics that are determined by a single pair of genes.

Let us analyze several such traits, such as eye color and the blood factor Rh.

Example 1   It is known that in human eyes, the color brown is dominant and the color blue is recessive. Let B represent a gene for brown eyes and let b represent a gene for blue eyes. Thus, a person who has BB genes (that is, brown from each parent) or Bb genes (brown from one parent and blue from the other) will have brown eyes, since brown is dominant. On the other hand, a person who has bb genes (blue from both parents) will have blue eyes. Arlene who is known to have Bb genes

Arlene

Gene *B*   Gene *b*

	Gene *B*	Gene *b*
Gene *B*	*BB*	*Bb*
Gene *b*	*bB*	*bb*

Tom

Solution

marries Tom who is also known to have Bb genes. What is the probability that their child will have blue eyes?

We set up a chart indicating how the genes can be matched and the resulting child's eye color.

In this chart, every time B is paired with any other gene, then the child will have brown eyes. Only the pairing bb will result in blue eyes for the baby. Out of the 4 possible outcomes BB, Bb, bB, and bb, only 1 is favorable. Thus the probability that the child will have blue eyes is 1/4.

Example 2   Another inherited trait that is known to be determined by a single gene from each parent is the Rh blood factor. When a person has this factor in his blood, we say that he is Rh positive. Otherwise he is Rh negative. Eighty-five percent of American Caucasians, and 93 percent of blacks are born with this inherited substance. Thus if the parents of the husband and of the wife are pure Rh positive, then all the children will be Rh positive. On the other hand, if both husband and wife had one Rh-positive and one Rh-negative parent, then such a couple would produce, in every 4 children, one child who is pure Rh positive, 2 children who are partial Rh positive, and one child who is Rh negative.

In a recent court suit, Jane, who was known to be Rh negative, accused Bill, who was also Rh negative, of being the father of her Rh-positive baby. Find the probability that Bill actually was the father.

Solution   Since Jane is known to be Rh negative, it is impossible for Bill to be the father, since he is also Rh negative. Two Rh-negative parents can only produce an Rh-negative child. Thus, the probability that Bill is the father of Jane's child is 0.

Comment   Certain traits in human beings are known to be determined by the genes received from each parent. For example, albinism (no skin color, eye color, or hair color), muscular atrophy, Tay-Sach's disease, hemophilia, sickle-cell anemia, etc., are known to be determined by genes. With appropriate medical care and advice, the probability of these occurring can be reduced.

EXERCISES   Another example of a recessive trait is albinism, where a child is born with no skin color, eye color, or hair color. Albinism is known to be a recessive trait. Suppose two normal parents have an albino child named Peter.

1. Find the probability that Peter's sister is also an albino.
2. Peter plans to marry Susan or Carol. Both Susan and Carol have normal parents. However, Susan is an albino whereas Carol is normal.
   a) If Peter marries Susan, find the probability that they will have an albino child.
   b) If Peter marries Susan, find the probability that they will have a normal child.
   c) If Peter marries Carol, find the probability that they will have an albino child.
   d) If Peter marries Carol, find the probability that they will have a normal child.
   e) If Carol has an albino brother, how does this affect the answers to parts (c) and (d)?

STUDY GUIDE   In this chapter the following ideas of probability were discussed.

Counting (p. 242)   Relative frequency (p. 265)
Tree diagram (p. 246)   Statistical probability (p. 265)
Permutations (p. 248)   Axiomatic probability (p. 265)
Factorial notation (p. 249)   Mutually exclusive events (p. 268)
$_nP_r$ (p. 249)   Odds (p. 275)
Combinations (p. 254)   Mathematical expectation (p. 275)
$_nC_r$ (p. 254)   Random numbers (p. 279)
Pascal's triangle (p. 255)   Table of random digits (p. 281)
Sample space (p. 261)   Genetics (p. 282)
Event (p. 261)   Genes (p. 283)
Probability (p. 261)   Dominant traits (p. 283)
Null event (p. 264)   Recessive traits (p. 283)
Definite event (p. 264)

Formulas   1. $_nP_r = \dfrac{n!}{(n-r)!}$ (p. 249)

2. The number of permutations of n things when p are alike, q are alike, r are alike, etc., is

$$\frac{n!}{p!\,q!\,r!}. \text{ (p. 251)}$$

3. $_nC_r = \dfrac{n!}{(n-r)!\,r!}$ (p. 254)

4. Prob $(A$ or $B)$ = prob (A) + prob (B) if A and B are mutually exclusive. (p. 269)

5. Prob $(A$ or $B)$ = prob (A) + prob (B) − prob $(A$ and $B)$ for any events. (p. 271)

SUGGESTED FURTHER READINGS

Adler, I., *Probability and Statistics for Everyman.* New York: New American Library, 1966.

Bell, E. T., *Men of Mathematics.* New York: Simon and Schuster, 1961. Chapter 5 contains a biography of Pascal and chapter 4 contains a biography of Fermat.

Bergamini, D., et al., *Life. Mathematics* (*Life* Science Library). New York: Time-Life Books, 1970. Pages 126–147 discuss figuring the odds in an uncertain world and the fascinating game of probability and chance.

David, F. N., *Games, Gods and Gambling.* New York: Hafner, 1962.

Epstein, R. A., *Theory of Gambling and Statistical Logic.* New York: Academic Press, 1967. Contains an interesting discussion on the fairness of coins.

Gallup, G., *The Sophisticated Poll Watcher's Guide.* Princeton, N.J.: Princeton Opinion Press, 1972. Contains an interesting discussion on taking and interpreting polls.

Gardner, M., *New Mathematical Diversions From The Scientific American.* New York: Simon and Schuster, 1966. See Chapter 19.

Havermann, E., "Wonderful Wizard of Odds." *Life* 51(14): 30ff. (6 October 1961).

Huff, Darrell, *How to Take a Chance.* New York: Norton, 1959.

Kac, M., "Probability." *Scientific American* (September 1964).

Kalb, M., and B. Kalb, "Twenty Days in October." *New York Times Magazine* 23 June 1974.

Kasner, E., and J. Newman. *Mathematics and the Imagination.* New York: Simon and Schuster, 1940. See the chapter on chance and probability.

Kline, M., *Mathematics for Liberal Arts.* Reading, Mass.: Addison-Wesley, 1967. Chapter 23 discusses probability.

Mathematics in the Modern World (Readings from *Scientific American*). San Francisco: W. H. Freeman, 1968. Article 22 discusses chance; articles 23 and 24 discuss probability.

Newmark, J., *Statistics and Probability in Modern Life.* New York: Holt, Rinehart and Winston, 1975. Chapters 5 through 7 discuss the elementary rules for probability and probability functions.

Ore, Øystern, *Cardano, The Gambling Scholar.* New York: Dover, 1965.

Polya, G., *Mathematics and Plausible Reasoning.* Princeton, N.J.: Princeton University Press, 1957.

Weaver, W., *Lady Luck.* New York; Anchor Books, 1963.

The following articles, which all discuss some of the aspects of probability and its applications, appear in James R. Newman, ed., *The World of Mathematics.* New York: Simon and Schuster, 1956.

Haldane, J. B. S., "The Mathematics of Natural Selection." Volume 2, part 5, article 14.

Haldane, J. B. S., "On Being the Right Size." Volume 2, part 5, article 15.

Keynes, J. M., "The Application of Probability to Conduct." Volume 2, part 5, article 4.

Laplace, P. S., "Concerning Probability." Volume 2, part 7, article 1.

Mendel, G., "The Mathematics of Heredity." Volume 2, part 5, article 13.

Peirce, C. S., "The Probability of Induction." Volume 2, part 7, article 3.

Peirce, C. S., "The Red and the Black." Volume 2, part 7, article 2.

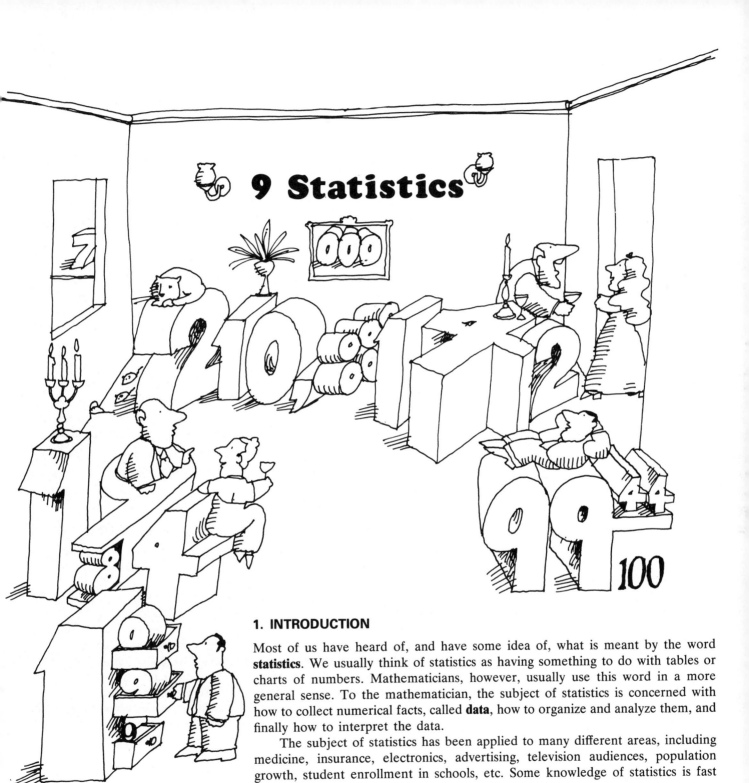

9 Statistics

1. INTRODUCTION

Most of us have heard of, and have some idea of, what is meant by the word **statistics**. We usually think of statistics as having something to do with tables or charts of numbers. Mathematicians, however, usually use this word in a more general sense. To the mathematician, the subject of statistics is concerned with how to collect numerical facts, called **data**, how to organize and analyze them, and finally how to interpret the data.

The subject of statistics has been applied to many different areas, including medicine, insurance, electronics, advertising, television audiences, population growth, student enrollment in schools, etc. Some knowledge of statistics is fast becoming an important tool for everyone. The following examples of the use of statistics should be quite familiar.

1. Statistics show that male drivers under the age of 25 have more accidents than other drivers. Based on this information, insurance companies charge higher premiums for these drivers.

2. On election-day television newscasts, computers are used to "project" the winners based on only very early returns. Samples from representative districts are collected and the predictions are based on these statistics.

3. The latest statistics released by the FBI indicate that serious crime in a large northeastern city increased by 18% during the year 1975.

4. Statistics indicate that due to the extremely high tax rate in New York City, many citizens and businesses are moving out of the city into neighboring states. Consequently, the statistics indicate that an increase in taxes will result in an exodus of people and businesses from the city.

5. When the Social Security system was first put into effect, the rate of contribution was determined by statistics that predicted how long a person would live. In recent years, statistics show that people have been living longer. As a result, the rate of contribution has been steadily increasing.

6. The Nielsen television ratings show that one network has 20% more viewers than another between the hours of 7:00 P.M. and 8:00 P.M. on Monday night.

The study of statistics was really begun by an Englishman, John Graunt (1620–1674). Graunt studied death records in various cities and noticed that the percentages of deaths from different causes were about the same, and did not change much from year to year. Graunt was also the first to discover (using statistics) that there were more boys born than girls. Because, at the time, men were more subject to death from occupational accidents and diseases and from war, it turned out that at the age suitable for marriage, the number of men and women was about equal. Graunt believed that this was a natural way of guaranteeing monogamy.

In 1662, Graunt published *Natural and Political Observations . . . upon the Bills of Mortality*, which has been said to have founded the science of statistics.

The work begun by Graunt was continued by others, who wanted to make the social sciences more "quantitative," that is, based more on mathematics. In the late seventeenth century, life insurance companies were formed and they, of course, were also interested in the information to be obtained from statistics, such as death rates and life expectancy. The Industrial Revolution increased interest in statistics even further. Government agencies and social reformers wanted statistics on births and deaths, national and individual incomes, unemployment, occurrence of disease, etc. By the nineteenth century, statistics was also accepted as an important tool in the physical sciences.

What mathematical knowledge is needed to understand and use statistics? The collection and organization of the data require little or no mathematical background. On the other hand, the interpretation of the data is another story. The statistician should have some mathematical knowledge if he or she is to interpret the data in a meaningful way.

If the data are not interpreted properly, very wrong and sometimes ridiculous conclusions can be drawn from them. As an example of how data can be misinterpreted, consider the following statistics. During 1976, in a southern state, there were 1246 accidents involving drunken pedestrians. Also there were 723 accidents involving drunken drivers. You could conclude that it is more dangerous to be a drunken pedestrian than a drunken driver. Do you agree that this is a reasonable way of interpreting the data, that is, do you agree with the conclusion?

In this chapter, we will discuss how statistical data can be organized and tabulated so that meaningful results can be drawn from them.

---

2. SAMPLING   Suppose a producer is interested in knowing what percentage of the television audience enjoys watching a new show. He or she obviously will not (or cannot) ask every individual who watches television for a reaction to the new show. What he or she will do is take a **sample**. A relatively small group of TV viewers will be selected and asked for their reactions. From their comments, a generalization will be made for *all* television viewers.

Before indicating some of the difficulties involved in taking a sample, we first state formally what we mean by a sample.

Definition 2.1   *A **sample** is a small group of individuals (or objects) selected to stand for a larger group usually called the **population**.*

In taking a sample, a number of problems can arise. The first problem is that of sample size. If a sample is too small, then the individuals used may not be truly typical (or representative) of the population. A sample that is too large is usually very costly. A good sample should be large enough to be typical of the population it represents, but not so large that its cost is ridiculously expensive. How large a sample to select varies from situation to situation. Only by applying statistical procedures can one be reasonably confident of the correct sample size.

The most important thing in taking a sample is the requirement that it be a **random** one. This means that each individual in the population should have an equally likely chance (or probability) of being selected. Unless the sample is a random one, it may not give a true picture of the population it represents.

In 1948 the pollsters predicted that Harry Truman would lose the presidential election. This prediction was based on a sample of the population. It turned out that the pollsters were wrong. This was because their sample was not random and was not truly representative of the population.

Statisticians have devised various techniques for obtaining random samples. In the previous chapter (pp. 280–282), we discussed various ways of obtaining a random sample. Although the *table of random digits* technique discussed in that section is the simplest way of obtaining a random sample, there are other techniques that can be used. A discussion of these is beyond the scope of this text.

In this chapter we will assume that all data have been collected from random samples.

3. MEASURES OF CENTRAL TENDENCY

In this section, we shall be concerned with several methods of interpreting data. To understand what we mean by this, consider a used-car dealer, who, due to economic conditions, has decided to fire one of two salesmen, Crazy Eddie or Mad Mike. Obviously, the dealer wants to keep the better salesman. To help him decide which employee to fire, he has made a chart as shown below, indicating the number of cars sold by each man over the last seven weeks.

	Crazy Eddie	Mad Mike
Week 1	8	10
Week 2	6	12
Week 3	12	12
Week 4	8	11
Week 5	6	12
Week 6	38	12
Week 7	6	8
Total	84	77

At first glance, one would claim that Crazy Eddie is a better salesman, since he sold a total of 84 cars, while Mad Mike sold only 77 cars. However, let us analyze the situation a bit more carefully. Notice that Mad Mike sold more cars than, or the same number of cars as Crazy Eddie during every week but the sixth.

Let us compute the average number of cars sold by both salesmen by dividing each total by the number of weeks (which is 7). We get

Crazy Eddie's average: $\dfrac{84}{7} = 12$;

Mad Mike's average: $\dfrac{77}{7} = 11$.

Thus Crazy Eddie sold 12 cars on the average, whereas Mad Mike sold only 11 on the average. It would again appear that Crazy Eddie is a better salesman.

Another look at the data, however, shows that Mad Mike sold 12 cars most often (on 4 weeks). Crazy Eddie sold 12 cars only once. Crazy Eddie sold 6 cars most often. One might now say that Mad Mike is a better salesman in terms of consistent performance.

Suppose we were to arrange the number of cars sold by each (per week) in order from smallest to largest. We get the following chart.

Crazy Eddie	Mad Mike
6	8
6	10
6	11
(8)	(12)
8	12
12	12
38	12

Note that two numbers have been circled. These are the numbers that are in the middle. For Mad Mike this number is 12 and for Crazy Eddie this number is 8. This example leads us to the following definitions.

Definition 3.1   *The **mean** or **average** of a set of numbers is found by adding them together and dividing the total by the number of numbers added.*

Definition 3.2   *The **mode** of a set of numbers is the number that occurs most often. If every number occurs only once, then we say that there is no mode. A set of numbers may have more than one mode.*

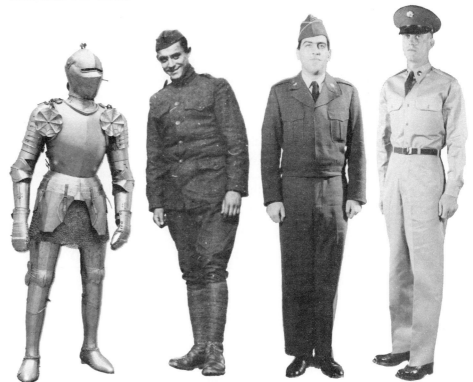

The median height of a medieval knight in armor was 5 ft 6 in. In World War I, the median height of a soldier was 5 ft $7\frac{3}{4}$ in.; in World War II, 5 ft $8\frac{1}{2}$ in.; and after 1958, 5 ft $8\frac{9}{10}$ in. What is the value of knowing these statistics?

Medieval armor: courtesy of Museum of Fine Arts, Boston; soldiers of World War I, World War II, and 1958: E. P. Jones.

Definition 3.3   *If a set of numbers is arranged in order (from smallest to largest), then the number that is in the middle is called the **median**. This is only when there is an odd number of numbers. If there is an even number of them, then the **median** is the average of the middle two numbers (when arranged in order).*

Let us now apply these definitions to our two salesmen. We have the following:

	Crazy Eddie	Mad Mike
Mean	12	11
Median	8	12
Mode	6	12

Who is a better salesman, Crazy Eddie or Mad Mike? One would probably say Mad Mike, even though his mean (or average) is less than Crazy Eddie's. Eddie's average is 12 only because of the sixth-week sales when he sold 38 cars. Mike, on the other hand, consistently sold 11 or 12 cars. In this case, the mean does not tell us as much about the salesmen as does the median or the mode.

The above ideas will further be illustrated by several examples.

Example 1   A professor recently gave a test to his statistics class of eleven students. The following results (grades) were obtained: 78, 53, 100, 27, 94, 88, 98, 93, 98, 91, and 89. Find the mean, median, and mode for this class.

Solution   We first arrange the grades in order from lowest to highest. We get 27, 53, 78, 88, 89, 91, 93, 94, 98, 98, and 100.

The grade that occurred most often is 98. Thus the mode is 98.

The grade that is in the middle (now that we have arranged them in order) is 91, so the median is 91.

To find the mean, we first add all the numbers. The sum is 909. We then divide 909 by the total number of grades (which is 11), getting $909/11 = 82.64$ (rounded off to two decimal places). The mean is then 82.64.

Which is a better indication of class performance in this particular example, the mean, median, or mode?

Example 2   The mathematics department at a state university consists of 8 members whose salaries are given in the following chart.

Rank	Salary
Dave, chairman	$21,000
Arthur, professor	19,000
Roger, assoc. professor	18,500
Betsy, assoc. professor	17,000
Nancy, assist. professor	16,000
Alice, assist. professor	15,810
Bob, instructor	13,120
Jim, lecturer	12,810

Find the mean, median, and modal salary for the members of the mathematics department at this particular university.

Solution   We notice that the salaries are already arranged in order, from highest to lowest. We can then read up the list.

To find the median, we look for the number that is in the middle. In this case, there is an even number of salaries, so that no number is in the middle. The median is somewhere between $17,000 and 16,000. Definition 3.3 tells us that the median is the average of these two numbers, or that the median is

$$\frac{17,000 + 16,000}{2} = \frac{33,000}{2} = \$16,500.$$

What about the mode? Notice that no salary occurred more than once. Definition 3.2 tells us that there is no mode.

To find the mean, we first add all the salaries. The total is $133,240. We now divide the total by 8, getting

$$\frac{133,240}{8} = \$16,655.$$

For the salaries in the mathematics department of this university we have

Mean $16,655,
Median $16,500,
Mode   none.

Which is a better indication of the teachers' salary, the mean or the median?

Comment   The mean, median, and mode are called **measures of central tendency**. The reason for this name should be obvious. Each of these (mean, median, or mode) measures some central or general trend of the data. Depending on the situation, one will usually prove to be more meaningful than the other.

EXERCISES   1. Richard has recently bought a new car and is interested in knowing how many miles he gets per gallon of gas. On 8 successive weeks he has computed the miles per gallon. The results of his calculations are

Week	1	2	3	4	5	6	7	8
Miles per gal	14	21	18	14	19	18	15	17

Find the mean, median, and mode for these numbers. Which is a better indication of performance?

2. A health insurance company is interested in determining the cost for maternity care in a large city. (They assume a 4-day hospital stay in a semiprivate room.) Eleven hospitals have responded to a questionnaire sent by the company. At these hospitals, the cost is $920, $480, $630, $810, $1050, $630, $725, $890, $970, $500, and $770. What is the average maternity cost in these hospitals?

3. A car manufacturer is interested in knowing how often customers return to the dealer for repairs on their new cars. He sends a questionnaire to 20 owners and finds that the number of times each returned to the dealer was 7, 3, 0, 1, 10, 19, 4, 5, 8, 3, 10, 4, 3, 5, 2, 0, 1, 3, 4, and 2. Find the mean, median, and mode for these numbers. Which is a better indication of the frequency of repair?

4. Rosemary has taken seven math tests during the term. Her grades were 93, 67, 75, 100, 31, 78, and 95. Find her mean, median, and mode. Which would you say is most important to her?

5. An IRS agent is reviewing the deductions for medical expenses by families in the $15,000–$20,000 income bracket for the year 1975. She has selected 15 returns at random. In these returns, the following medical deductions were claimed: $490, $350, $178, $210, $952, $103, $215, $95, $164, $300, $208, $0, $1320, $175, and $38. Find the average medical deduction for this income group.

6. The number of minutes spent by several students in the biology laboratory performing their experiments is as follows:

Student	Time spent in lab (min)
Pedro	100
Agnes	95
Rosalie	125
Liz	60
Cathy	75
Juanita	100
George	85
Phil	80

Using what you consider to be the most appropriate measure of central tendency, find the average amount of time spent by a student in the lab. Which measure of central tendency did you use and why?

ACME	A-1
$13,000	$15,000
12,000	16,000
18,000	13,000
10,000	17,000
22,000	14,000

7. John is interested in learning how to drive a truck and is considering two possible driving schools, the ACME Training School and the A-1 Truck Driving School. Both claim that their graduates have an average first-year salary of $15,000. Judging by these claims, John believes that both are equally good. His friend persuades him to speak to some recent graduates of each school and find out their individual salaries, rather than their average salaries. The information he gets is shown on the left. How is this information useful in helping John make his decision? Explain.

8. A company pays its workers the following salaries depending on title.

Title	Number of employees	Weekly salary
Manager	1	$500
Supervisor	3	300
Machinists	8	275
Other workers	14	200
Secretaries	4	150

Two students were asked to calculate the average. One student did as follows:

$$\frac{500 + 300 + 275 + 200 + 150}{5} = \frac{1425}{5} = \$285.$$

This student then claimed that the average salary was $285 per week.

A second student computed a **weighted arithmetic average** as follows:

$$\frac{1 \cdot 500 + 3 \cdot 300 + 8 \cdot 275 + 14 \cdot 200 + 4 \cdot 150}{30} = \frac{7000}{30} = \$233.33.$$

He claimed the average salary was $233.33. Which student is correct and why?

9. If a dress manufacturer claims that the average size dress sold is 10, which average is he referring to, the mean, median, or mode?

10. If the mean IQ in one sixth-grade class is 105 and if the mean IQ in another sixth-grade class is 115, can we conclude that the mean IQ for both classes together is 110? Explain.

11. Mary keeps a record of the number of hours that she spends studying math.

Day	Number of hours spent
Monday	3
Tuesday	5
Wednesday	4
Thursday	2
Friday	4
Saturday	7

a) Calculate the mean, median, and mode.

b) How would the mean, median, and mode be affected if she had studied one hour less each day?

Chain A	Chain B
43	42
38	40
51	38
37	39
31	41

12. A state consumers' group is investigating the milk prices charged by two large supermarket chains in various parts of the city. Both chains claim that their average milk price is 40¢ a quart. Investigators find that the prices shown on the left are charged by these chains in five different neighborhoods of the city.

a) Verify that the average price charged by each chain is 40¢.

b) What difference do you notice in the prices charged by both chains?

---

*4. PERCENTILES   In the preceding section, we discussed different ways of analyzing data. Quite often, one is interested in knowing the position of a score in a list of numbers. Thus, if Mary Ruth gets a score of 83 on a civil service exam that she has just taken, she undoubtedly would be interested in knowing how this score compares with others who have taken the exam. In such a situation, she would probably be more interested in her **percentile rank**, rather than in the mean, median, or mode.

Let us analyze the results of the civil service exam that Mary Ruth took. The exam was given to 200 people, including Mary Ruth. She finds that 70% of the people who took the exam got below 83, 10% of the people got 83, and the remaining 20% scored above 83. Since 70% of the people got below 83 and 20% got above 83, her percentile rank should be between 70 and 80 (why?). We use 75, which is halfway between 70 and 80. What we have done is find the percentage of scores that are below her score and add one-half of the scores that are the same as her score. The result is her percentile rank. In our case, Mary Ruth's percentile rank is 75. This means that *approximately* 25% of the people who took the exam scored higher than she did and 75% of the people scored lower than she did. (Why is it only approximate?) Most civil service tests are graded using such a procedure.

This leads us to the following.

Definition 4.1   *The **percentile rank** of a score is found by adding the percentage of scores below it to one-half of the percentage of scores equal to it.*

Although the above definition enables us to find the percentile rank of an individual score, in practice we can use a convenient formula. Let X be a given score, let B represent the *number* of scores below the given score X, and let E represent the *number* of scores equal to the given score X. If the total number of scores is n, then the percentile rank of the given score is given by the following formula.

Formula 4.1

$$\text{Percentile rank of } X = \frac{B + \frac{1}{2}E}{n} \cdot 100$$

Let us illustrate the use of Formula 4.1 with an example.

Example 1   There are 23 students in a statistics class. On the midterm exam, the grades of the students were 79, 63, 94, 100, 83, 92, 78, 62, 53, 84, 76, 22, 17, 52, 57, 66, 83, 72, 81, 70, 69, 46, and 97. Douglas got an 83 on the exam. Find his percentile rank.

Solution   Since there were 23 people who took the exam, $n = 23$. Analyzing the individual scores, we find that there are 16 scores that are below 83 and exactly 2 scores (including Douglas's) that are equal to 83. Thus $B = 16$ and $E = 2$. We now apply Formula 4.1 to find Douglas's percentile rank. We have

$$\text{Douglas's percentile rank} = \frac{B + \frac{1}{2}E}{n} \cdot 100$$

$$= \frac{16 + \frac{1}{2}(2)}{23} \cdot 100$$

$$= \frac{16 + 1}{23} \cdot 100$$

$$= \frac{17}{23} \cdot 100 = 73.91.$$

Therefore, Douglas's percentile rank is 73.91. This means that Douglas did better than approximately 73.91% of the students and only about 26.09% of the students did better than Douglas.

Comment   In the above example, Douglas's percentile rank was 73.91. We sometimes say that Douglas was in the 73.91st percentile.

Statisticians have special names for certain percentiles. The 25th percentile is called the *lower or first quartile*. The 50th percentile is called the *median or middle quartile*. The 75th percentile is called the *upper or third quartile*.

EXERCISES   1. Professor Greenberg has just graded his statistics exams. The following are the grades of his students.

Mark:	98	Mary Ann:	68	Arthur:	100	Laurie:	58
Denise:	72	Charles:	75	Gregory:	76	Jack:	60
Ernest:	31	Morton:	58	Rocco:	85	Isaac:	91
Alma:	82	Marilyn:	82	John:	74	Janet:	75
Angel:	91	Hilda:	46	Ann:	66	Helen:	74

Find the percentile rank of (a) Hilda; (b) Isaac; and (c) Helen.

2. In Kathleen's English class there are 25 students. On the midterm exam, 14 students got lower grades than Kathleen, 3 others got the same grade as Kathleen, and 7 got higher grades than Kathleen.

a) Find Kathleen's percentile rank.

b) If the teacher decides to curve the exam by adding five points to everyone's grade, how is Kathleen's percentile rank affected?

---

5. MEASURES OF VARIATION   It is very hard to find two things of any type that are identically the same. Two people of the same age and sex may differ a great deal in height, weight, etc. Every cook knows that the same recipe will not necessarily result in the same quality of cake. Even things that are mass-produced are not really exactly the same. There is some variation or difference among them. In this section, we will discuss some of the methods used to measure variation.

To get started, let us again look at the two supermarket chains of Exercise 12 of Section 3 (p. 295). The prices charged by the supermarket chains are as follows.

Chain A	43	38	51	37	31
Chain B	42	40	38	39	41

The average (mean) price of a quart of milk for each chain is 40¢. Yet for chain A, the price of a quart of milk varies from 31 to 51 cents. This gives a *range* of 51–31 or 20 cents. For chain B, the prices vary from 38 to 42 cents. Their range is 42–38, or 4 cents.

This leads us to the following definition.

Definition 5.1   *The **range** of a set of numbers is found by subtracting the smallest number from the largest.*

Comment   No matter how many numbers are in the original data, only two of them (the smallest and the largest) are needed to compute the range.

Unfortunately, the range does not tell us anything about how the other numbers vary. For this reason we need another measure of variation called the **standard deviation.** This tells us how "spread out" the numbers are. To find the standard deviation for each of the supermarket chains is a simple procedure. We first find the mean. (In our case, we already know that it is 40¢.) We subtract the mean from each price and square the result. We then find the average of these squares. Finally, we take the square root of this average. The result is called the standard deviation.

We will now calculate the standard deviation for each supermarket chain.

Supermarket chain A

Price (in cents)	Difference from mean	Square of difference
43	$43 - 40 = 3$	$(3)(3) = +9$
38	$38 - 40 = -2$	$(-2)(-2) = +4$
51	$51 - 40 = 11$	$(11)(11) = +121$
37	$37 - 40 = -3$	$(-3)(-3) = +9$
31	$31 - 40 = -9$	$(-9)(-9) = +81$
Total 200		Total 224

$$\text{Mean} = \frac{200}{5} = 40 \qquad\qquad \text{Average} = \frac{224}{5} = 44.8.$$

Thus, the standard deviation for supermarket chain A is $\sqrt{44.8}$ or 6.69.[1]

What about supermarket chain B? Let us compute its standard deviation.

Supermarket chain B

Price (in cents)	Difference from mean	Square of difference
42	$42 - 40 = 2$	$(2)(2) = +4$
40	$40 - 40 = 0$	$(0)(0) = 0$
38	$38 - 40 = -2$	$(-2)(-2) = +4$
39	$39 - 40 = -1$	$(-1)(-1) = +1$
41	$41 - 40 = +1$	$(1)(1) = 1$
Total 200		Total 10

$$\text{Mean} = \frac{200}{5} = 40 \qquad\qquad \text{Average} = \frac{10}{5} = 2$$

Therefore, for supermarket chain B, the standard deviation is $\sqrt{2}$ or 1.41.

---

1. A knowledge of how to compute square roots is not assumed. These values can be obtained by using a calculator or from square root tables.

We can summarize the procedure to be used in finding the standard deviation by the following rule.

Rule *The standard deviation of a set of numbers is the result obtained by finding (in order)*

a) *the mean (or average) of the numbers,*
b) *the difference between each number and the mean,*
c) *the squares of each of these differences,*
d) *the average of these squares,*
e) *the square root of the average of these squares.*

We will illustrate this rule further with another example.

Example 1 A large city in the south requires that its policemen be at least 5 ft 7 in. tall (67 in.). Seven policemen are selected and their heights recorded. What is the standard deviation if their heights are 69, 72, 74, 67, 68, 70, and 70 in.?

Solution We arrange their heights in order in the following chart and then calculate the standard deviation as indicated.

Height	Difference from mean	Square of difference
67	$67 - 70 = -3$	$(-3)(-3) = +9$
68	$68 - 70 = -2$	$(-2)(-2) = 4$
69	$69 - 70 = -1$	$(-1)(-1) = 1$
70	$70 - 70 = 0$	$(0)(0) = 0$
70	$70 - 70 = 0$	$(0)(0) = 0$
72	$72 - 70 = 2$	$(2)(2) = 4$
74	$74 - 70 = 4$	$(4)(4) = 16$
490		34

$$\text{Mean} = \frac{490}{7} = 70 \qquad\qquad \text{Mean} = \frac{34}{7} = 4.857 \text{ (rounded off)}$$

The standard deviation of their heights is thus $\sqrt{4.857}$, or 2.2.

Comment   You may feel that the standard deviation is a rather complicated number to calculate, so why bother. However, it is an extremely important and useful number to the mathematician. A detailed discussion of how it is used is beyond the scope of this text.

Another measure of variation that is often used is the *variance*. It is computed in a manner similar to that for the standard deviation with one difference. We omit step (e) of the rule on p. 299. Thus for Example 1, the standard deviation is $\sqrt{4.857}$ and the variance is 4.857.

Example 2   A Women's Lib group is interested in determining the number of fulltime women professors in the mathematics departments of several western colleges. It has obtained the following information on 9 such colleges. For every 100 faculty members, these schools employed 21, 7, 10, 2, 43, 15, 22, 14, and 1 women. Find the standard deviation and the variance.

Solution   We arrange the numbers in the following chart and then calculate the variance and standard deviation as shown.

Number of female employees	Difference from mean	Square of difference
21	$21 - 15 = 6$	$(6)(6) = 36$
7	$7 - 15 = -8$	$(-8)(-8) = 64$
10	$10 - 15 = -5$	$(-5)(-5) = 25$
2	$2 - 15 = -13$	$(-13)(-13) = 169$
43	$43 - 15 = 28$	$(28)(28) = 784$
15	$15 - 15 = 0$	$(0)(0) = 0$
22	$22 - 15 = 7$	$(7)(7) = 49$
14	$14 - 15 = -1$	$(-1)(-1) = 1$
1	$1 - 15 = -14$	$(-14)(-14) = 196$
Total 135		Total 1324

$$\text{Mean} = \frac{135}{9} = 15 \qquad\qquad \text{Mean} = \frac{1324}{9} = 147.11$$

Thus, the variation is 147.11 and the standard deviation is $\sqrt{147.11}$ or 12.13.

EXERCISES

1. During the week of July 4–9, a Texas highway patrolman issued the number of speeding tickets shown in the table below. Find the range, standard deviation, and variance for the number of tickets issued.

Date	Number of tickets issued	Date	Number of tickets issued
July 4	16	July 7	10
July 5	8	July 8	11
July 6	12	July 9	3

2. Tom Edison recently tested a dozen light bulbs. They were used for 81, 63, 48, 75, 59, 93, 72, 61, 88, 33, 56, and 63 hours before burning out. Find the range, standard deviation, and variance for the life of a bulb.

3. Justin Case is an insurance salesman. During 1976, he sold the number of policies indicated in the chart below.

Jan.	Feb.	Mar.	Apr.	May	June	July	Aug.	Sept.	Oct.	Nov.	Dec.
23	34	29	19	49	8	18	20	53	46	33	28

Find the range, standard deviation, and variance for the number of policies sold.

4. A local clean-air committee keeps records on the numbers of summonses issued by the health department to industrial firms for air pollution violations. Over the last six months, the department has issued the following numbers of summonses.

Nov.	Dec.	Jan.	Feb.	Mar.	Apr.
12	3	11	19	6	9

Find the range, standard deviation, and variance for the number of summonses issued.

5. In Professor Cupid's "Marriage and the Family" course, there are 9 women students. The age at which each had her first child is 25, 19, 34, 21, 13, 41, 22, 27, and 14 years. Find the range, standard deviation, and variance for the age at which these women had their first child.

6. Gertrude is interested in buying a 21-in. color television set. She has shopped around and has come up with the list of prices given below.

Store	Price (including installation)
Store A	$425
Store B	379
Store C	465
Store D	399
Store E	439
Store F	455

a) Calculate the mean, median, and mode price for the set.

b) How are the mean, median, and mode affected if there is a delivery charge of $10 for each set?

c) Calculate the range, standard deviation, and variance for the prices of the TV set.

d) How are these affected by the $10 delivery charge?

7. What effect, if any, would doubling each number of the data have on the mean and the standard deviation?

8. Calculate the range, standard deviation, and variance for Exercises 1 through 6 of the preceding section.

6. FREQUENCY DISTRIBUTIONS

There are many situations in which the data may be so numerous that it would be difficult (if not impossible) to come up with any meaningful interpretation of them. To see how this can happen, consider Michael who is late for work many times. His boss tells him that for the past 6 weeks he has been late the number of minutes per day shown below.

7	6	1	4	0	13
15	10	12	3	3	12
10	12	11	13	2	11
2	3	8	5	3	14
5	14	6	7	7	9

All one can say definitely at first glance is that he was on time only once and one day he was as much as 15 minutes late. Since the numbers are not arranged in order, it is somewhat difficult to conclude anything else from them. For this reason we use a frequency distribution. First we have the following definition.

Definition 6.1   *A **frequency distribution** is a convenient way of organizing data so that we may see what patterns they have. The word **frequency** will be interpreted to mean how often the number occurred.*

A frequency distribution is made very easily. We first make a list of numbers from 0 to 15 in a column to show how many minutes Michael was late. Then we make a second column for tally marks. We go through the original numbers, and each time he was late we put a tally mark in the appropriate space. Finally, we add the tally marks per line and indicate this sum in the frequency column. When we apply this to our problem, we get the following.

Minutes late to work	Tally	Frequency	Minutes late to work	Tally	Frequency
0	l	1	8	l	1
1	l	1	9	l	1
2	ll	2	10	ll	2
3	llll	4	11	ll	2
4	l	1	12	lll	3
5	ll	2	13	ll	2
6	ll	2	14	ll	2
7	lll	3	15	l	1

Once we have done this, we can come up with meaningful interpretations. We see that most latenesses were 3-minute ones (there were four of them). Michael was also late more than 5 minutes 19 times. This would represent 19 out of 30 times, or approximately 60% of the time. Still other interpretations can be given to the above data.

One may want to draw a *bar graph* for these numbers.

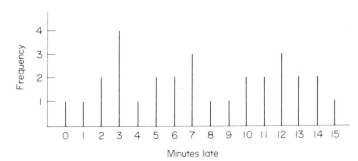

To construct the bar graph we first draw two lines, one horizontal (across) and one vertical (up-down). The horizontal line we will label "Minutes late" and the vertical line we will label "Frequency."

Once we have the frequency distribution, we can draw the bar graph very easily. The height of each bar will represent the frequency. The bar graph will also tell us at a glance that the most latenesses were the 3-minute ones.

We could shorten the above frequency distribution as shown in the following chart.

Minutes late	Tally	Frequency		Minutes late	Tally	Frequency
0–1	ll	2		8–9	ll	2
2–3	ЖНГ l	6		10–11	llll	4
4–5	lll	3		12–13	ЖНГ	5
6–7	ЖНГ	5		14–15	lll	3

This chart is more compact, *but* some of the information is lost in this version. For example, we can see that Michael is late 2–3 minutes 6 times. But we cannot tell exactly how often he is 2 minutes late or 3 minutes late.

When there are many numbers, listing them separately, as we did in the first chart, may make it difficult to look at the data and draw meaningful conclusions. If the data are grouped, as in the second chart, they may be easier to interpret.

To further illustrate the idea of a frequency distribution and a bar graph, consider the following example.

Example 1   A large midwestern university is reviewing the performance of its star basketball player. During the past season, he has scored the following number of points per game.

27	16	19	24	18	23	24	18	24	25
23	16	23	24	19	22	17	25	19	27
19	29	24	25	24	18	32	23	21	30

Find the frequency distribution and draw the bar graph for the above numbers.

Solution We make three columns. The first column will contain the number of points scored, the second will have the tally, and the third will give the frequency.

Number of points	Tally	Frequency		Number of points	Tally	Frequency
16	ll	2		25	lll	3
17	l	1		26		0
18	lll	3		27	ll	2
19	llll	4		28		0
20		0		29	l	1
21	l	1		30	l	1
22	l	1		31		0
23	llll	4		32	l	1
24	JHt l	6				

The bar graph for this distribution is given on the left. One thing should be immediately obvious. The player scored 24 points most often.

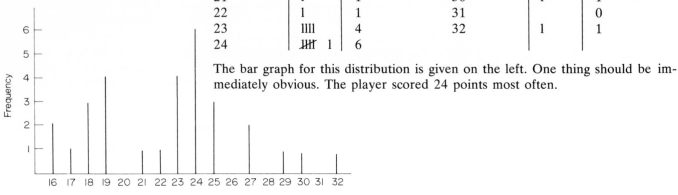

Number of points scored

Example 2 George is a maintenance man in a large office building. He has compiled the following list of numbers that indicate the life length (in hours) of 25 bulbs.

50	50	45	45	60	50	40
45	40	35	55	50	45	
55	55	40	50	45	60	
60	50	50	55	65	55	

Find the frequency distribution and draw the bar graph for these numbers.

Solution Again we make three columns. The first column is for the life length, the second for the tally, and the third for the frequency. We have the chart shown below.

Life length	Tally	Frequency
35	l	1
40	lll	3
45	JHt	5
50	JHt ll	7
55	JHt	5
60	lll	3
65	l	1

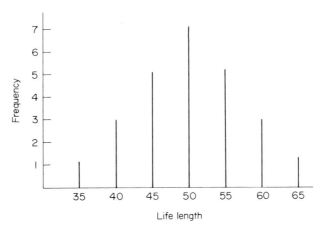

The bar graph for this distribution is shown here.

EXERCISES   Find the frequency distribution and draw the bar graph for each of the following lists of numbers.

1. The number of rapes reported in a certain state during the month of February 1975 was as follows.

12	2	4	10	4	5	17	0
5	11	4	11	3	9	8	
3	11	9	11	3	3	7	
4	3	4	4	9	9	8	

2. The heights of individuals in a statistics course of 30 students were (in inches):

61	64	62	66	62	63
72	73	65	64	65	65
65	71	72	69	63	68
67	68	70	68	67	67
63	66	65	65	71	69

3. The number of people returning merchandise to the lingerie department of the J. C. Nickel Department Store over a 4-week period was as follows.

12	9	8	18	15
17	15	13	17	5
10	17	15	16	17
2	12	17	14	7

4. The number of performers interviewed by the manager of the Dingling Brothers Circus in a 5-week period was

5	4	8	8	7
7	7	6	9	1
1	9	1	4	4
2	3	0	5	3
7	2	7	3	2

5. The number of smugglers arrested during the month of July was

6	12	9	7	6	1	7
4	4	8	9	4	10	
11	6	2	14	3	8	
0	10	11	12	8	12	
5	3	12	9	13	9	

6. The number of rooms rented daily at the Waldorf Hysteria Hotel during the month of June 1976 was as follows.

43	60	57	60	44
57	45	49	45	57
62	57	52	57	60
49	51	43	58	62
51	52	51	60	44

7. GRAPHS

In the last section, we saw how a bar graph can be used to picture information. In statistics we use not only bar graphs, but also other kinds of graphs to show all the information given in a situation. This is often of great help in arriving at meaningful conclusions.

The following examples will illustrate how bar graphs and other kinds of graphs can be used.

Example 1   The number of vehicles passing through a toll booth on a turnpike during a 24-hour period is given in the table on the left.

Time	Number of cars arriving
12 midnight–2 A.M.	120
2 A.M.–4 A.M.	640
4 A.M.–6 A.M.	1790
6 A.M.–8 A.M.	5780
8 A.M.–10 A.M.	3460
10 A.M.–12 noon	2010
12 noon–2 P.M.	1860
2 P.M.–4 P.M.	2000
4 P.M.–6 P.M.	4030
6 P.M.–8 P.M.	5640
8 P.M.–10 P.M.	2440
10 P.M.–12 midnight	560

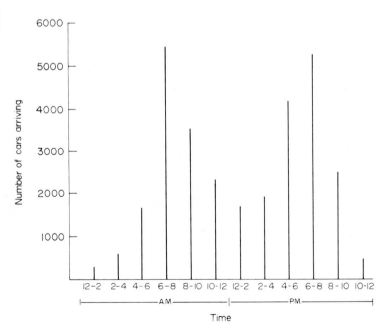

In order to be able to draw any meaningful conclusion from this data, we construct a bar graph that contains all the given information. From this graph we can see that the greatest number of cars passed through the toll booth during the hours of 6 to 8 A.M. Traffic then began dropping off until after 2 P.M., at which time the number of cars increased until about 8 P.M. and then dropped off again.

Information of this type is needed by the authorities to determine the number of toll collectors to hire and the hours to hire them for, so that motorists will not have to wait in long lines. It is easier to determine this information from a graph than from the table of numbers.

Example 2   The latest statistics showing the incidence of heart attacks among various age groups in a certain community in the southwest are pictured in the following bar graph.

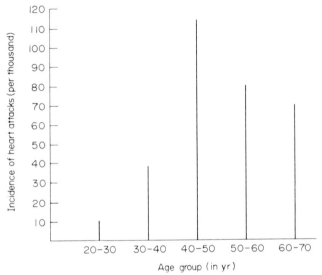

a)   For this particular community, in what age group are there more than 60 heart attacks per thousand?

b)   In what age group are there more than 100 heart attacks per thousand?

Solution   a)   We see from the graph that in the age groups between 40 and 70 yr there were more than 60 heart attacks per thousand people.

b)   From the graph we see that only the age group of 40–50 yr had more than 100 heart attacks per thousand people.

Example 3   A large supermarket chain is interested in knowing which flavor of soda is in greatest demand and during which months. It needs this information so that it can adequately stock its warehouses in advance. For its three most popular flavors, it has available last year's statistics, which have been recorded in the form of a bar graph.

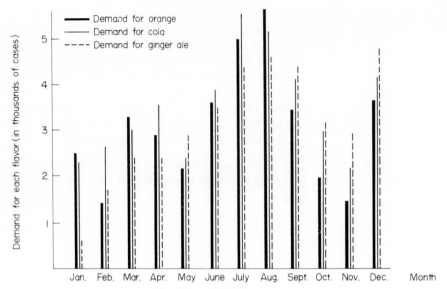

a) In which months is orange most popular?

b) Which flavor is most popular in December?

c) Which flavor is least popular in February?

Example 4   The budget of the Wegetum Crime Syndicate has just been approved by the council members. The following *circle graph* indicates how the syndicate plans to spend its money.

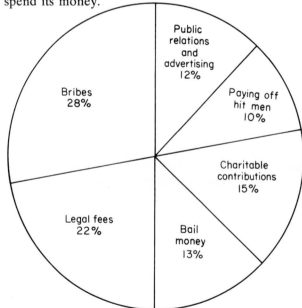

If the total budget is $2,500,000, how much money will be spent for

a) bribes?

b) legal fees?

c) paying off hit men?

d) charitable contributions?

Solution a) 28% of $2,500,000 will be spent for bribes. (We write 28% in decimal form as 0.28.) Thus they will spend 0.28 × $2,500,000 or $700,000 for bribes.

b) They will spend 22% of $2,500,000 for legal expenses. (We write 22% in decimal form as 0.22.) Thus they will spend 0.22 × $2,500,000 or $550,000 for legal expenses.

c) They will spend 0.10 × $2,500,000 or $250,000 for paying off hit men.

d) They will spend 0.15 × $2,500,000 or $375,000 for charitable contributions.

Example 5 An electric company wants to know during which hours electrical supply is in greatest demand on a typical summer day. It needs this information so that it can prepare itself to satisfy consumer demands adequately. Information for one typical day has been gathered and is pictured in the *line graph* below.

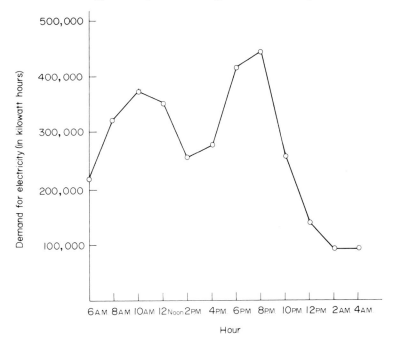

a) During which hour(s) is electricity in greatest demand?

b) During which hours is the demand for electricity decreasing?

c) During which hour(s) is electricity in least demand?

EXERCISES

1. The graph below shows the number of students enrolled in the adult education program of a city school system.

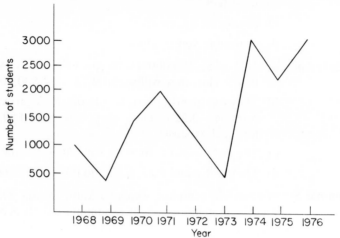

a) In what year(s) was enrollment highest?

b) In what year(s) was enrollment lowest?

c) Between which two years did enrollment increase most?

2. The graph below shows the number of unemployed workers in a large state during 1975.

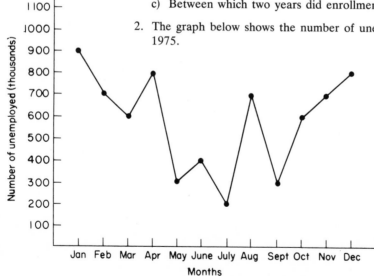

a) In what month was unemployment highest?

b) In what month was unemployment lowest?

c) Between which two months did unemployment decrease most?

3. The number of Jolly Giants sold by the Happy Toy Company during 1976 was as follows.

January	5,600	July	6,500
February	10,000	August	6,900
March	8,400	September	8,000
April	9,000	October	7,500
May	4,400	November	12,300
June	3,300	December	14,000

a) Make a line graph to show the sales.

b) Between which two months was the increase in sales greatest?

c) In which month were sales for this toy the lowest?

4. The number of violent crimes in a western city over the past 15 years was as follows.

Year	Number of violent crimes	Year	Number of violent crimes
1962	450	1970	1190
1963	540	1971	1220
1964	1000	1972	1500
1965	1110	1973	1670
1966	1090	1974	2040
1967	1230	1975	2200
1968	1210	1976	2350
1969	1200		

a) Draw a line graph to represent the above information.

b) How would you describe the crime trend in this city?

5. The causes of fatal automobile accidents during 1975 were the following:

Inattention	28%
Speeding	23%
Intoxication	9%
Mechanical failure	6%
Other traffic violations	19%
Poor road and weather conditions	11%
Miscellaneous	4%

Make a circle graph to picture this information.

8. NORMAL DISTRIBUTION

If you connect the endpoints of each line of a bar graph, you get a **curve** for the given frequency distribution. Let us do this for Example 2 of Section 6 (p. 304). We get a figure such as the one on the left.

This curve looks rather like a bell. There are many frequency distributions whose graphs are bell-shaped curves, as shown in Fig. 2. Such a graph is called a **normal curve** and the frequency distribution is called a **normal distribution**.

For any normal distribution, the mean, median, and mode are always at the center, which is the highest point. Thus, in Fig. 1, the mean, median, and mode are all 50 years. It is rather obvious why the mode is in the middle. It is the most

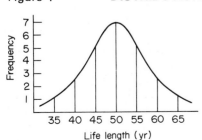

Figure 1

Life length (yr)

frequently occurring score, that is, it is the highest point on the frequency distribution. Can you see why the median and mean are at the middle?

One of the important properties of a normal distribution is that it is symmetrical. This means that if we draw a vertical line as we have done in Fig. 2, and then fold the graph along the dotted line, then one side will fall exactly on top of the other. Thus 50% of the data will be to the left of the dotted line and 50% of the data will be to the right of the dotted line. This line, of course, is drawn at the point that represents the mean, median, and mode.

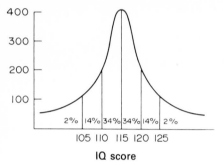

Figure 2

We will give several examples of normal distributions and point out what information can be obtained from the normal curve.

Example 1   An intelligence test was given at a school to 1000 students. The graph of their IQ's is shown on the left.

The vertical line in the middle represents the average IQ. It is 115.

The standard deviation is known to be 5. What percent of the students have an IQ within 1 standard deviation of the mean? This really means, what percent of the students have an IQ between 110 and 120? From the graph, we see that it is 68%.

What percent of the students have an IQ within 2 standard deviations of the mean? Again this means, what percent of the students have an IQ between 105 and 125? From the graph, we see that is 96%.

What percent of the students have an IQ within 3 standard deviations of the mean?

What percent of the students have an IQ below 115? above 115?

Example 2   The weight of the 200 members of the Jack La Lean Health Club are normally distributed as shown in the following figure.

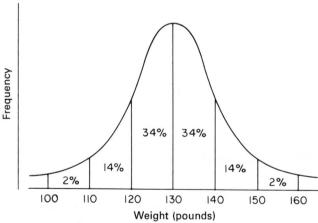

Weight (pounds)

a) What percentage of the women weigh between 120 and 140 lb?

b) What percentage of the women weigh between 110 and 150 lb?

c) What percentage of the women weigh between 140 and 160 lb?

Solution   We read the answers from the normal curve. We find that 68% of the women weigh between 120 and 140 lb, 96% of the women weigh between 110 and 150 lb, and 16% of the women weigh between 140 and 160 lb.

Comment   If a frequency distribution has a normal curve as its graph, then it is known that approximately 68% of the population will be within 1 standard deviation of the mean. Also approximately 95% (more accurately 95.4%) of the population will be within 2 standard deviations of the mean. Approximately all (more accurately 99.7%) of the population will be within 3 standard deviations of the mean.

Example 3   A teacher believes that the scores of his students on the final exam are normally distributed with a mean of 75 and a standard deviation of 5. He wants 2% of the class to get A or F, 14% of the class to get B or D, and 68% of the class to get C. What range of grades on the final should he use to accomplish this?

Solution   We first draw the normal curve as shown on the left.

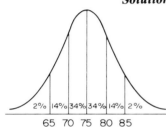

Since we are told that the standard deviation is 5, we conclude that the range of 75 to 80 will contain 34% of the class. Similarly, the range of 70 to 75 will contain 34% of the class.

Between 80 and 85 (between 1 and 2 standard deviations of the mean), we have 14% of the class. Similarly, between 65 and 70, there is 14% of the class.

Finally, only 2% of the class will score above 85 and the same is true below 65. Thus the teacher should use the following table.

Grade on final	Grade assigned	Percentage of class
Above 85	A	2%
80–85	B	14%
70–79	C	68%
65–69	D	14%
Below 65	F	2%

EXERCISES

1. Which of the following would be likely to be normally distributed? (In each case assume a sample size of 300 randomly selected people.)

   a) age of victims of heart attacks

   b) age of unemployed person during 1975

   c) height of statisticians

   d) intelligence of math teachers

   e) useful life of a car

   f) number of cavities in 10-year-old children

   g) age at which American men marry for the first time

   h) unemployment rate during 1975

2. The following distribution represents the number of accidents (per year) occurring in a factory that employs 10,000 workers. (The mean is 10; the standard deviation is 2.)

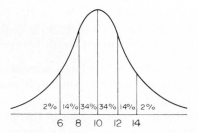

a) What percentage of the employees had 8–12 accidents per yr?

b) What percentage of the employees had more than 8 accidents per yr?

c) What percentage of the employees has less than 8 accidents per yr?

d) Approximately how many employees had between 6 and 14 accidents per yr?

3. The age at which individuals die of a certain disease is found to be normally distributed with a mean of 30 and a standard deviation of 4, as shown at the left.

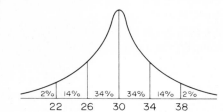

a) What percentage of the population affected with this disease die between the ages of 22 and 38 yr?

b) What percentage of the population die between the ages of 34 and 38 yr?

c) What percentage of the population die under the age of 30 yr?

4. In a large hospital in Los Angeles, the length of stay for routine maternity cases is found to be normally distributed with a mean of 5 days and a standard deviation of 1 day, as shown below.

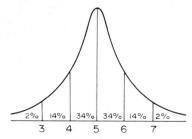

a) What percentage of the mothers remain in the hospital between 4 and 6 days?

b) What percentage of the mothers remain in the hospital more than 6 days?

c) What percentage of the mothers remain in the hospital less than 7 days?

5. A recent survey found that the life (in hours) of television picture tubes is normally distributed with a mean of 1000 and a standard deviation of 100, as shown at the left.

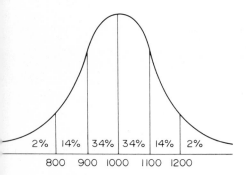

a) What percentage of picture tubes lasted more than 800 hr?

b) What percentage of picture tubes lasted less than 1100 hr?

c) What percentage of picture tubes lasted between 1000 and 1100 hr?

d) What percentage of picture tubes lasted between 900 and 1100 hr?

e) How do the answers to parts (c) and (d) compare? Explain.

9. CORRELATION   Many colleges administer exams to entering freshmen. One reason for the exams is to place the student in the appropriate course. Naturally a college would be interested in determining whether its tests actually do succeed in placing the student in the right course. The college wants to know, therefore, what relationship there is between performance on the exam and performance in the course the student is placed in as a result of the exam. Such a relationship is known as **correlation**.

To be specific, imagine that we have administered a mathematics placement exam to ten students. We now want to compare these results with the students' numerical grades in the college math course in which they are placed. The results are given in the table below: x represents the placement score, and y represents the grade in the college math course. We graph this information as follows:

x	y
88	92
91	98
79	78
68	71
94	100
73	72
65	69
77	80
84	90
80	85

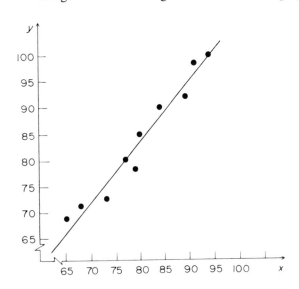

This graph is called a **scatter diagram**. The straight line through the scattered points is used so that we can note if any trend exists. Just exactly how this line is drawn is beyond the scope of this book, but the line is known as the **line of regression**.

In any case, once this line is drawn, the following facts will be true. If the points all lie close to this line, we say that there is a **high correlation** between x and y. If the points are widely scattered, then there is a **low correlation** between them. Roughly speaking, high correlation means that x and y are closely connected, and low correlation means that x and y are not necessarily connected.

In our example, since all the points are relatively close to the line, there is a high correlation between the placement test score and the college math course score. This tells us that the test did its job fairly well for these ten students.

Instead of drawing a scatter diagram, there is also a purely algebraic formula one can use that gives what is known as the **coefficient of correlation**.[2] Using this formula, one can get results similar to those we obtained using the scatter diagram.

One must, however, be very careful in interpreting results obtained when using correlation. For example, if either of the above methods were applied in order to determine if there is a correlation between the rise in teachers' salaries and the rise in alcoholic consumption between 1950 and 1972, one might conclude that there is a definite relationship. Other factors besides the scatter diagram or the formula have to be considered.

Sometimes, instead of having a *line* of regression, we have a *curve* of regression. The scatter diagrams for some possibilities are shown here.

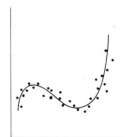  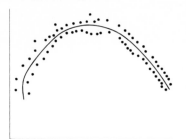

Again we point out that other factors must be considered.

EXERCISES   Draw the scatter diagram for each of the following and try to determine if any relationship exists between x and y. If you believe that there is a correlation, try to draw the line of regression.

1.

x	10	4	2	5	7	3	8	8
y	24	12	6	15	16	10	19	21

2. The following table shows the average weights of boys in different age groups at a certain school.

Age (yr) (x)	5	6	7	8	9	10	11	12	13	14
Weight (lb) (y)	53	56	60	62	69	75	83	90	103	120

3. The following table shows the height (in inches) of men at a certain college, and their average shoe sizes.

Height (x)	62	63	64	65	66	67	68	69	70	71	72
Shoe size (y)	8	$7\frac{1}{2}$	7	8	$8\frac{1}{2}$	9	$9\frac{1}{2}$	8	$9\frac{1}{2}$	$10\frac{1}{2}$	10

---

2. The formula for coefficient of correlation is $\sqrt{(x - \bar{x})(y - \bar{y})}/s_x s_y$, where s_x and s_y are the standard deviations of x and y, and $\bar{x}$ and $\bar{y}$ are the means of x and y.

4. The following table shows the age of drivers in a certain town and the average number of automobile accidents in which they were involved in a given month.

Age of driver (x)	18	19	20	21	22	23	24	25	26	27
Average number of accidents (y)		2	3	1	1	0	2	1	2	0	1

5. The following table shows the relationship between the salary of an individual and the number of years of schooling beyond the eighth grade in the town of Camelot.

Number of years schooling beyond eighth grade	1	2	3	4	5	6	7	8	9
Average salary	$11,400	$12,309	$12,908	$13,800	$14,700	$15,201	$16,378	$16,981	$18,871

6. The following table shows the number of children in a family in a certain community, and the average salary of a family that has that number of children.

Number of children	8	7	6	5	4	3	2	1
Income	$5000 and under	$5000– 6000	$6000– 8000	$8000– 10,000	$10,000– 14,000	$14,000– 18,000	$18,000– 21,000	Above $21,000

10. HYPOTHESIS TESTING

One of the important uses of statistics is in hypothesis testing. In this situation, we are concerned with an assumption called the **hypothesis**, which we must either accept or reject. For example, if we were to flip a coin 1000 times, then we would expect heads to come up about 500 times if the coin is fair. Thus if we are tossing a coin to test the hypothesis that the coin is fair, and if we obtain only 17 heads in 1000 flips of the coin, then it is very unlikely that we would accept this hypothesis.

All of us do hypothesis testing in one form or another in our daily lives. Thus if we are trying to decide which box of strawberries to purchase at the supermarket, we are really involved in hypothesis testing. What we would usually do in this case is look at the top of the strawberry box and make a decision to either buy or reject it. Suppose the strawberries are rotten and we decide not to buy them. Then we have rejected the hypothesis that the strawberries are good, and our decision is correct. If they are rotten, but we buy them, then we have accepted the hypothesis that the strawberries are good and our decision is not correct.

Similarly, suppose the strawberries are good and we decide to buy them. Then we have accepted the hypothesis that the strawberries are good and it is correct. If we reject the strawberries, we are actually rejecting the hypothesis that they are good and our decision is incorrect. In either case, our decision about the strawberries is based on our past experience (statistics) with strawberries.

In the following examples, we indicate several different areas in which statistics can be applied to hypothesis-testing situations.

Example 1   A large auto manufacturer employs a quality-control engineer whose job is to sample and inspect 10 cars from each day's production line. If the engineer inspects these cars and finds them defective, he will reject the entire day's production. Thus the manufacturer will have to inspect and repair all defective cars. This can be costly for the manufacturer.

On the other hand, if the engineer inspects these cars and says that they are satisfactory, then it may turn out that only these cars are not defective. All the others may be defective. If they are then sold to the public, the manufacturer can incur financial loss through recalls or lawsuits.

This is a hypothesis testing situation because the decision to reject or accept is based on past statistics. Statistics are used in determining the sample size, the probability of obtaining a defective car, the chances of making a wrong decision, etc.

Example 2   When a student who has a low high-school average is accepted into a college on the basis of freshman placement tests, this decision is based on past statistics. If the college decides to admit such a student, it may have to deny other students admission. Then if the student does not succeed, it is at the expense of other students who might have succeeded. On the other hand, if the college does not admit this student, then this person might actually be a genius, but be denied the opportunity to attend college. The statistics used in making this decision would be based on past experience with students with similar backgrounds.

Example 3   A large dress manufacturer has recently installed new lighting, carpeting, etc., in his factory. Also music is piped in to improve the working conditions for the employees. These improvements were quite costly. Naturally, the manufacturer wants to know whether they have actually increased productivity. If yes, then he will do the same in his other factories. If no, then he has wasted his money. Thus, the manufacturer is testing the hypothesis that improved working conditions will increase productivity. Statistics are needed to help arrive at a decision. Thus, this is also a problem in hypothesis testing.

Comment   As we mentioned earlier, the above examples were given merely to indicate the nature of hypothesis testing. A typical hypothesis-testing problem is actually much more complicated. Nevertheless, the basic ideas are the same. For a complete discussion of this topic, the reader should consult any basic statistics and probability text, such as *Statistics and Probability in Modern Life*, 2nd ed., by Joseph Newmark (New York: Holt, Rinehart, and Winston, 1977).

---

11. CONCLUDING REMARKS

Statistics can be very meaningful and useful when applied properly, but great care must be taken to make sure that we do not read too much into them. This is the job of the statistician. It is important to know the size and extent of the sample, whether it was selected randomly, the kinds of analysis used, etc.

An interesting example of how statistics can be misused is the following: A university in Texas has three female faculty members in the mathematics depart-

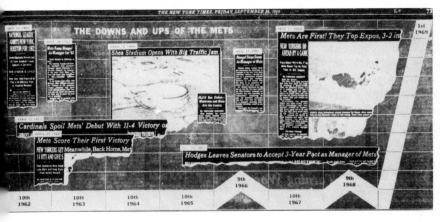

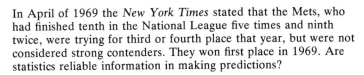

In April of 1969 the *New York Times* stated that the Mets, who had finished tenth in the National League five times and ninth twice, were trying for third or fourth place that year, but were not considered strong contenders. They won first place in 1969. Are statistics reliable information in making predictions?

Wide World Photos, Inc.

ment. Recently, one of them married one of her students. The student newspaper then printed an article under the following headline.

$33\frac{1}{3}$% OF OUR FEMALE FACULTY MEMBERS MARRY THEIR STUDENTS

What is wrong, if anything, with this headline?

There are many other examples of how statistics can be misused. For example, consider the following argument. There were at least two automobile accidents for every driver per 10,000 miles driven during 1976. Also, there were two electrical storms during the year, but there were no airplane accidents during these storms. Can we conclude that it is safer to be in an airplane during an electrical storm than it is to drive a car at any time?

As another example, a large urban university began its open admissions program five years ago. A study recently completed indicated that the number of A's given in English courses at this university increased by 30%. Can we conclude that, as a result of open admissions, students are performing 30% better in English than they did before?

Sometimes, statistics can lead to contradictory conclusions. For example, in 1970 the Nobel Prize winner Dr. Linus Pauling claimed that, based on the statistical data that he had collected, large doses of vitamin C are quite effective in preventing the occurrence of the common cold and also in reducing its severity. To test this claim, many other studies have been made. Specifically, several doctors at the University of Toronto's School of Hygiene conducted such a study during the winter of 1971–1972. They found that vitamin C had no significant effect in preventing colds, but did seem to reduce the severity of colds. So in one respect, the findings of the Toronto group contradicted those of Dr. Pauling, while in another respect, they confirmed them.

Other studies have both supported and contradicted Dr. Pauling's results. Thus, whether or not you believe that vitamin C is a "cure" for the common cold may depend on how you collect and interpret your statistics.

For further examples of how statistics can be misused or misinterpreted, see *How to Lie with Statistics* by Darrell Huff (New York: W. W. Norton, 1954).

STUDY GUIDE   In this chapter, the following ideas about statistics were discussed.

Data (p. 287)	Variance (p. 300)
Sample (p. 289)	Frequency distributions (p. 302)
Population (p. 289)	Frequency (p. 302)
Random sample (p. 289)	Bar graph (p. 302)
Measures of central tendency (p. 290)	Graphs (p. 306)
Mean (p. 291)	Circle graph (p. 308)
Median (p. 292)	Line graph (p. 309)
Mode (p. 291)	Normal distribution (p. 311)
Percentiles (p. 295)	Correlation (p. 315)
Percentile rank (p. 296)	Scatter diagram (p. 315)
Lower or first quartile (p. 297)	Line of regression (p. 315)
Middle quartile (p. 297)	High correlation (p. 315)
Upper or third quartile (p. 297)	Low correlation (p. 315)
Measures of variation (p. 297)	Curve of regression (p. 316)
Range (p. 298)	Hypothesis testing (p. 317)
Standard deviation (p. 298)	Misuse of statistics (p. 318)

SELECTED FURTHER READINGS

Adler, I., *Probability and Statistics For Everyone.* New York: New American Library, 1966.

Cooke, W. P., "Beginning Statistics at the Track." *Mathematics Magazine* 46: 250–255 (November–December 1973).

David, F. N., *Games, Gods and Gambling.* New York: Hafner, 1962.

Huff, D., *How to Lie with Statistics.* New York: W. W. Norton, 1954. Discusses how statistics can be misused.

Los Angeles Times, "Nielsen raters' views decide your TV fare." 9 June 1974.

Kline, M., *Mathematics For Liberal Arts.* Reading, Mass.: Addison-Wesley, 1967. Chapter 22 discusses statistics.

Mathematics in the Modern World (Readings from *Scientific American*). San Francisco: W. H. Freeman, 1968. Article 25 discusses statistics.

New York Times, "On the Nielsen families." 15 September 1974.

Newmark, J., *Statistics and Probability in Modern Life*, 2nd ed., New York: Holt, Rinehart, and Winston, 1977. Chapters 2–4 contain a thorough but easy discussion on how to analyze data by graphical and arithmetic techniques.

Senter, R. J., *Analysis of Data: Introductory Statistics for the Behavioral Sciences.* Glenview, Ill.: Scott Foresman, 1969.

Tanur, J. M., ed., *Statistics; A Guide to the Unknown.* San Francisco: Holden Day, 1972. Contains a discussion on the uses of statistics.

Weaver, W., *Lady Luck.* New York: Anchor Books, 1963.

The following articles, which all discuss some of the uses of statistics appear in James Newman, ed., *The World of Mathematics.* New York: Simon and Schuster, 1956.

Bernoulli, J., "The Law of Large Numbers." Volume 3, part 8, article 3.

Fisher, R. A., "Mathematics of a Lady Tasting Tea." Volume 3, part 8, article 6.

Graunt, J., "Foundations of Vital Statistics." Volume 3, part 8, article 1.

Halley, E., "First Life Insurance Tables." Volume 3, part 8, article 2.

Malthus, T., "Mathematics of Population and Food." Volume 2, part 6, article 3.

Moroney, M. J., "Average and Scatter." Volume 3, part 8, article 5.

Richardson, L. F., "The Statistics of Deadly Quarrels." Volume 2, part 6, article 7.

Shaw, G. B., "The Vice of Gambling and the Virtue of Insurance." Volume 3, part 8, article 7.

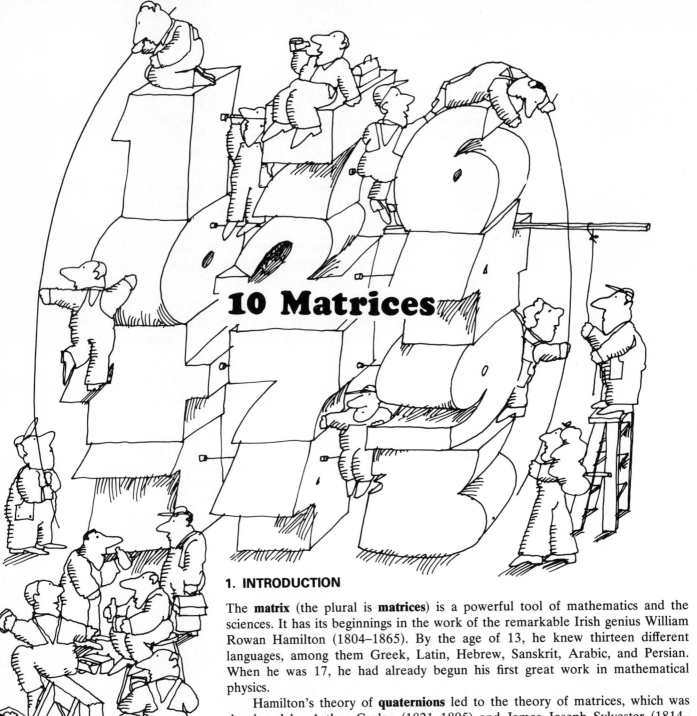

10 Matrices

1. INTRODUCTION

The **matrix** (the plural is **matrices**) is a powerful tool of mathematics and the sciences. It has its beginnings in the work of the remarkable Irish genius William Rowan Hamilton (1804–1865). By the age of 13, he knew thirteen different languages, among them Greek, Latin, Hebrew, Sanskrit, Arabic, and Persian. When he was 17, he had already begun his first great work in mathematical physics.

Hamilton's theory of **quaternions** led to the theory of matrices, which was developed by Arthur Cayley (1821–1895) and James Joseph Sylvester (1814–1897). Cayley and Sylvester were lifelong friends, although their personalities were quite different. Sylvester was hot-tempered, whereas Cayley was calm and rarely lost his temper. Each demonstrated his exceptional mathematical talent as a child, and each continued his mathematical activities until practically the day of his death.

Cayley and Sylvester were both victims of some form of religious prejudice. Cayley received a teaching position at Cambridge University. At the time, all teachers at Cambridge were required to take religious vows. Although Cayley was a member of the Church of England, he refused to take holy orders just to keep his position. As a result, his teaching position was not renewed. Since he could not work as a mathematics professor because of his refusal to take these religious vows, he became a lawyer. After fourteen years as a successful lawyer, he was again offered a professorship at Cambridge, which he accepted.

James Joseph Sylvester was born in London in 1814 to an orthodox Jewish family. He attended the University of London at the young age of 14, where he studied under the famous mathematician Augustus DeMorgan. At the age of 15, he entered the Royal Institution of Liverpool. Here he was subject to religious persecution and at one point ran away to Dublin. At 17, he entered Cambridge, where he was unable to compete in certain mathematical contests because he was not a Christian. In fact, Cambridge would not even grant him his degree because he refused to accept the religious beliefs of the Church of England. In 1871, when the religious requirements were removed, Sylvester finally received his degree. This was 40 years after he first entered Cambridge.

Sylvester held many jobs including various teaching positions, two of them in the United States. He taught at the University of Virginia where he resigned his position when the university refused to discipline a student who had insulted him. He also taught at Johns Hopkins University. In London, he worked for an insurance company as an actuary, preparing statistical charts. While working as an actuary, he gave private lessons in mathematics and one of his students was the famous nurse Florence Nightingale.

As we have said, Cayley and Sylvester continued the work begun by Hamilton. Although Hamilton's work was concerned with geometric situations, it has been found that the theory of matrices can be applied to almost all areas of mathematics and physics. As we shall see, it has many interesting applications in business and economics. Of course, matrices are also used extensively in pure mathematics.

You are probably wondering what a matrix is. It is just any list of numbers such as the following:

$$\begin{pmatrix} 1 & 7 \\ 6 & 3 \end{pmatrix} \qquad \begin{pmatrix} 1 & 7 & 6 \\ 0 & -2 & 4 \\ 4 & 7 & 9 \\ 1 & 3 & 3 \end{pmatrix} \qquad \begin{pmatrix} 1 \\ 4 \\ 3 \\ -7 \end{pmatrix} \qquad (3 \quad 2 \quad 9 \quad 10).$$

We enclose each matrix within parentheses, as shown.

In the next section we will restate this definition and discuss it in detail. We will also investigate some of the properties of matrices and their applications.

2. DEFINITION AND NOTATION

We begin our discussion with the definition of a matrix.

Definition 2.1   *A matrix is a rectangular array (table) of numbers. We enclose the matrix within parentheses.*

Some examples of matrices are as follows:

a) $\begin{pmatrix} 1 & 2 \\ 4 & -7 \end{pmatrix}$ b) $\begin{pmatrix} -1 & 0 & 3 \\ 5 & 0 & -9 \\ 1 & 3 & 8 \end{pmatrix}$ c) $\begin{pmatrix} 1 & 0 \\ 0 & 1 \end{pmatrix}$ d) $(0 \quad 0 \quad 0)$

e) $\begin{pmatrix} 1 \\ 2 \\ -5 \\ 7 \end{pmatrix}$ f) $\begin{pmatrix} 1 & 7 & 8 \\ -2 & -1 & -3 \\ 1 & -4 & 9 \\ -8 & 0 & 1 \end{pmatrix}$ g) (4).

The matrix of (a) has 2 rows and 2 columns. We call it a 2×2 matrix. The matrix of (b) has 3 rows and 3 columns. We call it a 3×3 matrix. The matrix of (f) has 4 rows and 3 columns. We call it a 4×3 matrix. The matrix of (c) is a 2×2 matrix; (d) is a 1×3 matrix; (e) is a 4×1 matrix; and (g) is a 1×1 matrix.

In general, *a matrix that has m rows and n columns is called an m × n matrix,* read as "*m by n matrix.*" We say that the **dimension** of the matrix is $m \times n$.

If a matrix has the same number of rows and columns, we call it a **square matrix**. Both (a) and (c) above are 2×2 square matrices, and (b) is a 3×3

Much information that is put into a computer can be fed in and processed in matrix form. Computers do addition and subtraction, as well as multiplication, of matrices. The computer can also provide results in matrix form, as shown here in a section of a printout.

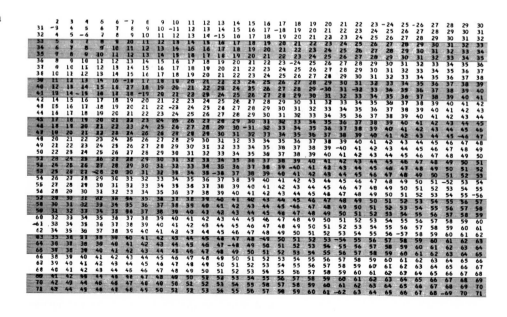

square matrix. On the other hand, (d), (e), and (f) are not square matrices. Is (g) a square matrix?

In our discussion, we will denote matrices by capital letters, such as A, B, C, ...

---

3. ADDITION AND SUBTRACTION OF MATRICES

Adding two matrices is a simple procedure. We just add corresponding elements. Let $A = \begin{pmatrix} 1 & 3 \\ -2 & 0 \end{pmatrix}$, and let $B = \begin{pmatrix} 5 & 4 \\ 1 & -7 \end{pmatrix}$. Then $A + B$ is

$$\begin{pmatrix} 1 & 3 \\ -2 & 0 \end{pmatrix} + \begin{pmatrix} 5 & 4 \\ 1 & -7 \end{pmatrix} = \begin{pmatrix} 1+5 & 3+4 \\ -2+1 & 0+(-7) \end{pmatrix} = \begin{pmatrix} 6 & 7 \\ -1 & -7 \end{pmatrix}.$$

Example 1   Add the matrices $\begin{pmatrix} 2 & 4 \\ -1 & 0 \\ 6 & 6 \end{pmatrix}$ and $\begin{pmatrix} 1 & 1 \\ 10 & 8 \\ 12 & -3 \end{pmatrix}$.

Solution $\begin{pmatrix} 2 & 4 \\ -1 & 0 \\ 6 & 6 \end{pmatrix} + \begin{pmatrix} 1 & 1 \\ 10 & 8 \\ 12 & -3 \end{pmatrix} = \begin{pmatrix} 2+1 & 4+1 \\ -1+10 & 0+8 \\ 6+12 & 6+(-3) \end{pmatrix} = \begin{pmatrix} 3 & 5 \\ 9 & 8 \\ 18 & 3 \end{pmatrix}$

Example 2   Add the matrices $(1 \quad 7 \quad 9)$ and $(2 \quad -3 \quad 0)$.

Solution $(1 \quad 7 \quad 9) + (2 \quad -3 \quad 0) = (1+2 \quad 7+(-3) \quad 9+0)$

$$= (3 \quad 4 \quad 9).$$

Example 3   Add the matrices $\begin{pmatrix} 1 & -2 & 4 \\ 8 & 0 & -1 \\ 0 & 3 & 4 \end{pmatrix}$ and $\begin{pmatrix} 7 & 7 & -7 \\ -5 & 2 & 1 \\ 4 & 3 & 4 \end{pmatrix}$.

Solution $\begin{pmatrix} 1 & -2 & 4 \\ 8 & 0 & -1 \\ 0 & 3 & 4 \end{pmatrix} + \begin{pmatrix} 7 & 7 & -7 \\ -5 & 2 & 1 \\ 4 & 3 & 4 \end{pmatrix} = \begin{pmatrix} 1+7 & -2+7 & 4+(-7) \\ 8+(-5) & 0+2 & -1+1 \\ 0+4 & 3+3 & 4+4 \end{pmatrix}$

$$= \begin{pmatrix} 8 & 5 & -3 \\ 3 & 2 & 0 \\ 4 & 6 & 8 \end{pmatrix}.$$

Notice that you can add together two matrices *only* if they are of the same *dimension*. This means that the number of rows in each is the same, and the number of columns in each is the same.

Example 4   The matrices $\begin{pmatrix} 1 & 4 \\ -3 & 9 \end{pmatrix}$ and $\begin{pmatrix} 4 & 9 & -3 \\ -2 & 0 & 1 \end{pmatrix}$ *cannot* be added, because the first matrix has 2 columns and the second matrix has 3 columns.

Subtraction of matrices is similar to addition. However, we subtract the corresponding elements, as the following examples will illustrate.

Example 5   Subtract the matrix $\begin{pmatrix} 2 & 0 \\ -3 & -5 \end{pmatrix}$ from $\begin{pmatrix} 4 & 8 \\ -3 & 5 \end{pmatrix}$.

Solution $\begin{pmatrix} 4 & 8 \\ -3 & 5 \end{pmatrix} - \begin{pmatrix} 2 & 0 \\ -3 & -5 \end{pmatrix} = \begin{pmatrix} 4-2 & 8-0 \\ -3-(-3) & 5-(-5) \end{pmatrix} = \begin{pmatrix} 2 & 8 \\ 0 & 10 \end{pmatrix}.$

Example 6   Perform the subtraction $\begin{pmatrix} 4 & 1 & 1 \\ -2 & 0 & -8 \end{pmatrix} - \begin{pmatrix} 1 & 5 & 9 \\ 8 & 3 & -5 \end{pmatrix}$.

Solution $\begin{pmatrix} 4 & 1 & 1 \\ -2 & 0 & -8 \end{pmatrix} - \begin{pmatrix} 1 & 5 & 9 \\ 8 & 3 & -5 \end{pmatrix} = \begin{pmatrix} 4-1 & 1-5 & 1-9 \\ -2-8 & 0-3 & -8-(-5) \end{pmatrix}$

$$= \begin{pmatrix} 3 & -4 & -8 \\ -10 & -3 & -3 \end{pmatrix}.$$

Comment   Wherever addition is possible, the commutative and associative laws hold. This means that if A, B, and C are matrices, then

$$A + B = B + A \qquad \text{(Commutative law)}$$
$$A + (B + C) = (A + B) + C. \qquad \text{(Associative law)}$$

EXERCISES   1. Perform the indicated operations (where possible).

a) $\begin{pmatrix} -3 & 8 \\ 4 & 1 \end{pmatrix} + \begin{pmatrix} -2 & 0 \\ 2 & 7 \end{pmatrix}$

b) $\begin{pmatrix} -4 & 2 \\ -1 & 6 \end{pmatrix} - \begin{pmatrix} 9 & 3 \\ -7 & 5 \end{pmatrix}$

c) $\begin{pmatrix} -3 & 2 \\ 3 & -4 \\ 0 & 1 \\ 1 & 5 \end{pmatrix} + \begin{pmatrix} 1 & -4 \\ 2 & -3 \\ 3 & -2 \\ 4 & -1 \end{pmatrix}$

d) $(-1 \quad -2) + (2 \quad 1)$

e) $\begin{pmatrix} 4 & 3 & 8 \\ 1 & -1 & 6 \\ 2 & 3 & -4 \end{pmatrix} + \begin{pmatrix} 1 & 0 \\ 1 & 2 \\ -1 & 5 \end{pmatrix}$

f) $\begin{pmatrix} 8 & 5 & 0 \\ -2 & -5 & -2 \\ 7 & 0 & -1 \end{pmatrix} - \begin{pmatrix} 1 & 4 & 7 \\ 2 & 5 & 8 \\ 3 & 6 & 9 \end{pmatrix}$

g) $\begin{pmatrix} 0 & -6 & 8 \\ -1 & -5 & 9 \end{pmatrix} - \begin{pmatrix} 10 & 1 \\ 9 & 1 \end{pmatrix}$

h) $\begin{pmatrix} 2 \\ 1 \\ -4 \\ 3 \\ 6 \end{pmatrix} + \begin{pmatrix} 1 \\ 0 \\ -1 \\ 5 \\ 3 \end{pmatrix}$

i) $\begin{pmatrix} 4 & 2 \\ 6 & -3 \\ 5 & -1 \\ 3 & 2 \\ 1 & 7 \end{pmatrix} - \begin{pmatrix} 3 & 5 \\ -3 & 2 \\ 8 & 6 \\ 1 & 0 \\ 5 & 3 \end{pmatrix}$

j) $\begin{pmatrix} 4 & 2 & 5 \\ -1 & 0 & 6 \end{pmatrix} - \begin{pmatrix} -3 & 6 \\ -5 & 4 \\ 7 & -11 \end{pmatrix}$

k) $\begin{pmatrix} 1 \\ 2 \end{pmatrix} - \begin{pmatrix} 5 \\ -6 \end{pmatrix}$

l) $\begin{pmatrix} 5 & 7 \\ 3 & 2 \end{pmatrix} - \begin{pmatrix} 5 & 7 \\ 3 & 2 \end{pmatrix}$

m) $\begin{pmatrix} 0 & 0 \\ 0 & 0 \end{pmatrix} - \begin{pmatrix} 5 & 7 \\ 3 & 9 \end{pmatrix}$

2. Verify that the commutative law holds for Examples 1 and 2 on p. 325.

3. Let $A = \begin{pmatrix} 1 & 4 \\ 5 & 3 \end{pmatrix}$, $B = \begin{pmatrix} -3 & 2 \\ 5 & 4 \end{pmatrix}$, $C = \begin{pmatrix} 1 & 0 \\ 0 & 2 \end{pmatrix}$.

a) Calculate $A + (B + C)$.

b) Calculate $(A + B) + C$.

The answers in parts (a) and (b) should be the same, thereby verifying the associative law for this case.

---

4. MULTIPLICATION OF MATRICES

It may seem that multiplication of matrices should be done in a manner similar to that of addition of matrices. However, if multiplication is done this way, then the range of applications is extremely limited. Since our main interest in matrices is in their applications, we will multiply matrices in a way that turns out to be useful. The procedure for multiplication will at first appear strange and unnecessarily complicated. However, you will find after working with it that it is very useful.

We will illustrate the method by multiplying

$$\begin{pmatrix} 2 & 4 \\ 3 & -2 \\ 9 & 0 \end{pmatrix} \cdot \begin{pmatrix} 3 & 0 & 2 & 7 \\ 2 & 4 & -1 & 2 \end{pmatrix}.$$

The first matrix has 3 rows and 2 columns. It is a 3×2 matrix. The second matrix is 2×4. Notice that to multiply matrices, it is *not* necessary that they have the same dimension. *What is necessary is that the number of columns of the first matrix be exactly the same as the number of rows of the second matrix.* In our case, this number is 2. Our answer will be a 3×4 matrix.

The following diagram illustrates the relationship between the number of rows and columns of the matrices to be multiplied, and the answer.

$3 \times ②$ and $② \times 4$

Do they match?

If yes, our answer is

3×4

Now how do we obtain the numbers for the 3×4 matrix that is our answer? Let us rewrite the problem with blanks in the answer.

$$\begin{pmatrix} 2 & 4 \\ 3 & -2 \\ 9 & 0 \end{pmatrix} \begin{pmatrix} 3 & 0 & 2 & 7 \\ 2 & 4 & -1 & 2 \end{pmatrix} = \begin{pmatrix} ? & ? & ? & ? \\ ? & ? & ? & ? \\ ? & ? & ? & ? \end{pmatrix}.$$

William Rowan Hamilton

To find out what belongs in the first row, first column (the circled one), we multiply each element of row 1 of the first matrix by the *corresponding* element of column 1 of the second matrix, and then add. This gives

$$2 \cdot 3 + 4 \cdot 2 = 6 + 8 = 14.$$

To find what belongs in the third row, fourth column, we multiply row 3 of the first matrix by the corresponding elements of column 4 of the second matrix, and then add. This gives

$$9 \cdot 7 + 0 \cdot 2 = 63 + 0 = 63.$$

In a similar manner, we find the number that belongs in the second row, third column. We multiply row 2 of the first matrix by column 3 of the second matrix. We get

$$3 \cdot 2 + (-2)(-1) = 6 + 2 = 8.$$

Proceeding in the same way for all the other blanks, we get

$$\begin{pmatrix} 2 & 4 \\ 3 & -2 \\ 9 & 0 \end{pmatrix} \cdot \begin{pmatrix} 3 & 0 & 2 & 7 \\ 2 & 4 & -1 & 2 \end{pmatrix}$$

$$= \begin{pmatrix} 2 \cdot 3 + 4 \cdot 2 & 2 \cdot 0 + 4 \cdot 4 & 2 \cdot 2 + 4(-1) & 2 \cdot 7 + 4 \cdot 2 \\ 3 \cdot 3 + (-2)2 & 3 \cdot 0 + (-2) \cdot 4 & 3 \cdot 2 + (-2)(-1) & 3 \cdot 7 + (-2)(2) \\ 9 \cdot 3 + 0 \cdot 2 & 9 \cdot 0 + 0 \cdot 4 & 9 \cdot 2 + 0(-1) & 9 \cdot 7 + 0 \cdot 2 \end{pmatrix}$$

$$= \begin{pmatrix} 14 & 16 & 0 & 22 \\ 5 & -8 & 8 & 17 \\ 27 & 0 & 18 & 63 \end{pmatrix}.$$

Notice that if we try to multiply

$$\begin{pmatrix} 3 & 0 & 2 & 7 \\ 2 & 4 & -1 & 2 \end{pmatrix} \cdot \begin{pmatrix} 2 & 4 \\ 3 & -2 \\ 9 & 0 \end{pmatrix},$$

it cannot be done, since the first matrix is 2×4 and the second matrix is 3×2. The number of columns of the first matrix is *not* the same as the number of rows of the second.

Example 1   Multiply $\begin{pmatrix} 1 & 4 \\ 3 & -2 \end{pmatrix}$ by $\begin{pmatrix} 1 & 0 & 2 \\ -1 & 3 & 5 \end{pmatrix}.$

Solution   The matrices are 2×2 and 2×3, respectively. We draw a diagram.

$$2 \times ② \quad \text{and} \quad ② \times 3$$
Do they match?

Yes!

Our answer is 2×3.

Computing the entries for the 2×3 matrix, we get

$$\begin{pmatrix} 1 & 4 \\ 3 & -2 \end{pmatrix} \cdot \begin{pmatrix} 1 & 0 & 2 \\ -1 & 3 & 5 \end{pmatrix} = \begin{pmatrix} 1 \cdot 1 + 4(-1) & 1 \cdot 0 + 4 \cdot 3 & 1 \cdot 2 + 4 \cdot 5 \\ 3 \cdot 1 + (-2)(-1) & 3 \cdot 0 + (-2) \cdot 3 & 3 \cdot 2 + (-2) \cdot 5 \end{pmatrix}$$

$$= \begin{pmatrix} -3 & 12 & 22 \\ 5 & -6 & -4 \end{pmatrix}$$

Example 2   Multiply $(1 \quad 3 \quad -1)$ by $\begin{pmatrix} 7 \\ -8 \\ 0 \end{pmatrix}$.

Solution   The matrices are

$$1 \times ③ \quad \text{and} \quad ③ \times 1.$$
Do they match?

Yes!

Our answer is 1×1.

Computing the answer, we get

$$(1 \cdot 7 + 3 \cdot (-8) + (-1) \cdot 0) = (-17).$$

It would be wrong to write the answer as -17. Since the answer is a matrix, the parentheses are necessary.

Example 3   Multiply $\begin{pmatrix} 7 \\ -8 \\ 0 \end{pmatrix}$ by $(1 \quad 3 \quad -1)$.

The matrices are

$$3 \times ① \quad \text{and} \quad ① \times 3.$$
Do they match?

Yes!

Our answer is 3×3.

Computing the answer, we have

$$\begin{pmatrix} 7 \\ -8 \\ 0 \end{pmatrix} \cdot (1 \quad 3 \quad -1) = \begin{pmatrix} 7 \cdot 1 & 7 \cdot 3 & 7(-1) \\ (-8) \cdot 1 & (-8) \cdot 3 & (-8) \cdot (-1) \\ 0 \cdot 1 & 0 \cdot 3 & 0 \cdot (-1) \end{pmatrix}$$

$$= \begin{pmatrix} 7 & 21 & -7 \\ -8 & -24 & 8 \\ 0 & 0 & 0 \end{pmatrix}.$$

Compare this problem with Example 2. What does it tell us about the commutative law for multiplication of matrices?

We can obviously conclude that the commutative law does not hold. When Hamilton discovered this, he was so excited about it that he scratched the result on a bridge in Dublin, where he happened to be walking at the time.

Example 4   Multiply $\begin{pmatrix} 5 & 7 \\ -3 & 2 \end{pmatrix}$ by $\begin{pmatrix} 4 & 3 & 8 \\ -2 & 5 & 9 \\ 1 & 0 & 1 \end{pmatrix}$.

Solution   The matrices are

$2 \times ②$ and $③ \times 3$.

   Do they match?

No!

Since they do not match, this multiplication is impossible.

Example 5   Multiply $\begin{pmatrix} 5 & 7 \\ -3 & 2 \end{pmatrix}$ by $\begin{pmatrix} 1 & 0 \\ 0 & 1 \end{pmatrix}$.

Solution   The matrices are

$2 \times ②$ and $② \times 2$.

Do they match?

Yes!

Our answer is 2×2.

The product is

$$\begin{pmatrix} 5 & 7 \\ -3 & 2 \end{pmatrix} \cdot \begin{pmatrix} 1 & 0 \\ 0 & 1 \end{pmatrix} = \begin{pmatrix} 5 \cdot 1 + 7 \cdot 0 & 5 \cdot 0 + 7 \cdot 1 \\ (-3) \cdot 1 + 2 \cdot 0 & (-3) \cdot 0 + 2 + 1 \end{pmatrix}$$

$$= \begin{pmatrix} 5 & 7 \\ -3 & 2 \end{pmatrix}.$$

In the last example, the answer was the same matrix as the one we started with. Multiplying by $\begin{pmatrix} 1 & 0 \\ 0 & 1 \end{pmatrix}$ did nothing. If you multiply $\begin{pmatrix} 1 & 0 \\ 0 & 1 \end{pmatrix}$ by $\begin{pmatrix} 5 & 7 \\ -3 & 2 \end{pmatrix}$, you will find that the answer is also $\begin{pmatrix} 5 & 7 \\ -3 & 2 \end{pmatrix}$. Verify this. We say that $\begin{pmatrix} 1 & 0 \\ 0 & 1 \end{pmatrix}$ is the **identity** for multiplication of 2×2 matrices. Multiplying by the matrix $\begin{pmatrix} 1 & 0 \\ 0 & 1 \end{pmatrix}$ does not change anything.

Example 6   a) Multiply $\begin{pmatrix} 4 & 3 \\ 5 & 7 \\ -1 & 0 \end{pmatrix} \cdot \begin{pmatrix} 1 & 0 \\ 0 & 1 \end{pmatrix}$.

b) Multiply $\begin{pmatrix} 1 & 0 \\ 0 & 1 \end{pmatrix} \cdot \begin{pmatrix} 4 & 3 \\ 5 & 7 \\ -1 & 0 \end{pmatrix}$.

Solution   a) $\begin{pmatrix} 4 & 3 \\ 5 & 7 \\ -1 & 0 \end{pmatrix} \cdot \begin{pmatrix} 1 & 0 \\ 0 & 1 \end{pmatrix} = \begin{pmatrix} 4 & 3 \\ 5 & 7 \\ -1 & 0 \end{pmatrix}$

b) This multiplication cannot be done. Why not?

In Example 5, $\begin{pmatrix} 1 & 0 \\ 0 & 1 \end{pmatrix}$ was the identity matrix, because no matter which way we performed the multiplication, our answer was always $\begin{pmatrix} 5 & 7 \\ -3 & 2 \end{pmatrix}$. In Example 6, $\begin{pmatrix} 1 & 0 \\ 0 & 1 \end{pmatrix}$ is *not* the identity matrix since we can multiply only from one side. Thus, we define an identity matrix in the following way.

Definition 4.1   *An **identity** matrix is a **square** matrix which, when multiplied by another square matrix A on the left side or on the right side, leaves the matrix A unchanged. Thus, $\begin{pmatrix} 1 & 0 \\ 0 & 1 \end{pmatrix}$ is the 2×2 identity matrix and it can be shown that*

$$\begin{pmatrix} 1 & 0 & 0 \\ 0 & 1 & 0 \\ 0 & 0 & 1 \end{pmatrix}$$

is the 3×3 identity matrix.

Sometimes we want to multiply a matrix by an ordinary number. For example, suppose we want to multiply the matrix $\begin{pmatrix} 3 & 5 \\ -1 & 7 \end{pmatrix}$ by the number 4. This is done by multiplying each entry of the matrix by 4. Thus we have the following.

$$4\begin{pmatrix} 3 & 5 \\ -1 & 7 \end{pmatrix} = \begin{pmatrix} 4 \cdot 3 & 4 \cdot 5 \\ 4(-1) & 4(7) \end{pmatrix}$$

$$= \begin{pmatrix} 12 & 20 \\ -4 & 28 \end{pmatrix}.$$

In general terms, if k is any number and A is any matrix, then kA means "multiply each entry of A by the number k."

Example 7   a) $3\begin{pmatrix} 4 & 7 \\ 2 & -2 \\ 7 & 4 \end{pmatrix} = \begin{pmatrix} 3 \cdot 4 & 3 \cdot 7 \\ 3 \cdot 2 & 3(-2) \\ 3 \cdot 7 & 3 \cdot 4 \end{pmatrix} = \begin{pmatrix} 12 & 21 \\ 6 & -6 \\ 21 & 12 \end{pmatrix}$

   b) $-2\begin{pmatrix} 5 \\ 0 \\ -4 \\ 9 \end{pmatrix} = \begin{pmatrix} (-2)(5) \\ (-2)0 \\ (-2)(-4) \\ (-2)(9) \end{pmatrix} = \begin{pmatrix} -10 \\ 0 \\ +8 \\ -18 \end{pmatrix}$

EXERCISES   1. Perform the following multiplications, if possible.

a) $\begin{pmatrix} 1 & 5 \\ 3 & -6 \end{pmatrix} \cdot \begin{pmatrix} 4 & 2 \\ 0 & -3 \end{pmatrix}$

b) $\begin{pmatrix} 1 & 4 & 3 \\ 6 & 2 & 0 \\ -1 & -2 & 5 \end{pmatrix} \cdot \begin{pmatrix} 2 \\ 3 \\ -1 \end{pmatrix}$

c) $\begin{pmatrix} 6 \\ 0 \\ 0 \\ 5 \end{pmatrix} \cdot (4 \quad 3 \quad 7 \quad -1)$

d) $(4 \quad 3 \quad 7 \quad -1)\begin{pmatrix} 6 \\ 0 \\ 0 \\ 5 \end{pmatrix}$

e) $\begin{pmatrix} 5 \\ 6 \end{pmatrix} \cdot \begin{pmatrix} -3 & -1 \\ 0 & 4 \end{pmatrix}$

f) $(1 \quad 1 \quad 1)\begin{pmatrix} 7 \\ 6 \\ -3 \end{pmatrix}$

g) $(7) \cdot (-3)$

h) $\begin{pmatrix} 4 & -3 & 7 \\ 1 & 0 & 0 \\ 0 & -4 & 1 \end{pmatrix} \cdot \begin{pmatrix} 2 & 4 \\ 1 & 3 \\ -3 & -5 \end{pmatrix}$

i) $\begin{pmatrix} 5 & -3 & 7 \\ 2 & -21 & 17 \\ 4 & 3 & -4 \end{pmatrix} \cdot \begin{pmatrix} 1 & 0 & 0 \\ 0 & 1 & 0 \\ 0 & 0 & 1 \end{pmatrix}$

j) $(1 \quad 2 \quad -3 \quad 4 \quad -5 \quad 6) \cdot \begin{pmatrix} 6 \\ 5 \\ -4 \\ 3 \\ -2 \\ 1 \end{pmatrix}$

2. What is the 4×4 identity matrix?

*3. Can you find two 2×2 matrices M and N such that $MN = \begin{pmatrix} 0 & 0 \\ 0 & 0 \end{pmatrix}$ but M is not $\begin{pmatrix} 0 & 0 \\ 0 & 0 \end{pmatrix}$ and N is not $\begin{pmatrix} 0 & 0 \\ 0 & 0 \end{pmatrix}$?

4. Find each of the following.

a) $2\begin{pmatrix} -1 & 4 \\ 3 & 0 \end{pmatrix}$

b) $3\begin{pmatrix} -1 & 5 & 6 \\ 2 & 0 & 2 \\ 5 & 0 & 7 \end{pmatrix}$

c) $2\begin{pmatrix} -2 & 0 & 1 \\ 10 & 3 & 0 \\ -1 & 0 & -4 \end{pmatrix}\begin{pmatrix} 5 & 1 & 3 \\ 2 & -1 & 0 \\ 2 & 8 & 1 \end{pmatrix}$

d) $2\begin{pmatrix} 3 & 4 \\ -10 & 20 \end{pmatrix} + 3\begin{pmatrix} 1 & -5 \\ 1 & 5 \end{pmatrix}$

e) $5\begin{pmatrix} 1 & -7 & -2 \\ 1 & 0 & -3 \\ 4 & 2 & 6 \end{pmatrix} - 6\begin{pmatrix} 1 & 3 & 6 \\ -1 & 5 & 2 \\ 2 & 8 & 4 \end{pmatrix}$

*5. Find a 2×2 matrix A, such that $A = 2A$.

*6. Find a 2×2 matrix B, such that $B \cdot B = B$.

5. APPLICATIONS TO BUSINESS PROBLEMS

Matrices can be very useful in analyzing business problems such as determining cost, profit, amounts of materials needed, etc. The following examples illustrate how this can be done.

A drug and cosmetics company markets the following products: Easy-Off hair tonic, Vanish toothpaste, and Left-Guard deodorant. The following table indicates the number of cases of each product ordered by four different supermarkets.

	Hair tonic	Toothpaste	Deodorant
Market I	5	4	7
Market II	8	0	1
Market III	13	3	9
Market IV	2	6	3

The hair tonic costs $12 per case, the toothpaste costs $9 a case, and the deodorant costs $15 a case.

There are many things that may be of interest to both the manufacturer and store managers. Each of these can be expressed by some combination of matrices. Suppose the manufacturer wants to know the total income from each supermarket and also the total income from all supermarkets together.

From market I, the income is $5(12) + 4(9) + 7(15) \quad = \$201.$

From market II, the income is $8(12) + 0(9) + 1(15) \quad = \ 111.$

From market III, the income is $13(12) + 3(9) + 9(15) = 318$.

From market IV, the income is $2(12) + 6(9) + 3(15) = 123$.

We can put the final result in matrix form as

$$\begin{pmatrix} 201 \\ 111 \\ 318 \\ 123 \end{pmatrix}.$$

The entire computation can be done in a much more efficient way using matrices as follows.

Let A represent the amounts ordered of each product. We then have

$$A = \begin{pmatrix} 5 & 4 & 7 \\ 8 & 0 & 1 \\ 13 & 3 & 9 \\ 2 & 6 & 3 \end{pmatrix}.$$

Let C represent the cost of each item. Then

$$C = \begin{pmatrix} 12 \\ 9 \\ 15 \end{pmatrix},$$

$$A \cdot C = \begin{pmatrix} 5 & 4 & 7 \\ 8 & 0 & 1 \\ 13 & 3 & 9 \\ 2 & 6 & 3 \end{pmatrix} \cdot \begin{pmatrix} 12 \\ 9 \\ 15 \end{pmatrix} = \begin{pmatrix} 5(12) + 4(9) + 7(15) \\ 8(12) + 0(9) + 1(15) \\ 13(12) + 3(9) + 9(15) \\ 2(12) + 6(9) + 3(15) \end{pmatrix} = \begin{pmatrix} 201 \\ 111 \\ 318 \\ 123 \end{pmatrix}.$$

The matrix AC represents the total income from each supermarket. Market I paid \$201 for all three products; \$111 is the total income from market II. Similarly, \$318 and \$123 are the incomes from markets III and IV, respectively.

The total income from all the supermarkets is obtained by adding the numbers down. This gives \$753.

Some further examples will illustrate how matrices can be used in business problems.

Example 1 Suppose the student council of a school sponsors a benefit concert, the proceeds of which are to go to the local drug rehabilitation clinic. There are two performances, one on Friday and one on Saturday. Seat prices are as follows.

\$3.00 Front orchestra
 2.50 Back orchestra
 2.00 Mezzanine
 1.50 Lower balcony
 1.00 Upper balcony

In a concert hall, the location in the orchestra or balconies determines the price of seats.

© Lincoln Center for the Performing Arts, Inc. Photograph by Bob Serating

The table below shows the number of tickets sold and the amount of money collected on Friday.

Type of seat	Number of seats		Price per seat	Total collected
Front orchestra	250	×	$3.00	$ 750.00
Back orchestra	120	×	2.50	300.00
Mezzanine	73	×	2.00	146.00
Lower balcony	208	×	1.50	312.00
Upper balcony	124	×	1.00	124.00
				$1632.00

On Saturday night, the numbers of seats sold were: 300 front orchestra, 110 back orchestra, 78 mezzanine, 220 lower balcony, and 113 upper balcony.

The treasurer wants to figure out how much money has been collected altogether for the two concerts. He can do this in two ways.

1. He can calculate the amount collected on Saturday by setting up a table similar to the one above. He would get $1774.00. He would now add the totals together, getting $1632.00 plus $1774.00, or $3406.00.

2. He can also calculate the total number of each kind of ticket sold for both nights, multiply each total by the price of that ticket, and then add. Such a calculation would look like this.

Type of seat	Seats sold Friday	Seats sold Saturday	Total seats sold	Price per seat	Total collected
Front orchestra	250	300	550	× 3.00	$1650.00
Back orchestra	120	110	230	× 2.50	575.00
Mezzanine	73	78	151	× 2.00	302.00
Lower balcony	208	220	428	× 1.50	642.00
Upper balcony	124	113	237	× 1.00	237.00
					$3406.00

The above information can be expressed using matrices as follows: Let the matrix A represent the number of tickets sold on Friday night. Let the matrix B stand for the number of tickets sold on Saturday. Let C be the cost matrix. Then

$$A = \begin{pmatrix} 250 \\ 120 \\ 73 \\ 208 \\ 124 \end{pmatrix}, \ B = \begin{pmatrix} 300 \\ 110 \\ 78 \\ 220 \\ 113 \end{pmatrix}, \ C = (3.00 \ \ 2.50 \ \ 2.00 \ \ 1.50 \ \ 1.00),$$

$$A + B = \begin{pmatrix} 250 + 300 \\ 120 + 110 \\ 73 + 78 \\ 208 + 220 \\ 124 + 113 \end{pmatrix} = \begin{pmatrix} 550 \\ 230 \\ 151 \\ 428 \\ 237 \end{pmatrix} = \text{Total seats sold of each type}$$
Friday and Saturday,

$$C \cdot (A + B) = (3.00 \quad 2.50 \quad 2.00 \quad 1.50 \quad 1.00) \cdot \begin{pmatrix} 550 \\ 230 \\ 151 \\ 428 \\ 237 \end{pmatrix} = (3406).$$

This represents the total income from both performances.

Example 2   The following table represents the number of summonses issued for various traffic violations for the first three months of 1976 in one midwestern town.

	Speeding	Illegal U-turn	Double parking	Drunken driving
January	2064	210	5314	206
February	3018	342	3709	421
March	1997	78	4112	308

Furthermore, the fines for these offenses are \$50, \$10, \$20, and \$30, respectively.

Mayor Pothole wants to know how much money the town can expect from these summonses. He may also be interested in determining how much money will be collected each month. Using matrices he can proceed in the following way.

Let the matrix A represent the number of summonses, and let the matrix C be the fine for each offense. We have

$$A = \begin{pmatrix} 2064 & 210 & 5314 & 206 \\ 3018 & 342 & 3709 & 421 \\ 1997 & 78 & 4112 & 308 \end{pmatrix}, \qquad C = \begin{pmatrix} 50 \\ 10 \\ 20 \\ 30 \end{pmatrix},$$

$$A \cdot C = \begin{pmatrix} 217{,}760 \\ 241{,}130 \\ 192{,}110 \end{pmatrix} \begin{matrix} \leftarrow \text{amount collected in January.} \\ \leftarrow \text{amount collected in February.} \\ \leftarrow \text{amount collected in March.} \end{matrix}$$

The total amount collected for all three months is found by adding the monthly totals together. It is \$651,000.

Example 3   Three salesmen for a dress manufacturer submit the following orders for the month of January. The results are coded in the form of matrices where the rows stand for the dress sizes: 8, 10, 12, and 14. The columns represent the colors: black, red, green, and beige.

MAUDE'S DRESS SHOP

SALESMAN'S ORDERING PATTERN

DRESS SIZE	12	14	16
BLUE	6	10	4
GREEN	4	12	8
BROWN	3	18	3

Salesman 1

$$\begin{pmatrix} 12 & 10 & 5 & 1 \\ 8 & 2 & 8 & 5 \\ 7 & 0 & 17 & 9 \\ 5 & 4 & 19 & 0 \end{pmatrix}$$

Salesman 2

$$\begin{pmatrix} 2 & 3 & 4 & 1 \\ 5 & 0 & 9 & 6 \\ 3 & 14 & 17 & 8 \\ 10 & 7 & 3 & 2 \end{pmatrix}$$

Salesman 3

$$\begin{pmatrix} 0 & 4 & 8 & 12 \\ 5 & 4 & 3 & 9 \\ 8 & 7 & 12 & 6 \\ 0 & 1 & 5 & 0 \end{pmatrix}$$

From these reports the manufacturer can obtain much information. Some of these results are summarized below. In each case, you should verify the results given.

108 size 12 dresses were sold,
 59 beige dresses were sold,
 65 black dresses were sold,
 84 dresses were sold by salesman 3,
290 dresses were sold altogether.

Comment   In doing these examples, it may be possible to write a matrix in two different ways. For instance, in Example 2 the matrix C was written as a 4×1 matrix,

$$C = \begin{pmatrix} 50 \\ 10 \\ 20 \\ 30 \end{pmatrix}.$$

We could have written it as a 1×4 matrix $(50 \quad 10 \quad 20 \quad 30)$. However, note that if we had written it as a 1×4, it would not have been possible to multiply it by A. (Why not?) We would then have been unable to solve this problem using matrices. Care is needed in deciding which matrices are to be used and how they are to be written. You have to consider what is to be done with the matrix when making these decisions.

Perhaps you are wondering why we use matrices to solve these problems when general arithmetic would work just as well. One reason is that the same type of problem may have to be solved many times with different numbers. For example, a salesman may send in monthly sales reports, all of which have different figures and all of which have to be analyzed by the same procedure. Matrices provide a *mechanical* procedure for doing this. All we have to do is "plug in" the new numbers each time and work out the answer mechanically. We do not have to rethink the method each time.

Another reason for using matrices is that the lists may be very long. There may be many different things that we want to analyze. In such situations, we could use a computer to help us with the calculations. Matrix operations can be very easily performed by a computer, and most machines have programs prepared to do such operations.

EXERCISES

1. A large computer dating service receives applications from a number of female college students. Some basic information is summarized in the following matrix.

	Redhead	Brunette	Blonde
Freshman	25	20	36
Sophomore	37	16	11
Junior	20	15	50
Senior	21	23	44

a) Bill is interested in a blonde. How many different blondes can the dating service offer him?

b) Michael will date only freshmen or sophomores. How many different dates can the dating service arrange for him?

c) Romeo will date anybody! How many different girls can he be offered by the dating service?

Vitamins

$$\begin{array}{l} \\ \text{Ice cream} \\ \text{Cereal} \\ \text{Nuts} \\ \text{Orange juice} \end{array} \begin{pmatrix} \text{A} & \text{B} & \text{C} & \text{D} \\ 4 & 5 & 1 & 3 \\ 0 & 1 & 5 & 3 \\ 2 & 2 & 9 & 11 \\ 4 & 7 & 15 & 2 \end{pmatrix}$$

2. Ima Hog is on a diet. Her doctor has given her the list at the left showing the vitamin content (expressed in convenient units) of one portion of some of her favorite foods.

a) If she has cereal and orange juice for breakfast, how much vitamin B will she get?

b) How much vitamin D does she get if she eats all four items?

c) She plans to eat two of these foods for lunch. Which two foods will give her the greatest amount of vitamin C?

3. A television manufacturer produces many different television sets in varying sizes, some of which are listed below (the numbers are expressed in thousands).

Size

$$Models\begin{cases} \text{Black and white table} \\ \text{Color table} \\ \text{Black and white portable} \\ \text{Color portable} \\ \text{Black and white console} \\ \text{Color console} \end{cases} \begin{pmatrix} \text{17-in.} & \text{19-in.} & \text{21-in.} & \text{23-in.} \\ 10 & 28 & 7 & 12 \\ 15 & 17 & 13 & 15 \\ 6 & 3 & 2 & 0 \\ 11 & 13 & 3 & 0 \\ 3 & 10 & 11 & 13 \\ 0 & 3 & 10 & 18 \end{pmatrix}$$

a) How many black and white televisions of all types are sold?

b) How many color 23-in. sets are made?

c) How many 17-in. sets are made?

d) How many color table models are sold?

e) How many black and white console models are made?

f) How many portable sets larger than 19-in. are made?

The cost for black and white table models is as follows: $80 for a 17-in., $110 for a 19-in., $120 for a 21-in., and $150 for a 23-in.

g) Write the above costs as a matrix.

h) Using matrix multiplication, find the income to the manufacturer from the black and white table models.

4. A mathematics professor teaches four classes and at the end of the semester has submitted the following grades:

Math   1   10 A's,   3 B's,   18 C's,   7 D's,   9 F's
Math 14   2 A's, 15 B's,   19 C's, 10 D's,   0 F's
Math 20   7 A's, 10 B's,   15 C's, 14 D's,   0 F's
Math 100   0 A's, 12 B's,   10 C's, 14 D's,   8 F's

Number of problems solved correctly	Grade
34	A
20	B
11	C
6	D
0	F

a) Rewrite the above information in matrix form.

b) How many students did the professor have this semester in all his classes?

c) How many A's did he give altogether?

d) How many students failed?

e) How many students were in Math 100?

f) In Math 20 the professor gives assignments, and grades are given according to the number of problems solved correctly, as indicated on the left. If each problem submitted is correctly solved, how many problems does the professor have to grade?

5. Bill, Phil, Will, Gil, and Jill ate lunch at the school cafeteria. Bill had 2 franks, a salad, and a coke. Phil had a veal cutlet and 2 cokes. Will had a frank, a coke, and a salad. Gil shared a veal cutlet with Jill. He also had a frank, a salad, and a coke. In addition, Jill had 3 franks and 2 cokes. The costs were: franks 50¢, cokes 35¢, salads 55¢, and veal cutlets 90¢.

a) Write the order as a matrix.

b) Write the cost of the items as a matrix.

c) How much did Will spend?

d) How many franks were ordered?

e) Using matrix multiplication, find the amount of money spent by each person.

f) How much money did the entire group spend?

6. The following chart indicates the prices of stocks of five companies on the first of the months of October, November, and December.

Stock	October 1	November 1	December 1
The Hot Seat Co.	$23	$27	$26
The Double-Header Hat Corp.	$6	$5	$12
American Distress Travel Co.	$32	$32	$31
Matrix Corp.	$10	$15	$12
International Drug Imports of Texas (IDIOT)	$51	$58	$46

a) Write a matrix to picture the above data.

b) On the 1st of each month, George bought 30 shares of Hot Seat, 20 shares of Double Header, 50 shares of American Distress, 65 shares of Matrix, and 100 IDIOT shares. Represent this information in matrix form.

c) Using matrix multiplication, find the total cost of the stocks to George over the 3-month period.

---

6. APPLICATIONS TO COMMUNICATION NETWORKS

We spend a good part of our lives communicating with other people in many different ways. When you call a friend, write a letter, or watch television, you are communicating or being communicated with. In writing this book, we hope we are communicating some ideas to you.

Communication also takes place when a teacher gives the class a homework assignment or when the students give the teacher a dirty look. Such activities as playing games, fighting, shaking hands, or kissing are also ways of communicating.

Some communication links are quite complicated. For example, suppose Steve wants to contact Florence, whom he met at a party last year. He doesn't have her phone number, so he calls his friend John who gave the party. John does

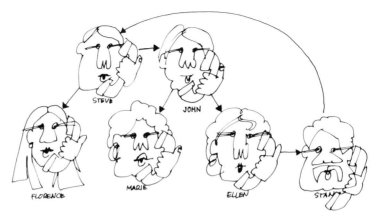

not have her number either, but he calls Ellen and Marie who came with Florence to the party. Neither has her number, but Ellen calls Stan, who does. He calls Steve and gives him the number. Through this *network*, Steve finally succeeds in getting Florence's number. We can illustrate this situation with the diagram shown above.

Another example of a complicated communications network is a publishing company that has a president, 2 vice-presidents, 8 office managers, 3 division heads, and various supervisory personnel (in addition to those who actually do the work). Some of these people have authority over others, that is, they give orders to those who work under their supervision. The president can give orders to everyone; the vice-presidents can give orders to everyone but the president; and so forth. An involved network such as this can become quite messy, and it may be very difficult to determine who can give orders to whom at certain levels.

Matrices can be helpful in analyzing communication networks. The following examples will illustrate how this is done. In our discussion, we will assume that *no one communicates with himself.*

Example 1   On a battlefield, four soldiers—Bob, Ed, Mike, and Al—have walkie-talkies with which they communicate with each other. Some of the walkie-talkies are broken. Bob's is completely broken. He can neither receive nor transmit any information. Ed's walkie-talkie can only receive but not transmit. Mike's is working properly. Al's can only transmit but not receive any information.

We form a matrix with rows and columns for each soldier. All entries are either 1 or 0. An entry of 1 means that the soldier in that row can transmit to the soldier in that column. Otherwise we put a 0. The matrix would then be the following:

$$
\begin{array}{c}
\\
\text{Bob} \\
\text{Ed} \\
\text{Mike} \\
\text{Al}
\end{array}
\begin{array}{cccc}
\text{Bob} & \text{Ed} & \text{Mike} & \text{Al} \\
\left(\begin{array}{cccc}
0 & 0 & 0 & 0 \\
0 & 0 & 0 & 0 \\
0 & 1 & 0 & 0 \\
0 & 1 & 1 & 0
\end{array}\right).
\end{array}
$$

a)

b)

c)

d)

e)

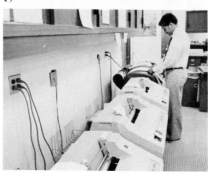

f)

Figure (a) and Figs. (c) through (h) are by courtesy of Merrill Lynch, Pierce, Fenner & Smith, Inc. Figure (b) is by courtesy of Bunker-Ramo Corporation.

Buying or selling stock sets a communication network in motion.

a) An account executive receives an order from a customer.

b) A telequote desk unit gives instantaneous data from the exchanges.

c) The order is sent via a communications terminal at the branch office to a computer in New York.

d) The computer switches the order directly to the floor of the New York Stock Exchange.

e) The firm's floor broker executes the order.

f) The local office receives the confirmation.

Example 2   Suppose in the situation of Example 1 the walkie-talkies are not completely broken, but are not working properly. There is also some interference. The following matrix represents the communications that are possible:

$$\begin{array}{c} \\ \text{Bob} \\ \text{Ed} \\ \text{Mike} \\ \text{Al} \end{array} \begin{array}{cccc} \text{Bob} & \text{Ed} & \text{Mike} & \text{Al} \\ \left(\begin{array}{cccc} 0 & 1 & 0 & 0 \\ 1 & 0 & 1 & 0 \\ 0 & 0 & 0 & 1 \\ 0 & 1 & 0 & 0 \end{array}\right). \end{array}$$

The matrix gives us the following information:

Bob can transmit only to Ed.
Ed can transmit to both Bob and Mike.
Mike can transmit only to Al.
Al can transmit only to Ed.

If Al wants to send a message to Bob, then he could transmit it by way of Ed. Is there any way for Mike to send a message to Bob?

You may be wondering why one would use matrices at all for such a situation. The matrix provides a quick and easy-to-read diagram of the entire network. By looking at the appropriate row and column, we can tell immediately who can communicate with whom.

Furthermore, by studying these matrices in greater detail it is possible to learn a lot about a network. For example, in a large corporation with many levels of supervision (some of them overlapping), one can find who has the most power, who has the least power, who is the second most powerful, and so on.

Example 3   In Chelem, there are four neighbors, Mrs. Yenta, Ms. Babble, Miss Tellall, and Mrs. Goody. Mrs. Yenta passes on all the gossip she hears to Ms. Babble and Mrs. Goody. Ms. Babble repeats gossip only to Miss Tellall. Miss Tellall tells all to Mrs. Yenta. Mrs. Goody never repeats any gossip. The following matrix represents the neighborhood gossip network:

$$\begin{array}{c} \\ \text{Mrs. Yenta} \\ \text{Ms. Babble} \\ \text{Miss Tellall} \\ \text{Mrs. Goody} \end{array} \begin{array}{cccc} \text{Mrs. Yenta} & \text{Ms. Babble} & \text{Miss Tellall} & \text{Mrs. Goody} \\ \left(\begin{array}{cccc} 0 & 1 & 0 & 1 \\ 0 & 0 & 1 & 0 \\ 1 & 0 & 0 & 0 \\ 0 & 0 & 0 & 0 \end{array}\right). \end{array}$$

EXERCISES   1. In any communications matrix, the diagonal from upper left to lower right has only 0's. Why?

2. a) In Example 3, will Miss Tellall hear all the gossip?

b) Will everybody in the system hear the gossip, no matter where it enters?

3. Ed Popp's wife, Lolly, decides to surprise him by inviting four of his old school chums, Ted, Fred, Ned, and Jed, to his birthday party. Since Ed hasn't seen them for a long time, she does not have all of their addresses. She does have Jed's address and she knows that Ted has some of the others. The following matrix shows who has which addresses:

	Lolly	Ted	Fred	Ned	Jed
Lolly	0	0	0	0	1
Ted	1	0	1	1	0
Fred	1	1	0	0	0
Ned	1	0	1	0	1
Jed	0	1	1	1	0

a) Which addresses does Ned have?

b) How can Lolly contact Fred?

c) What is the smallest number of letters that have to be written in order for Lolly to contact all four friends?

4. A doctor examines a patient and discovers that the patient has smallpox. It is vital that the local health service find out the names of all people who have had contact with the patient within the last 24 hours. Questioning him they discover the following: The patient has seen his wife and son. The wife has seen her parents. The son has seen his girlfriend. The girlfriend and the wife's parents have seen no one else. Make up a matrix showing who could have transmitted the disease to whom, using the above information only.

5. In a tennis match there are the following results:

Bill beats Sam and Fred,
Sam beats Sidney,
Sidney beats Bill, and
Fred beats Sidney and Sam.

Make a matrix showing who beat whom.

6. The following matrix shows who can control whom in the publishing company mentioned earlier:

	President	Vice president	Advertising manager	Sales manager	Personnel director
President	0	1	1	1	1
Vice president	0	0	1	0	1
Advertising manager	0	0	0	0	1
Sales manager	0	0	1	0	1
Personnel director	0	0	0	0	0

a) How many people control the sales manager?

b) How many people control the advertising manager?

c) From whom does the personnel director take orders?

†7. APPLICATIONS TO SYSTEMS OF EQUATIONS

If we are asked to solve the system of equations

$$\left\{\begin{array}{l} 2x + 3y = 11 \\ 7x + 5y = 33 \end{array}\right\}$$

for x and y, then we are looking for one value of x and one value of y that satisfy both equations. The solution for these equations is $x = 4$ and $y = 1$. This can be verified by substituting $x = 4$ and $y = 1$ into each of the equations. This answer can be obtained using matrices in the following way.

We first write the numbers on the left side, in matrix form. Call this matrix A:

$$A = \begin{pmatrix} 2 & 3 \\ 7 & 5 \end{pmatrix}.$$

The numbers on the righthand side can be written as a matrix C:

$$C = \begin{pmatrix} 11 \\ 33 \end{pmatrix}.$$

The unknowns x and y can be written as a matrix M:

$$M = \begin{pmatrix} x \\ y \end{pmatrix}.$$

If we multiply matrices A and M (by matrix multiplication), we get

$$AM = \begin{pmatrix} 2 & 3 \\ 7 & 5 \end{pmatrix} \cdot \begin{pmatrix} x \\ y \end{pmatrix} = \begin{pmatrix} 2x + 3y \\ 7x + 5y \end{pmatrix}.$$

Notice that the result is the same as C since $2x + 3y = 11$ and $7x + 5y = 33$. Therefore

$$AM = C.$$

It can be shown that a 2×2 matrix, such as A, may have an **inverse**. This means that there may exist another matrix which when multiplied by A will give the identity matrix (see Definition 4.1, p. 331). Such an inverse will be denoted by A^{-1}. We can verify (see Example 3, p. 348) by direct matrix multiplication that if matrix $T = \begin{pmatrix} a & b \\ c & d \end{pmatrix}$, then the inverse, T^{-1}, is $\begin{pmatrix} d & -b \\ -c & a \end{pmatrix}$ multiplied by the number $\dfrac{1}{ad - bc}$, provided that $ad - bc$ is not 0.[1] Using this result, we find that the inverse of $A = \begin{pmatrix} 2 & 3 \\ 7 & 5 \end{pmatrix}$ is $A^{-1} = \begin{pmatrix} 5 & -3 \\ -7 & 2 \end{pmatrix}$ multiplied by $\dfrac{1}{2(5) - 3(7)}$, or

---

1. If $ad - bc = 0$, then the number $1/(ad - bc)$ would be $1/0$. This, of course, is undefined.

$$A^{-1} = \begin{pmatrix} 5 & -3 \\ -7 & 2 \end{pmatrix} \text{ multiplied by } -\frac{1}{11}. \text{ This gives}$$

$$A^{-1} = \begin{pmatrix} \dfrac{-5}{11} & \dfrac{+3}{11} \\ \dfrac{+7}{11} & \dfrac{-2}{11} \end{pmatrix}.$$

(If you have forgotten how this type of multiplication is performed, see p. 332). The reader should verify that this is the inverse by multiplying it with A. You should get

$$\begin{pmatrix} 2 & 3 \\ 7 & 5 \end{pmatrix} \cdot \begin{pmatrix} \dfrac{-5}{11} & \dfrac{+3}{11} \\ \dfrac{+7}{11} & \dfrac{-2}{11} \end{pmatrix} = \begin{pmatrix} 1 & 0 \\ 0 & 1 \end{pmatrix}.$$

Let us go back to our equation $AM = C$. Multiplying both sides by A^{-1}, we get

$$A^{-1}(AM) = A^{-1}C$$

$$(A^{-1}A)M = A^{-1}C \qquad \text{(by the associative law)}.$$

Since $A^{-1}A$ is the 2×2 identity matrix, and the identity matrix (Definition 4.1) does nothing when you multiply with it, we get

$$M = A^{-1}C.$$

(Be careful! $A^{-1}C$ may not equal CA^{-1}, since matrix multiplication is not always commutative.)

This means that

$$\begin{pmatrix} x \\ y \end{pmatrix} = \begin{pmatrix} \dfrac{-5}{11} & \dfrac{+3}{11} \\ \dfrac{+7}{11} & \dfrac{-2}{11} \end{pmatrix} \cdot \begin{pmatrix} 11 \\ 33 \end{pmatrix}$$

$$= \begin{pmatrix} \dfrac{-5}{11} \cdot (11) + \dfrac{3}{11} \cdot (33) \\ \dfrac{+7}{11} \cdot (11) + \dfrac{-2}{11} \cdot (33) \end{pmatrix}$$

$$= \begin{pmatrix} -5 + 9 \\ 7 - 6 \end{pmatrix}$$

$$= \begin{pmatrix} 4 \\ 1 \end{pmatrix}.$$

Thus we have $x = 4$ and $y = 1$.

Example 1   Solve $\begin{Bmatrix} 3x - 2y = 0 \\ 2x + y = 7 \end{Bmatrix}$ for x and y.

Solution $A = \begin{pmatrix} 3 & -2 \\ 2 & 1 \end{pmatrix}, \qquad M = \begin{pmatrix} x \\ y \end{pmatrix}, \qquad C = \begin{pmatrix} 0 \\ 7 \end{pmatrix},$

$A^{-1} = \begin{pmatrix} 1 & 2 \\ -2 & 3 \end{pmatrix}$ multiplied by $\dfrac{1}{3 \cdot (1) - (-2)2}$,

$A^{-1} = \begin{pmatrix} 1 & 2 \\ -2 & 3 \end{pmatrix}$ multiplied by $\dfrac{1}{+7}$.

Therefore,

$$A^{-1} = \begin{pmatrix} \dfrac{1}{7} & \dfrac{2}{7} \\ \dfrac{-2}{7} & \dfrac{3}{7} \end{pmatrix}.$$

Our answer is then

$$M = A^{-1}C,$$

$$\begin{pmatrix} x \\ y \end{pmatrix} = \begin{pmatrix} \dfrac{1}{7} & \dfrac{2}{7} \\ \dfrac{-2}{7} & \dfrac{3}{7} \end{pmatrix} \cdot \begin{pmatrix} 0 \\ 7 \end{pmatrix}$$

$$= \begin{pmatrix} \dfrac{1}{7} \cdot (0) + \dfrac{2}{7} \cdot (7) \\ \dfrac{-2}{7} \cdot (0) + \dfrac{3}{7} \cdot (7) \end{pmatrix}$$

$$= \begin{pmatrix} 0 + 2 \\ 0 + 3 \end{pmatrix}$$

$$= \begin{pmatrix} 2 \\ 3 \end{pmatrix}.$$

Finally, we have $x = 2$ and $y = 3$. (The reader should check these answers by substituting $x = 2$ and $y = 3$ into the original equations.)

This technique can be extended to solve a system with any number of equations and the same number of unknowns.

Example 2   Solve $\begin{Bmatrix} 4x - 3y = 5 \\ x + 2y = 4 \end{Bmatrix}$ for x and y.

Solution $A = \begin{pmatrix} 4 & -3 \\ 1 & 2 \end{pmatrix}, \qquad M = \begin{pmatrix} x \\ y \end{pmatrix}, \qquad C = \begin{pmatrix} 5 \\ 4 \end{pmatrix}$

$A^{-1} = \begin{pmatrix} 2 & 3 \\ -1 & 4 \end{pmatrix}$ multiplied by $\dfrac{1}{4(2) - (-3)(1)}$,

$$A^{-1} = \begin{pmatrix} 2 & 3 \\ -1 & 4 \end{pmatrix} \quad \text{multiplied by } \frac{1}{11}.$$

Therefore,

$$A^{-1} = \begin{pmatrix} \dfrac{2}{11} & \dfrac{3}{11} \\ \dfrac{-1}{11} & \dfrac{4}{11} \end{pmatrix}.$$

Our answer is then

$$M = A^{-1}C,$$

$$\begin{pmatrix} x \\ y \end{pmatrix} = \begin{pmatrix} \dfrac{2}{11} & \dfrac{3}{11} \\ \dfrac{-1}{11} & \dfrac{4}{11} \end{pmatrix} \cdot \begin{pmatrix} 5 \\ 4 \end{pmatrix}$$

$$= \begin{pmatrix} \dfrac{2}{11} \cdot (5) + \dfrac{3}{11} \cdot (4) \\ \dfrac{-1}{11} \cdot (5) + \dfrac{4}{11} \cdot (4) \end{pmatrix}$$

$$= \begin{pmatrix} \dfrac{10}{11} + \dfrac{12}{11} \\ \dfrac{-5}{11} + \dfrac{16}{11} \end{pmatrix}$$

$$= \begin{pmatrix} \dfrac{22}{11} \\ \dfrac{11}{11} \end{pmatrix}$$

$$= \begin{pmatrix} 2 \\ 1 \end{pmatrix}.$$

Thus we have $x = 2$ and $y = 1$. (Again the reader should check these answers by substituting $x = 2$ and $y = 1$ into the original equations.)

Example 3   Verify, by matrix multiplication, that if matrix

$$T = \begin{pmatrix} a & b \\ c & d \end{pmatrix},$$

then its inverse is

$$\begin{pmatrix} d & -b \\ -c & a \end{pmatrix} \text{ multiplied by } \frac{1}{ad - bc}.$$ (We will assume that $ad - bc$ is not equal to 0. Otherwise the inverse does not exist.)

Solution   We first multiply $\begin{pmatrix} a & b \\ c & d \end{pmatrix}$ by $\begin{pmatrix} d & -b \\ -c & a \end{pmatrix}$, getting

$$\begin{pmatrix} a & b \\ c & d \end{pmatrix} \cdot \begin{pmatrix} d & -b \\ -c & a \end{pmatrix} = \begin{pmatrix} ad + b(-c) & a(-b) + ba \\ cd + d(-c) & c(-b) + da \end{pmatrix}$$

$$= \begin{pmatrix} ad - bc & -ab + ba \\ cd - dc & -bc + da \end{pmatrix}$$

$$= \begin{pmatrix} ad - bc & 0 \\ 0 & ad - bc \end{pmatrix}.$$

Finally, multiplying this matrix by $\dfrac{1}{ad - bc}$ gives

$$\begin{pmatrix} \dfrac{ad - bc}{ad - bc} & \dfrac{0}{ad - bc} \\ \dfrac{0}{ad - bc} & \dfrac{ad - bc}{ad - bc} \end{pmatrix} = \begin{pmatrix} 1 & 0 \\ 0 & 1 \end{pmatrix},$$

which is the 2×2 identity matrix.

Thus, the inverse of $\begin{pmatrix} a & b \\ c & d \end{pmatrix}$ is $\begin{pmatrix} d & -b \\ -c & a \end{pmatrix}$ multiplied by $\dfrac{1}{ad - bc}$.

EXERCISES

1. Verify by matrix multiplication that $\begin{pmatrix} 2 & 3 \\ 7 & 5 \end{pmatrix} \cdot \begin{pmatrix} -\dfrac{5}{11} & \dfrac{3}{11} \\ \dfrac{7}{11} & -\dfrac{2}{11} \end{pmatrix} = \begin{pmatrix} 1 & 0 \\ 0 & 1 \end{pmatrix}.$

2. Using matrices, solve the following systems of equations.

a) $3x - 4y = -5$ b) $5x + 3y = -1$ c) $-x - y = -7$
   $-4x + 3y = 2$ $3x - 4y = 11$ $3x + 5y = 9$

d) $4x + 3y = 9$ e) $2x - y = 1$ f) $5x + 4y = 26$
   $-5x + 7y = -22$ $-3x + 3y = 0$ $4x - 3y = -4$

g) $x + y = 10$ h) $7x + 3y = 21$ i) $2x - y = 4$
   $x - y = 4$ $-3x + 7y = -9$ $6x - 3y = 1$

3. The result of part (i) of Exercise 2 is different than the other parts. Explain.

†*4. Show that the inverse of matrix $\begin{pmatrix} a & b \\ c & d \end{pmatrix}$ is $\begin{pmatrix} d & -b \\ -c & a \end{pmatrix}$ multiplied by $\dfrac{1}{ad - bc}$, by using the following procedure. Consider

$$\begin{pmatrix} a & b \\ c & d \end{pmatrix} \cdot \begin{pmatrix} w & x \\ y & z \end{pmatrix} = \begin{pmatrix} 1 & 0 \\ 0 & 1 \end{pmatrix}.$$

Using matrix multiplication on the left side, we get

$$\begin{pmatrix} aw + by & ax + bz \\ cw + dy & cx + dz \end{pmatrix} = \begin{pmatrix} 1 & 0 \\ 0 & 1 \end{pmatrix}.$$

This means that

$aw + by = 1,$

$ax + bz = 0,$

$cw + dy = 0,$

$cx + dz = 1.$

Solve the first two equations and the last two equations in pairs for w, x, y, and z. You should get

$$w = \frac{d}{ad - bc}, \qquad y = \frac{-c}{ad - bc}, \qquad x = \frac{-b}{ad - bc}, \qquad z = \frac{a}{ad - bc}.$$

Compare this with the answer we got. They are the same.

8. CONCLUDING REMARKS

There are many other interesting and useful applications of matrix theory. Some of the ways in which matrices can be used are in

1. predicting population growth,
2. analyzing marriage rules of various societies (in an anthropology discussion),
3. various problems in economics,
4. studying heredity (genetics).

The interested reader can consult the suggested further readings.

STUDY GUIDE

In this chapter the basic ideas of matrices were introduced and several applications were given.

Basic ideas

Matrix (p. 322)
Addition and subtraction of matrices (p. 325)
Dimension (p. 324)
Square matrix (p. 324)
Multiplication of matrices (p. 327)
Identity matrix (p. 331)
Multiplication of a matrix by a number (p. 331)
Inverse of a matrix (p. 345)

Applications were given for:

a) business problems (p. 333),
b) communications networks (p. 340),
c) systems of equations (p. 345).

SUGGESTED FURTHER READINGS

Bell, E. T., *Men of Mathematics*. New York: Simon and Schuster, 1961. Chapter 19 contains a biography of Hamilton; chapter 21 discusses Sylvester and Cayley.

Campbell, H., *An Introduction to Matrices, Vectors and Linear Programming*, 2nd ed. New York: Appleton-Century-Crofts, 1971. An excellent introduction to matrix theory.

Kemeny, J., L. Snell, and J. Thompson, *Introduction to Finite Mathematics*, 3rd ed. Englewood Cliffs, N.J.: Prentice-Hall, 1974. Chapter 4 discusses matrices and their applications.

Mathematics in the Modern World (Readings from *Scientific American*). San Francisco: W. H. Freeman, 1968. Article 7 contains a biographical sketch of Hamilton.

Sawyer, W. W., *Prelude to Mathematics*. Baltimore, Md.: Penguin Books, 1959. Chapter 8 discusses matrix algebra.

Von Neumann, J., and O. Morgenstern, *Theory of Games and Economic Behavior*, 3rd ed. Princeton, N.J.: Princeton University Press, 1953. See chapter 1.

Whittaker, E., "William Rowan Hamilton." *Scientific American* (May 1954).

11 Game Theory

1. INTRODUCTION

Throughout our lives we play various games. Some of these games are played for fun, or to pass time, or for money. Card games, chess, dominoes, checkers, and tic-tac-toe are of this type. There are other situations (such as the stock market, politics, war, bargaining between a union and an employer, business activities, etc.) that so closely resemble games that in discussing them we frequently use the language of games. We say, for example, "I made a good deal," "That was a good move," "She plays the market," "He's bluffing." Because these activities are similar to games in many ways, we can use the mathematical theory of games in

352

analyzing them. For this reason, even nonmathematicians such as economists or sociologists find this branch of mathematics of great interest.

For example, when unions and employers sit down to negotiate a new contract, each side pushes for certain items that will result in benefits to itself. In trying to obtain the best terms, each side will use various strategies. By a combination of bluffing, arm-twisting, and "friendly persuasion," they will try to get the other side to agree to their terms. In doing so, not only must they plan their own moves, but they must also anticipate the other's moves and act accordingly.

This is the same type of situation that occurs when two or more people are playing a game such as chess, Monopoly, or football. Each player tries to figure out his or her opponents' moves and act accordingly.

Similar situations arise in many areas of business. For example, suppose a large hamburger chain is considering opening a new restaurant in one of two towns. In deciding which town to open in, it must consider the reactions of its competitors (if any).

There are certain games that have "best" moves for a player, no matter what his or her opponent does. Then again, there are other games that have no single best possible strategy. Most military situations are of the latter type. In this case, a player must first decide on one or more strategies. There is no "perfect" or "best" strategy for all situations. Obviously a player must hide his or her strategy so that the opponent does not figure it out. The first person to develop a method for finding best moves under these circumstances was the Hungarian mathematician John Von Neumann (1903–1957).

Although Von Neumann was born in Budapest, he left Europe just before World War I to come to the United States. For many years he worked at the Institute for Advanced Study at Princeton University, as well as for the United States government. He was one of the scientists who developed the atomic bomb. Much of the theory of games discussed in the chapter was developed by Von Neumann.

In mathematical game theory, we deal only with **games of strategy**. These are games in which there are two or more players competing against each other. Each player has several alternative "moves" available. He or she must decide which is the best move to make. All of the games mentioned above are games of strategy. On the other hand, such games as Russian roulette, dice, pitching pennies, or lotteries are not games of strategy. In them, no "plan" is involved. Some of them involve physical skill. Others are just "betting against the odds."

In this chapter, we deal only with two-person, zero-sum games. A **two-person game** is any game that involves only two players. A game is called a **zero-sum game** if on each move, one player wins what the other player loses. A game in which there is a bank or kitty is not a zero-sum game and will not be discussed in this book.

2. STRICTLY DETERMINED GAMES

Let us look in on Jack Row and Bill Column, who are playing a game. Each player has two cards. Row has a black 7 and a red 7. Column has a black 7 and a red 4. The rules of the game follow.

Each player puts out one card at the same time. If the colors match, then Row is the winner. If the colors do not match, then Column is the winner. In each case, the amount of money won is the difference of the numbers (the larger minus the smaller) appearing on the cards. All possible plays and the resulting payoffs are given in the following table, which is called the **payoff matrix**. Each entry represents Row's winnings or losses.

	Player Column Bk 7	Rd 4
Player Row Bk 7	0	−3
Rd 7	0	3

If both players play black 7, then the colors match. Row wins, and the amount of money won is the difference of the numbers appearing on the cards. So Row wins 7 − 7, or 0. We put 0 in the first row and first column.

If Row plays a black 7 and Column plays a red 4, then the colors do not match. Therefore Column wins. The amount of money won is the difference between the numbers, or 7 − 4 which is 3. Since Column wins 3, Row loses 3. This we indicate by −3 in the chart.

If Row plays red 7 and Column plays black 7, then Column wins 7 − 7, or 0.

If Row plays red 7 and Column plays red 4, then Row wins 7 − 4, or 3.

What is the best move for each player? Row reasons as follows: If he plays black 7, then he can either break even or lose 3, depending upon Column's play. He can never win. On the other hand, if he plays red 7, then he can either break even or win 3. (Remember that the entries in the chart always represent Row's winnings which are the same as Column's losses.) Clearly, the best move for Row is a red 7. He can never lose any money with this play, no matter what Column does, and he could win 3.

What should Column do? He knows that Row will always play red 7. If Column plays black 7, then he will break even. On the other hand, if he plays red 4, then he will lose 3. His best move would then be to play black 7. We circle each player's best move as shown on the left.

Notice that one number has been circled twice. It is 0. This number is called the **value** of the game. If the value of a game is zero, then we say that it is a **fair** game. This game is obviously a fair game. What this means is that in the long run, the amount of money won by either Row or Column is 0 (assuming that both players do not make any foolish moves).

Comment   Perhaps the thought has crossed your mind that there is no point in playing this game since no money can be won. In a sense you are right. However, you do not know that this is the case until you have analyzed it mathematically. In more complicated games, it is usually impossible to tell without mathematical analysis whether a game is worth playing at all. Some games are too complicated to be analyzed, even mathematically.

Let us refer back to the game above. The payoff matrix again is

	Bk 7	Rd 4
Bk 7	0	−3
Rd 7	0	3

What is the smallest number in the first row? It is −3. (Remember, −3 is smaller than 0.) Is −3 the largest number of the column that it is in? Clearly not, since 3 is larger.

In the second row, the smallest number is 0. Is 0 the maximum (largest number) in that column? Since there is no number that is larger than 0, then 0 is the maximum. In other words, *this number is at the same time the smallest in its row and the largest in its column.* Whenever a number like this exists in a game, then we say the game is **strictly determined**. We restate as definitions the important ideas given above.

Definition 2.1   *The **payoff matrix** of a game is a table of numbers representing the amount of money won or lost by each player as a result of any move.*

Definition 2.2   *A game is said to be **strictly determined** if there is a number in the payoff matrix which is **at the same time** the smallest number of the row it's in and the largest number of its column.*

Definition 2.3   *If we can find a number satisfying the conditions described by Definition 2.2, then this number is called the **value** of the game. We will refer to the value of a game as v.*

Definition 2.4   *A game is called a **fair game** if its value is zero.*

Definition 2.5   *By an **optimal strategy** for a player, we mean a best move. This will mean either winning the most amount of money or losing the least amount.*

In any game, a player is interested in finding his or her optimal strategy (best move). If we know that a game is strictly determined, then it is very easy to find the optimal strategy for each player. This is done as follows:

Player Row: Play the row that contains the value v. (If there is more than one such row, play either.)

Player Column: Play the column that contains the value v. (If there is more than one such column, play either.)

The following examples illustrate all of these ideas.

The game of *Go-Bang* in Japan is an ancient war game. A heavy block of wood is used as the board and the surface is divided into squares by cross lines. Smooth elliptically-shaped stones, called "Ishi" in Japanese, are used for "men."

(Peabody Museum of Salem, M. Sexton)

Example 1   Row and Column are playing a card game. Each player has only three cards, which are a 2, a 4, and a 5. The rules are as follows. At a given signal, each player puts out a card. If the numbers match, then nobody wins. If the numbers don't match, the person with the higher number wins. The amount won is the sum of the numbers shown. The payoff matrix is

		Player Column		
		2	4	5
Player Row	2	0	−6	−7
	4	6	0	−9
	5	7	9	0

The smallest number in the first row is −7. This is *not* the largest in its column. 0 is larger.

The smallest number in the second row is −9. This is *not* the largest in its column.

The smallest number in the third row is 0. This *is* the largest in its column.

This means that the game is strictly determined and its value is 0. Since the value, v, is 0, the game is *fair*. The value v is in the third row and third column. The optimal strategies are:

Player Row: Play row 3, or the card with 5 on it.

Player Column: Play column 3, or the card with 5 on it.

Example 2   In a card game, each player has three cards, a 2, a 4, and a 5. Each player puts out one card at the same time. If the numbers match, then player Row wins the sum of the numbers. If the numbers don't match, then player Column wins the sum of the numbers shown. The payoff matrix is the following one.

		Player Column		
		2	4	5
Player Row	2	4	−6	−7
	4	−6	8	−9
	5	−7	−9	10

Is this game strictly determined? To answer this, we notice that the smallest number in row 1 is −7. This is not the largest number in the third column.

The smallest number in row 2 is −9. This again is not the largest in the third column.

In row 3 the smallest number, −9, is not the largest in the second column. Since there is no number that is the smallest of the row and the largest of its column, the game is *not* strictly determined.

Example 3   Two players, Sue and Pete, are "choosing." Each player shows 1 or 2 fingers at the same time. If they both show the same number of fingers, then Sue wins the sum of the fingers shown. If they both do not show the same number of fingers, then Pete wins the sum of the fingers shown. The payoff matrix is on the left.

Is this game strictly determined? The smallest number in row 1 is −3. This is not the largest in the second column. Similarly −3, which is the smallest of the second row, is not the largest of the first column, so the game is not strictly determined.

		Pete	
		1	2
Sue	1	2	−3
	2	−3	4

Example 4   Suppose we have a game whose payoff matrix is given by the following.

		Player C			
		1	2	3	4
Player R	1	1	4	5	9
	2	0	−2	−7	−8
	3	1	3	2	8

The smallest number of the first row is 1. This is also the largest number in the column. Similarly, the minimum of the third row is 1. This is again the largest number in the column.

The game is strictly determined. The value of the game is 1. This means that the game is not fair. It is in favor of Row. The optimal strategies are:

Player R: Play *either* row 1 or row 3.

Player C: Play column 1.

Example 5 Mrs. Jones has told her two children Pat and Ned to play the game of stone, paper, and scissors. The game works as follows: Each player says the word "stone" or "scissors" or "paper." Scissors cuts paper, paper wraps stone, and stone breaks scissors. Therefore if one says "stone" and the other says "scissors," then "stone" wins a lollipop. If one says "scissors" and the other says "paper," then "scissors" wins a lollipop. Similarly, if one says "stone" and the other says "paper," then "paper" wins a lollipop. If they both say the same thing, then the game is a tie. The payoff matrix is

		Ned		
		Scissors	Stone	Paper
Pat	Scissors	0	−1	+1
	Stone	+1	0	−1
	Paper	−1	+1	0

Is this game strictly determined? If so, find the optimal strategies for each player.

Example 6 Two national car rental agencies offer weekly discounts to customers in certain weeks.

If they both offer the discount in the same week, nobody gets any points. If one company offers a discount and the other company offers this discount a week later, credit the first company with 5 points. If one company offers a discount and the other company offers it two weeks later, credit the first company with 3 points. The payoff matrix is

		Company 2 offers discount		
		Week 1	Week 2	Week 3
	Week 1	0	+5	+3
Company 1 offers discount	Week 2	−5	0	+5
	Week 3	−3	−5	0

This chart is read as follows: The entry in the second row and third column indicates that company 1 offers the discount in week 2 and company 2 offers the

MAS

During and after the period of the Crusades, the cultural exchange between Christians and Moslems included, along with such things as Arabic numerals and the works of Aristotle, the game of chess. From a thirteenth-century manuscript of rules comes this miniature in which a Christian and a Moslem play at chess.

In 1973, American Bobby Fischer and Russian Boris Spassky competed for the world chess championship.

discount in week 3. Thus we credit company 1 with 5 points. The +5 entry represents this.

The business contest between the two companies is a strictly determined game. The value is 0. The optimal strategy for each company is to offer the discount in week 1.

EXERCISES

1. If you were Sue, how would you play the game of Example 3? How would you play if you were Pete?

*2. In Example 2 on p. 357, what do you think are the optimal strategies for each player?

3. Why is the game of Example 4 (p. 357) in favor of Player R?

4. Is the game of Example 5 (p. 358) strictly determined?

5. In Example 6 (p. 358), why is the contest a strictly determined game?

6. Each payoff matrix given below represents some game. Determine whether each game is strictly determined or not. If the game is strictly determined, find the value and optimal strategies for each player. Are any of the games fair?

a)

           C

R
−4	1
5	0

b)

           C

R
2	−1	3
4	7	1
7	8	4

c)

           C

R
2	−2
−2	2

d)

           C

R
5	5
7	3

e)

           C

R
5	0
1	0

f)

           C

R
0	0
0	−7

g)

           C

R
1	3	−2	4
3	−1	2	0
0	1	−4	3
4	1	6	−6

h)

           C

R
8	3	1	6
1	5	−5	7
−3	11	−6	0
4	−2	−3	1

7. Mr. Rummy and Ms. Casino have each been given three cards as follows: Mr. Rummy has a black 3, a red 8, and a red 2. Ms. Casino has a black 6, a red 4, and a black 8. They each put out a card at the same time. If the colors match, then Rummy wins the sum of the numbers appearing on the cards. If they don't match, then Casino wins the sum of the numbers.

   a) Set up the payoff matrix for this game.   b) Is the game strictly determined?

8. Jean and Dave are "choosing" as described in Example 3. If they both show the same number of fingers, then Jean wins the sum. Otherwise, Dave wins the product of the number of fingers shown.

   a) Set up the payoff matrix for this game.   b) Is the game strictly determined?

9. Two competing companies, Shlep Trucking Corp. and Speedy Delivery Service, are each considering opening new branches in two nearby cities, Benton and Stonehill. Credit Shlep Trucking Corp. with 8 points if it decides on Benton and Speedy Delivery Service decides on Stonehill. Credit Speedy Delivery Service with 6 points if it opens in Benton and Shlep Trucking Corp. opens in Stonehill. Credit both companies with 0 if they open in the same city.

   a) Set up the payoff matrix.

   b) Is the game strictly determined?

   c) Find the value of the game (if it is strictly determined).

   d) Is it fair?

   e) Find the optimal strategies for each company.

3. NON-STRICTLY DETERMINED GAMES

In the previous section we discussed how to find the optimal strategies for strictly determined games. We have come across many games that are not strictly determined (that is, there exists no number that is the smallest of its row and also the largest of its column). In this section, we will consider how to find the optimal strategies for such games. For simplicity, we will consider only 2×2 games. These are games in which each player has only two possible moves.

Example 1

Consider a game whose payoff matrix is that shown at the left.

$$
\begin{array}{c}
 & \text{C} \\
 & \text{Bk 2} \quad \text{Rd 4} \\
\text{R} \quad
\begin{array}{c}
\text{Bk 1} \\
\text{Rd 3}
\end{array}
&
\begin{array}{|c|c|}
\hline
0 & 6 \\
\hline
6 & 0 \\
\hline
\end{array}
\end{array}
$$

This is also a non-strictly determined game. However, it is not obvious what R should do. If R plays black 1, then he either breaks even or wins 6. If R plays red 3, then he again will win 6 or break even. It would seem that R should play either black 1 or red 3, since in both cases, the amount of money won is the same. If R always plays black 1, then C will undoubtedly notice this. C will then play black 2 to break even. Similarly, if R always plays red 3, then C will again notice this and play red 4 to break even. R may decide to alternate his play between black 1 and red 3. This is also dangerous since C is sure to notice this and play accordingly. The best strategy for R is to play black 1 sometimes and red 3 at other times, but not according to any pattern that C will recognize. One way of doing this is to flip a coin (without C seeing what is being done) and play black 1 if heads comes up and play red 3 if tails comes up. In the long run, heads and tails will each come up approximately half the time. Thus R will end up playing black 1 and red 3, each half the time.

C should follow a strategy similar to that of R. He should toss a coin and play black 2 or red 4 depending upon whether heads or tails comes up. Why is this the best strategy for C?

Example 2

Mabel and Bob are playing the following game. Mabel hides either 1 penny or 2 pennies in her hand. Bob has to guess 1 or 2 and wins the number of pennies if he guesses correctly. Otherwise he wins nothing. The payoff matrix of this game is as follows:

$$
\begin{array}{c}
 & \text{Bob guesses} \\
 & 1 \qquad 2 \\
\text{Mabel hides} \quad
\begin{array}{c}
\text{1 penny} \\
\text{2 pennies}
\end{array}
&
\begin{array}{|c|c|}
\hline
-1 & 0 \\
\hline
0 & -2 \\
\hline
\end{array}
\end{array}
$$

This game is not strictly determined. If Mabel always hides 1 penny and Bob observes this pattern, then Bob will always say 1. Mabel will always lose 1 penny. Similarly, if Mabel always hides 2 pennies and Bob figures this out, then Bob will always say 2. Mabel will always lose 2 pennies.

It is obvious that Mabel should hide 1 penny sometimes and 2 pennies other times, but according to no particular pattern. She should *mix* her strategies. How is this to be done?

You may think that Mabel could do what R did in the last example. In that situation, no matter what R did, his payoff was either 0 or 6. This meant that any move he made would give him the same amount of money. In this example, Mabel loses different amounts of money, depending upon her move. Thus, just flipping a coin will not work. For this example and others like it, we need the following results.

It can be shown[1] that a non-strictly determined game whose payoff matrix is shown on the left has the following optimal strategies (the formulas may look frightening, but don't let them scare you off; they are really quite simple to use).

C

	Column 1	Column 2
Row 1	a	b
Row 2	c	d

R

Let $R_1 = \dfrac{d - c}{a + d - b - c}$, $R_2 = \dfrac{a - b}{a + d - b - c}$,

$C_1 = \dfrac{d - b}{a + d - b - c}$, $C_2 = \dfrac{a - c}{a + d - b - c}$.

Player R: Depending upon the total number of games played, play row 1, R_1 of the time and row 2, R_2 of the time.

Player C: Depending upon the total number of games played, play column 1, C_1 of the time and column 2, C_2 of the time.

The *value* of the game is

$$v = \frac{ad - bc}{a + d - b - c}.$$

Comment   The value for a non-strictly determined game is not defined as it was for strictly determined games. In this case the value is given by a formula.

Let us now return to Example 2. The payoff matrix is

Bob guesses

		1	2
Mabel hides	1 penny	-1	0
	2 pennies	0	-2

Here $a = -1$, $b = 0$, $c = 0$, and $d = -2$.

$$R_1 = \frac{d - c}{a + d - b - c} = \frac{(-2) - 0}{(-1) + (-2) - 0 - 0} = \frac{-2}{-3} = \frac{2}{3}$$

1. See, for example, J. Kemeny, J. Snell, and G. Thompson, *Introduction To Finite Mathematics.* Englewood Cliffs, N.J.: Prentice-Hall, 1966, pp. 343–350.

$$R_2 = \frac{a - b}{a + d - b - c} = \frac{(-1) - 0}{(-1) + (-2) - 0 - 0} = \frac{-1}{-3} = \frac{1}{3}$$

$$C_1 = \frac{d - b}{a + d - b - c} = \frac{(-2) - 0}{(-1) + (-2) - 0 - 0} = \frac{-2}{-3} = \frac{2}{3}$$

$$C_2 = \frac{a - c}{a + d - b - c} = \frac{(-1) - 0}{(-1) + (-2) - 0 - 0} = \frac{-1}{-3} = \frac{1}{3}$$

$$v = \frac{ad - bc}{a + d - b - c} = \frac{(-1)(-2) - 0 \cdot 0}{(-1) + (-2) - 0 - 0} = \frac{+2}{-3} = \frac{-2}{3}$$

Thus the optimal strategies are for R (or Mabel) to play row 1 (1 penny) $\frac{2}{3}$ of the time and to play row 2 (2 pennies) $\frac{1}{3}$ of the time.

Similarly, the optimal strategies for C (or Bob) are to play column 1 (guess 1) $\frac{2}{3}$ of the time and to play column 2 (guess 2) $\frac{1}{3}$ of the time.

The value of the game is $-\frac{2}{3}$. Since the value is negative, this means that the game is in favor of C (Bob). We could have guessed this, since the payoff matrix indicates that R can never win.

R now knows that she should play row 1, $\frac{2}{3}$ of the time, and row 2, $\frac{1}{3}$ of the time. Again she does not want to play row 1 twice and then row 2 once, constantly repeating this pattern, since C is bound to catch on and will play accordingly. She is therefore looking for some way of playing 1, $\frac{2}{3}$ of the time and row 2, $\frac{1}{3}$ of the time, but without C being able to figure out what R is going to do next. One way of doing this is to use a spinner such as the one pictured on the left.

It is obvious that the pointer will land in the shaded portion $\frac{2}{3}$ of the time and in the unshaded portion $\frac{1}{3}$ of the time. R should spin the pointer. If it lands in the shaded area, she should play row 1 (hide 1 penny). If the pointer stops in the unshaded area, then she should play row 2 (hide 2 pennies). By using this scheme, R will end up playing row 1, $\frac{2}{3}$ of the time and row 2, $\frac{1}{3}$ of the time.

C can use a similar spinner to determine his moves.

Example 3   Row and Column each have two cards. Row has a black 4 and a red 4. Column has a black 2 and a red 3. Each player puts out a card at the same time. If the colors match, Row wins the difference. Otherwise Column wins the difference. The payoff matrix is shown at the left.

Column

	Bk 2	Rd 3
Bk 4	2	−1
Rd 4	−2	1

Row

This game is not strictly determined. If Row plays black 4, then he can win 2 but lose only 1. If he plays red 4, then he can win 1 but lose 2. His best move would seem to be to play black 4. However, if Row always plays black 4, then Column will undoubtedly notice this. He will then play red 3 all the time, and Row will wind up losing 1 each time.

The formula tells us how often Row should play black 4 and red 4. It also will tell us Column's best strategy. We have the following.

$a = 2,\ b = -1,\ c = -2,\ d = 1.$

$$R_1 = \frac{d - c}{a + d - b - c} = \frac{1 - (-2)}{2 + 1 - (-1) - (-2)}$$

$$= \frac{1 + 2}{2 + 1 + 1 + 2} = \frac{3}{6} = \frac{1}{2}$$

$$R_2 = \frac{a - b}{a + d - b - c} = \frac{2 - (-1)}{2 + 1 - (-1) - (-2)}$$

$$= \frac{2 + 1}{2 + 1 + 1 + 2} = \frac{3}{6} = \frac{1}{2}$$

$$C_1 = \frac{d - b}{a + d - b - c} = \frac{1 - (-1)}{2 + 1 - (-1) - (-2)}$$

$$= \frac{1 + 1}{2 + 1 + 1 + 2} = \frac{2}{6} = \frac{1}{3}$$

$$C_2 = \frac{a - c}{a + d - b - c} = \frac{2 - (-2)}{2 + 1 - (-1) - (-2)}$$

$$= \frac{2 + 2}{2 + 1 + 1 + 2} = \frac{4}{6} = \frac{2}{3}$$

$$v = \frac{ad - bc}{a + d - b - c} = \frac{2(1) - (-1)(-2)}{2 + 1 - (-1) - (-2)}$$

$$= \frac{2 - 2}{2 + 1 + 1 + 2} = \frac{0}{6} = 0$$

The optimal strategies are

Player Row: Play row 1 (black 4) $\frac{1}{2}$ of the time.
Play row 2 (red 4) $\frac{1}{2}$ of the time.

Player Column: Play column 1 (black 2) $\frac{1}{3}$ of the time.
Play column 2 (red 3) $\frac{2}{3}$ of the time.

The value of the game is zero, so that the game is fair.

Row can flip a coin each time and play black 4 if heads comes up and red 4 if tails comes up. In the long run (assuming the coin is fair), approximately $\frac{1}{2}$ of the tosses will show heads, and $\frac{1}{2}$ will show tails. Thus Row will end up playing black 4, $\frac{1}{2}$ of the time and red 4, $\frac{1}{2}$ of the time.

In order for column to play black 2, $\frac{1}{3}$ of the time and red 3, $\frac{2}{3}$ of the time, he should use a spinner as in Example 2.

Comment   After you have computed the values of R_1, R_2, C_1, and C_2, you can check your answers by using the fact that $R_1 + R_2 = 1$ and $C_1 + C_2 = 1$.

EXERCISES

1. For each of the following non-strictly determined games, find R_1, R_2, C_1, C_2, and v.

a)

2	5
4	1

b)

5	−3
−4	−2

c)

−3	0
0	−4

d)

5	6
7	4

e)

0	7
1	−4

f)

9	−3
−2	−1

2. For each of the games of Exercise 1, find the optimal strategies for the players and also describe some device that the player can use to carry out this strategy.

3. Find the value of each of the games of Exercise 1. Are any fair games?

4. Virginia Slims and Chester Fields are playing the following game: Virginia hides either 15 or 20 cigarettes in her hand. Chester must guess how many she has hidden. If he guesses correctly, then he gets that number of cigarettes. Otherwise he gives Virginia the number of cigarettes he guessed.

a) Set up the payoff matrix.

b) Verify that the game is not strictly determined.

c) Find the value of the game.

d) Is the game fair?

e) Find the optimal strategies for each player.

f) What device should each player use to carry out this strategy?

5. Sam and Jim are choosing by each showing one or two fingers at the same time. If they both show the same number of fingers, then Sam wins an amount equal to the *product* of the number of fingers shown. Otherwise, Jim wins an amount equal to the *sum* of the number of fingers shown.

a) Set up the payoff matrix.

b) Find the value of the game.

c) Is the game fair? If not, in whose favor is the game?

d) Find the optimal strategies for each player.

6. Sam and Jim again are choosing by each showing one or two fingers. If the sum of the fingers shown is odd, Sam gets the sum, and if even, Jim gets the sum.

a) Set up the payoff matrix.

b) Find the value of the game.

c) Is it a fair game? If not, in whose favor is the game?

d) Find the optimal strategies for each player.

Alan Band Associates

A professional war-game specialist in London, England, prepares his moves, like a general, on his operations table.

7. The following is a war game. A navy ship is to shell a coastal town. If the ship comes in close to the shore, the shelling will of course be more accurate. It is also more dangerous for the ship, since the enemy has coastal gunners whose accuracy is greater

at short range. The coastal gums may be adjusted in one of two positions: short range and long range. Credit the ship with 8 points if it avoids enemy gunfire, and −9 points if it is within range of the guns. Also credit the ship with 5 extra points for accurate shelling if it comes in close.

a) Set up the payoff matrix.

b) Find the optimal strategies for each side.

c) What device should each side use to carry out this strategy?

8. Two competing soda companies, Croke and Seven-Down, decide to sell their drinks on either of two crowded beaches. The following payoff matrix indicates the differences in sales depending on which beach each chooses. (A + indicates that Croke sold more and a − indicates that Seven-Down sold more.)

| | Seven-Down | |
	Beach 1	Beach 2
Croke Beach 1	5	−7
Beach 2	−9	10

a) Find the optimal strategies for each company.

b) What device should each compamy use to carry out this strategy?

9. Why, in the above games, is it necessary for a player to use a spinner or a coin to decide on a move? Why can't he just decide himself what move to make the appropriate number of times?

4. CONCLUDING REMARKS

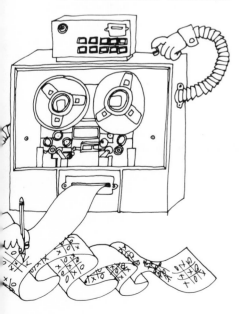

In this chapter we have barely scratched the surface of game theory. We have discussed very simple games only. However, far more complicated games can also be analyzed mathematically. This, of course, does not mean that game theory will provide optimal strategies for every game.

A simplified form of poker has been studied by mathematicians, and techniques have been devised for finding optimal strategies. However, even this version of the game (which is played with a deck of only three cards) requires a payoff matrix that has 1728 entries. To calculate the optimal strategies for each player with an accuracy of ten percent, we would have to perform about 2 billion multiplications and additions. (Of course, part of the strategy is that while you are doing these computations, your opponent will get tired of waiting and give up.)

Very complicated games, such as regular poker, chess, or bridge, involve so many computations that no formula exists that will give us the optimal strategies for each move in a reasonable amount of time. It has been estimated that even in a simple game where one player has 100 possible moves and his opponent has only 200 possible moves, it would take a modern high-speed computer a rather long time (many months) to compute the optimal strategies. For a human being to attempt this would therefore be ridiculous.

Computers have been programmed to play chess. One of the first people to do this was the mathematician A. M. Turing, who designed a chess-playing

Courtesy A. I. Laboratory,
Massachusetts Institute of Technology

Computers can not only write music and compose poetry; they can also play the intellectual game of chess.

machine called *MADAM*. His machine was a very poor chess-player and made foolish moves. After several moves, the machine would be forced to give up. Many advances have been made since Turing's machine. Today it is possible to play a fairly advanced game of chess with a computer. However, no machine has been designed that analyzes every possible strategy corresponding to any move. It would take 10^{108} years to play all the possible games if we had a machine that could play a million games a second.

Computers have been programmed to play such a game as tic-tac-toe. The computer can never lose because it has been programmed with the optimal strategies.

STUDY GUIDE   In this chapter we discussed the concepts of game theory. The following ideas were covered.

Games of strategy (p. 353)   Value of a game (p. 355)
Two-person games (p. 353)   Fair game (p. 355)
Zero-sum games (p. 353)   Optimal strategy (p. 355)
Strictly determined games (p. 354)   Non-strictly determined games (p. 361)
Payoff matrix (p. 355)

SUGGESTED FURTHER READINGS

Bellman, R., and D. Blackwell, "Red Dog, Blackjack Poker." *Scientific American* 184: 45–47 (1951). Discusses strategies for blackjack poker.

Computers and Computation (Readings from *Scientific American*). San Francisco: W. H. Freeman, 1971. See articles 9, 10, and 13.

Gardner, M., *New Mathematical Diversions from the Scientific American.* New York: Simon and Schuster, 1966. See chapter 6.

Kemeny, J., et al., *Finite Mathematics with Business Applications.* Englewood Cliffs, N. J.: Prentice-Hall, 1968. Chapter 8 discusses games in business situations.

Kemeny, J., J. L. Snell, and G. Thompson, *Introduction to Finite Mathematics*, 2nd ed. Englewood Cliffs, N. J.: Prentice-Hall, 1966. Chapter 6 discusses game theory in general.

Lindsey, G. R., "An Investigation of Strategies in Baseball." *Operation Research* 2: 477–501 (1963).

MacDonald, J., *Strategy in Poker, Business and War.* New York: W. W. Norton, 1950.

Mathematics in the Modern World (Readings from *Scientific American*). San Francisco: W. H. Freeman, 1968. Article 39 discusses the mathematics of communication, and articles 40 and 41 discuss the theory of games and the misuses of game theory.

Rapoport, A., *Fights, Games and Debates.* Ann Arbor: University of Michigan Press, 1960. Discusses debates and strategies to outwit opponents.

Shubik, M., *Strategy and Market Structure.* New York: Wiley, 1959. Discusses optimal strategies in economic and business situations.

Thompson, G. L., "Game Theory." In *Encyclopedia of Science and Technology*, New York: McGraw-Hill, 1971.

Von Neumann, J., and O. Morgenstern, *Theory of Games and Economic Behavior*, 3rd ed. Princeton, N.J.: Princeton University Press, 1953. See chapter 1.

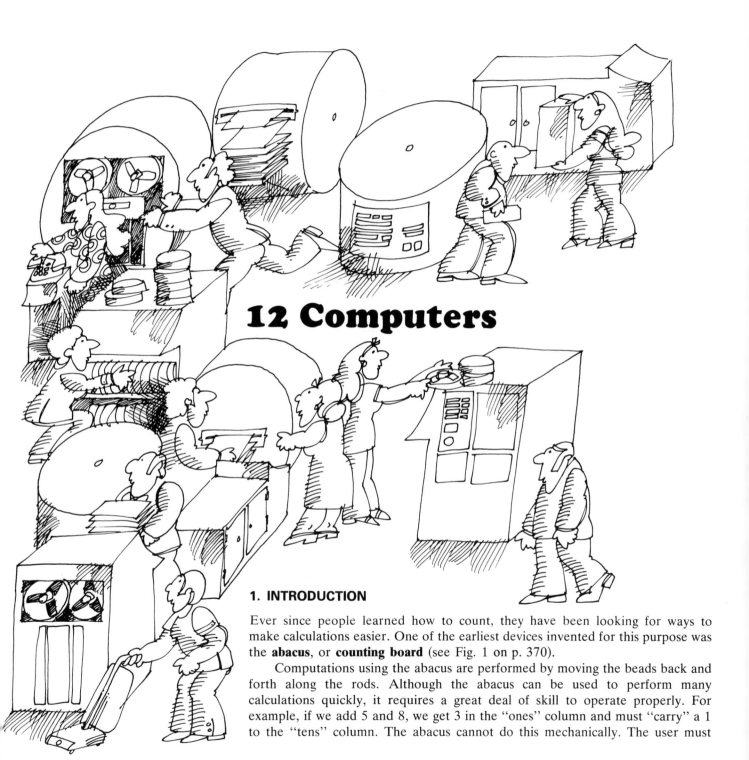

12 Computers

1. INTRODUCTION

Ever since people learned how to count, they have been looking for ways to make calculations easier. One of the earliest devices invented for this purpose was the **abacus**, or **counting board** (see Fig. 1 on p. 370).

Computations using the abacus are performed by moving the beads back and forth along the rods. Although the abacus can be used to perform many calculations quickly, it requires a great deal of skill to operate properly. For example, if we add 5 and 8, we get 3 in the "ones" column and must "carry" a 1 to the "tens" column. The abacus cannot do this mechanically. The user must

369

Fig. 1. An abacus.

(Courtesy IBM)

move a bead on the next (tens) rod by hand. This may be one reason why the abacus was not accepted by merchants who often had to add large columns of numbers.

The problems encountered in using the abacus led to the invention of other calculating devices in the seventeenth century. One such invention was Napier's bones, which were used to perform multiplication. The first mechanical adding machine was invented by Blaise Pascal. His adding machine was similar to many modern-day inexpensive desk calculators. The user had to enter each number on a dial by hand, and then pull a handle to register the number. This had to be repeated for each number entered. This was a slow and time-consuming process and various improvements were made on it in the years that followed.

About 150 years later, the Englishman Charles Babbage (1791–1871) invented his "difference engine," which was an automatic calculator capable of doing calculations with numbers of up to 20 decimal places. Although Babbage obtained the financial backing of the British government, his project was not completely successful because the technology of the time had not advanced sufficiently to produce the parts needed for his machine.

In 1833, Babbage began to work on the "analytical engine" which would have the capacity to read data from punched cards, and was to be powered by steam. It also was to have a "memory unit" and a unit for doing arithmetic operations. The results were to be printed out. These ideas are all part of the structure of our modern-day computers. Unfortunately, Babbage did not succeed because both the funds and technology to complete the project were not available.

Pascal's adding machine. In 1642 the French philosopher Blaise Pascal, 19 years old and tired of totaling up figures for his tax-collector father, invented this fancy machine for adding and subtracting. Its cylinders and gears were housed in a small box. The wheels on top of this box corresponded to units: 10s, 100s, and so on. Each wheel could register the digits 0 to 9.
(Courtesy IBM)

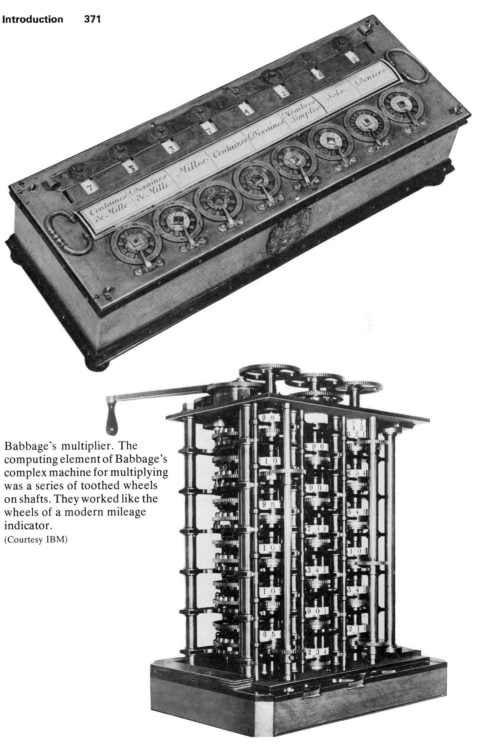

Charles Babbage, a nineteenth-century English inventor, designed the difference engine to calculate and print mathematical tables.
(Courtesy Bettman Archive)

Babbage's multiplier. The computing element of Babbage's complex machine for multiplying was a series of toothed wheels on shafts. They worked like the wheels of a modern mileage indicator.
(Courtesy IBM)

The first major advance in computers came about as a direct result of the 1890 United States Census. It had taken the U.S. Census Bureau almost 9 years to tabulate the results of the 1880 census. The population of the United States had grown considerably in the years from 1880 to 1890 and it was rather obvious that it would take considerably more time to process the 1890 results. At the time, Herman Hollerith (1860–1927) was working in the U.S. Patent Office. Hollerith knew that Joseph Marie Jacquard had invented a weaving loom that used punched cards. In 1804, Jacquard had built a loom capable of producing the most complicated designs and controlled by means of punched cards (see the figure on p. 104). "Instructions" to the machine were punched on a card which then wove the appropriate design. Hollerith was convinced that punched cards could be used to enter numbers into an adding machine. This would make the tallying of the census figures much easier.

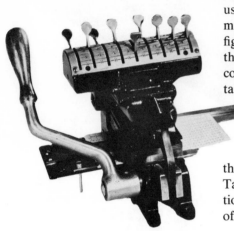

An early keypunch machine.

(Courtesy IBM)

Several years later, Hollerith did indeed invent such a machine. He invented the first **automatic data processing machine**. In 1896, Hollerith started the Tabulating Machine Corporation. This company was later to become the International Business Machines Corporation, commonly known as IBM. With the help of Hollerith's machine, the 1890 census was completed in three years.

In 1944, Howard Aiken and IBM completed the Mark I computer. It was the first completely automatic electromechanical computer. With it, computations were carried out by means of electrical impulses instead of by mechanical devices such as rotating wheels. In 1946, the ENIAC (Electronic Numerical Integrator and Calculator) computer was completed by the Remington Rand Corporation. Unlike the Mark I computer, the ENIAC used vacuum tubes to replace electric relays. This was the first purely electronic computer. It contained 18,000 vacuum tubes and was capable of adding 5000 numbers in one second. Unfortunately, neither the Mark I nor the ENIAC computer was capable of storing instructions internally. The instructions had to be read into the computer one at a time by means of punched cards, paper tapes, etc.

Historically, most of the rapid changes that took place in computer technology occurred around the time of World War II. The military demand of the Army and Navy required accurate and often time-consuming computations.

In 1946, John Von Neumann (1903–1957), who had worked on the atomic bomb, began work on the EDVAC computer. When completed in 1952, it became the first computer capable of operating at electronic speeds and of storing its programs internally. This important advance was a milestone in the development of the computer. It means that a problem can be placed into the computer and left there. Because the computer can store the information and instructions necessary to process the program, the programmer can do something else while the program is being run. The operation of the computer need not be supervised.

An example of a computer with less ability to store information internally is the hand-held calculator to be discussed later. Suppose you want to perform the calculation $[(2 + 3) - 1] \times 7$ using a calculator. You cannot just enter these numbers with one push of a button and expect the calculator to solve the problem. You must enter each number and operation by hand at the appropriate time.

Since World War II, technological advances have come so fast that computer professionals often classify computers as belonging to a particular "generation." All the early computers were called **first-generation computers**. They all used vacuum tubes and performed calculations in a few milliseconds (thousandths of a second). Examples of first-generation computers are the IBM 650 and the business-oriented version of the ENIAC, developed by Remington Rand and called the UNIVAC.

Although transistors were invented in 1948, the first transistorized computer did not appear on the market until 1954. Computers built with transistors instead of vacuum tubes are called **second-generation computers**. These computers perform calculations in a few microseconds (millionths of a second). The IBM 1410 is an example of a second-generation computer.

Beginning in 1965, **third-generation computers** appeared on the markets. These use groups of transistors and integrated circuits, and operate at such a speed (billionths of a second) that they are capable of working on several programs at the same time. The IBM 360 is an example of a third-generation computer.

A large high-speed computer is capable of doing far more work than most users actually need. Moreover, such computers are costly and often a company cannot afford to buy or rent such a machine. It is particularly costly and wasteful if the machine is left unused for large periods of time. Therefore, many companies do not buy or rent their own computer. Instead they buy time on another computer. This is called **time-sharing**. In time-sharing, a company has a *terminal* that looks like a large typewriter. The terminal is connected (by telephone or other means) to a computer that may be many miles away from the terminal. The user types programs on the terminal and receives results back the same way. The user pays only for the time that the computer is used, or the time that he or she wants it available for use. Because of the computer's tremendous capacity, many different terminals can be hooked up to the same computer. With such an arrangement the computer can work on several programs at the same time.

In the future, we can only expect increased use of the computer. It is even possible that every home will actually have its own terminal hooked up to a large computer.

2. APPLICATIONS OF COMPUTERS

Although computers have been on the market for about 30 years, nevertheless, in this short period of time they have already become a necessity for many businesses, schools, hospitals, government agencies, etc. This is because of the computer's wide-ranging capabilities, as we shall see shortly. Even small companies today are finding it harder and harder to compete in business unless they have access to a computer. They do not necessarily have to own a computer outright. They can have a terminal installed in their offices on some time-sharing plan. Present predictions are that there will be more than three billion time-sharing terminals in use by 1980 alone.

In the next few paragraphs, we will discuss some specific situations in which the computer has been applied.

Identifying the Hit and Run Driver

An old man was killed at the intersection of Broadway and Main Street by a car that went through a red light. The driver did not stop. Witnesses at the scene told the police that the car was either a 1972 or 1973 green Plymouth Valiant with a vinyl top. The last three digits of the license plate were KJP.

The police department relays this information to the State Motor Vehicle Bureau where all car registration records are stored. Through the computer, these records are searched to find all the registrations whose last license plate letters are KJP. Then this list is further narrowed down by car make, model, year, and color. Finally, the computer prints out a list of 45 addresses of owners whose cars fit the description given. The speed with which the Motor Vehicle Bureau computers can make such information available often allows the police department to track down the hit and run driver within hours of the accident. Without the aid of a computer, it would take days, perhaps even weeks, to search all the registrations in the state. By that time, the driver could easily have escaped or disposed of the car. Police departments, the FBI, and various governmental agencies use computers to search through their records to obtain essential information quickly.

Election Predictions

In November 1952, people watching the election-eve returns were informed shortly after the polls closed and long before all the ballots were counted that the next president would be Eisenhower. The television newscasters had used an IBM computer to project winners. This was the first time that a computer had been used to do this.

Based upon careful statistical analysis, samples of returns from key districts were fed into the computer. The computer had been programmed to analyze these returns and to make projections based upon them. This technique was so successful that it was not long before this became a standard procedure on election eve. In 1960, an IBM RAMAL computer predicted victory for Kennedy at 8:12 P.M. election night.

Computer-Assisted Instruction

Today there is considerable interest in computer-assisted instruction. In this method of teaching, the student works at a terminal and responds to questions asked by the machine. If the student answers the question correctly, the machine will tell him so and give him a slightly more difficult problem. Otherwise, the machine will give him another problem of the same type to try to solve again.

The computer gives the student the kind of individual attention that would be difficult or impossible for a teacher working with several students at the same time. Because the machine keeps an up-to-date and accurate record of the student's success or failure in handling each concept, it can decide at what rate to proceed. Different students, all learning the same material, can proceed at their own pace. Such individualized instruction would not otherwise be available unless each student had a private tutor.

The major disadvantage in using computers for this purpose is that personal communication between the teacher and student is lacking. The student cannot ask the machine any questions.

Simulation and Air Traffic Control

The most important part of any airport is the air traffic control center. From this center, the controllers direct air traffic into and out of the airport by means of radar, which is a kind of computer. Traffic controllers are trained for this demanding job by a process called **simulation**. Different traffic patterns and flight situations are randomly generated by computers to which the controller is trained to react.

The flow of automobile traffic on major roadways can also be controlled by means of computers which instantly analyze the traffic patterns and change the traffic signals accordingly.

The building of highways themselves (in fact, the entire analysis of urban transportation) often requires extensive computer analysis. Some of the important questions that can be analyzed by the computer are:

1. How many cars can be allowed on a particular street or bridge at any one time?
2. How many entrance and exit ramps to a highway are required at specific points?
3. What are the maximum and minimum speeds to be allowed on a highway?

Medical Care

In the intensive care units of hospitals, patients are "hooked up" to various devices that monitor their vital signs. This information is displayed visually on a screen either at the patient's bedside or in the nurse's station or both. This enables the nursing staff to see immediately if there has been any change in a patient's condition requiring immediate attention. The doctor can also recall this information from computer memory when it is needed.

Computers are also used to determine the amounts of medication to be administered in cases where extreme accuracy is necessary. For example, in radiation treatment of cancer patients, too much radiation could be dangerous, and too little could be ineffective. Computers are used to determine the exact amount needed.

Analysis of Genetic Defects

Many children are born with birth defects and disorders that can be traced back to defective chromosomes. The chromosomes carry the genes that determine which characteristics are inherited from parents. Thus, it is extremely important for researchers to analyze chromosomes and be able to recognize any that are abnormal. This kind of analysis takes time and requires a great deal of training. Because of the speed and accuracy with which computers operate, they are now being used to carry out this research in laboratories throughout the country.

The Internal Revenue Service

When an individual files his or her tax return with the Internal Revenue Service, the computer takes over. Some of the items that the computer will check are (1) the arithmetic calculations; (2) the use of the correct tax tables; and (3) the itemized deductions. If the computer finds irregularities in any of these, it will automatically "flag" the return for further review. In addition, the computer has been programmed to randomly select a certain percentage of returns for automatic review.

The Computer in the Arts

To apply the computer to the arts may seem strange. After all, the arts express human individuality and creativity. The computer cannot create and has no "mind of its own" at all; it does only what it is told to do. Yet the computer is being used with interesting results in several types of artistic activity. Possibly the most familiar is computer art. Everyone has seen "computer graphics" like those shown in Fig. 2.

A picture like this is created by programming the computer to print a character in certain places and leave blanks in others. Computers can also draw pictures on paper using automatic plotters or on TV screens using cathode-ray tubes to draw lines by means of an electron beam. The program controls the direction of the beam. Computers have also been programmed to design sculptures such as the one shown in Fig. 3.

Literature does not seem to be a promising field for the computer since of all the arts, it depends the most on the logical order of the words. To write a properly constructed and meaningful sentence is difficult enough for a human being. How can a computer, which cannot think, do it? For this reason, there has been little success up to now in getting computers to write intelligent prose. There has been more success with certain types of poetry where the meaning can be more unstructured. For example, Japanese haiku poems can be written by programming a basic form into the computer and letting it choose the words to fill in at random. For example, we might program in the following structure:

Figure 2

The ———— ———— ———— is not ————
———— is ———— ———— of the ————
All ———— is ————

and let the computer fill in the blanks at random. The results of this type of "poem" can be very pleasing to both the ear and eye. There are, of course, other methods of "writing" poetry by computer.

In music, computers have been programmed both to write music and to play it. In dance, they have even been programmed to do choreography.

Fig. 3. A computer graphic.
(Courtesy Bell Labs)

Cliometrics

In 1957, two young Harvard historians, Alfred H. Conrad and John R. Meyer, wrote two papers in which they tried to apply the methods of mathematics and statistics to the study of history, specifically economic history. This method of studying history is called **cliometrics**. In this approach, the historian searches through huge quantities of data relating to the topic under consideration. These data are then categorized and analyzed by means of computers. This process enables historians to examine numerical data, which because of its enormous quantity, could not have been completely evaluated before high-speed computers became available.

Many of the findings of cliometricians are very different from the traditional historical conclusions obtained by historians not using the cliometric approach. In a 1974 cliometric study of slavery,[1] the authors concluded that the long-held belief that the slave system destroyed the black family was not correct. They claimed in fact that the black family was essential to the practice of slavery and was in fact encouraged by southern planters.

Schoolchildren in the United States are taught that the building of railroads opened up the West. This too has been denied by some cliometricians.

While many of the findings of this "new" method of studying history are controversial (and cliometricians, like other historians and scientists, do not by any means always agree with each other), their computerized approach is definitely now an accepted and important technique in the study of certain aspects of history.

We have only scratched the surface of the vast range of computer applications. You can see from these that computers have become an essential part of our lives. Many people view a computer as the enemy—it is seen as an impersonal monster that sends incorrect bills and junk mail. It never answers when we write to it. However, computers are not monsters who are going to take over the world and reduce us all to numbers. They are actually very efficient servants of mankind and will perform only functions that human beings program them to do. When properly used, computers can make all of our lives easier and better.

EXERCISES   Give three applications of the computer to each of the following areas. (Do not mention any of the applications given in the text). Write a sentence or two about each application.

1. Business
2. The arts
3. Technology and science
4. Medicine
5. Crime control and prevention
6. Government
7. Ecology
8. Education
9. National defense

---

3. HOW A COMPUTER WORKS   Although a computer can perform difficult and time-consuming calculations at lightning speeds, we must realize that it is not capable of doing its own thinking. In this chapter, we will discuss only the type of computer called **digital**. Most of the computers produced in the United States today are of this type. There is another type of computer called an **analog computer** which we will not discuss at all. Such computers are widely used in science.

1. Robert W. Fogel and Stanley Engerman, *Time on the Cross.* Boston: Little, Brown, 1974.

How does a computer work? All computers must have an **input unit**. The input mechanism enters the data and instructions into the machine. "Input" into the machine can be accomplished by using punched cards, paper tapes, console typewriters, discs, and magnetic tapes. Input might even be at a terminal in an office miles away from the computer. If a punched card is used, it is placed into a card reader, and the machine will "read" the information from the card by noting the presence or absence of holes in certain positions. This information is punched on the card using one of the many possible programming languages. The information on the card is then transferred by means of electrical impulses into a **compiler**.

Although we perform all our arithmetical calculations in the decimal system, computers perform these calculations in the binary system (see Chapter 3). With the modern-day computer, it is not necessary for the programmer to convert decimal numbers into binary system numbers. All of this is done by the compiler or coder. The compiler converts numbers and instructions into binary numbers, so that the machine can understand them. Before the compiler became part of the computer, the programmer had to do all of the translation himself; that is, he had to type in instructions and data that were already written in the binary system.

After the input information has been translated into a language that the machine can understand, the information is passed on (by way of electrical signals) to the **control unit**. The control unit receives and interprets the instructions that it receives from the input unit. Then the control unit sends these signals to the various other units, called the **arithmetic unit**, the **memory unit**, and the **output unit**. The control unit determines which calculations are necessary for a given problem and in which order these calculations must be performed in order to solve the problem. The control unit also keeps track of the proper sequencing of the mathematical computations performed in the arithmetic unit.

The **arithmetic unit** performs all the calculations that are needed to solve a given problem. All computations that are performed here are done in the binary system.

The **memory unit** stores all the data, whether it be a number, an instruction for performing a calculation, or the results of the calculation itself. Information can be stored in several ways. Many computers store information by using magnetic cores. These are tiny doughnut-shaped pieces of metal that are no larger than a pinhead. The memory unit of a typical machine has thousands of such cores. These cores can be magnetized in one of two directions. When magnetized in one direction, the number 1 is represented, and when magnetized in the other direction, the number 0 is represented. (Remember that in the binary system, 0 and 1 are the only digits used.) With improvements in technology, the size of these magnetic cores is constantly decreasing. Many third-generation computers store information using cores grouped into small *chips* that are no larger than one-eighth of an inch on each side. There are 256 cores on a typical chip. Such storage ability greatly increases the capabilities of the machine. The more storage places a memory unit has, the more information the computer can store.

Some computers store information by using magnetic drums or discs. Such drums are usually about one foot in diameter. Information is stored on these drums by magnetizing certain spots on the drums or discs.

After the computer solves the problem, the control unit sends the solution to the **output unit**. Before reaching the output unit, the solution is decoded from machine language back into the programming language being used. The output unit may give the answer by typing the result on a typewriter, by punching holes in a card, by displaying it visually on a TV screen, or by plotting a graph. These all depend upon how the machine has been programmed.

In Fig. 4 we show the parts of a computer and how they interact with each other.

Fig. 4
The parts of a modern computer.

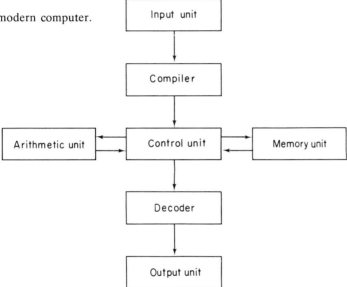

The physical parts of a computer are referred to as the **hardware**. This includes the machine itself, the terminals, the typewriters, card readers and printers, magnetic tapes and discs, etc. On the other hand, the programming techniques and the actual programs are referred to as the **software**. In recent years, there has been remarkable progress in both the hardware and software areas of the computer industry.

---

†4. PROGRAMMING A COMPUTER: FORTRAN

As we mentioned in the previous section, if you want a computer to do a job for you, it must be given detailed, step-by-step instructions. If one step is left out, then the machine cannot do the job properly. These instructions are contained in the **computer program** which is fed into the machine.

A computer program can be written in many different languages. Some of the more common programming languages are FORTRAN, BASIC, PL-1, COBOL, ALGOL, and APL. Each of these languages may be used for different applications. For example, COBOL is commonly used in business. In scientific applications, programs are often written in ALGOL. BASIC is a popular language frequently used with terminals. It is a very simple language to learn and is convenient for people who do not intend to use the computer extensively. It gives them quick access to the terminal without their having to spend a considerable amount of time learning one of the more complicated languages. FORTRAN (an abbreviation of FORmula TRANslation) is a general-purpose language and can be used in many different areas. It is perhaps the most widely used of computer languages and, for this reason, it is the language that we will discuss here.

Our discussion of FORTRAN is not intended to explain to you all the details of the language. We merely want to give you a feeling for what a computer program involves and the kind of detail that is necessary to get the program to run properly on the machine. For a complete discussion of FORTRAN, you should consult one of the many textbooks or manuals available on the subject.

Since many of the details of FORTRAN are based on concepts of elementary algebra, we suggest that the reader whose algebra is rusty consult Appendix B before continuing.

FORTRAN Symbols   In FORTRAN, the following symbols are used:

1. letters of the alphabet, always capitalized;
2. all numbers, with or without decimal points;
3. the arithmetic operation symbols + and −; and
4. special characters such as **/ , * ** . ( ) $**.

When writing FORTRAN programs, it is important that there be no possible misinterpretation of the symbols used. This is particularly important in programming, since we often use made-up names like IDIOT, which may have a different meaning or no meaning at all outside the program. Thus, to avoid confusion, we do the following:

Write the	as		Write the	as
number 0	0		letter I	I
letter o	Ø		number 2	2
number 1	1		letter Z	Z

Variables   In FORTRAN, we use the word **variable** to stand for any quantity whose value may change. As opposed to this, we also have **constants** which have a fixed value like 17, −2.3846, etc. There are two kinds of variables that we will use: the **integer** and the **real** variable. Integer variables are always treated by the machine as if they are whole numbers, with no decimal parts. Real variables may have decimal parts.

Integer variables consist of as many as six letters or digits. The first of these must be one of the letters **I, J, K, L, M,** or **N.**

Real variables also consist of as many as six letters or digits. However, the first of these must be a letter different from **I, J, K, L, M,** or **N.**

Example 1   Which of the following can be used for the names of integer variables, real variables, or neither?

a) **ATAX**   b) **I2MBBB**   c) **ELEPHANT**   d) **10Q**

e) **DØPEY**   f) **R*U*NUTS**   g) **RUNUTS**   h) **36*24*36**

i) **ILUVU**

Solution   a) **ATAX** is a real variable.

b) **I2MBBB** is an integer variable.

c) **ELEPHANT** is neither, since there are more than six characters.

d) **10Q** is neither, since it doesn't start with a letter of the alphabet.

e) **DØPEY** is a real variable.

f) **R*U*NUTS** is neither, since there are more than six characters.

g) **RUNUTS** is a real variable.

h) **36*24*36** is neither, since it has too many characters. Furthermore, it does not start with a letter of the alphabet.

i) **ILUVU** is an integer variable.

Arithmetic Operations   Some of the basic mathematical operations are written in FORTRAN exactly as they are in ordinary arithmetic. For the other operations, there are special symbols. The chart below lists the most important of these symbols used in FORTRAN.

Concept	Algebraic symbol	FORTRAN symbol	Example of an ordinary algebraic expression	Example of an equivalent FORTRAN expression
Addition	$+$	**+**	$a + b$	A+B
Subtraction	$-$	**−**	$a - b$	A−B
Multiplication	$\times$	*****	ab	A*B
Division	$\div$	**/**	$a \div b$	A/B
Exponent		******	a^b	A**B
Equals	$=$	**.EQ.**	$a = b$	A.EQ.B
Less than	$<$	**.LT.**	$4 < 5$	4.LT.5
Greater than	$>$	**.GT.**	$7 > 6$	7.GT.6
Not equal to	$\neq$	**.NE.**	$4 \neq 5$	4.NE.5

It is important to note that the symbol "=" is used in FORTRAN. However, it does not have the same meaning in FORTRAN as it does in algebra. In

FORTRAN, the symbol **=** means "replace by." Thus the statement **X=Y** tells the computer to replace the variable **X** with the variable **Y**. This statement does not say that **X** and **Y** have the same value. In general, **X** and **Y** will have different values when this statement occurs. In FORTRAN, if we wanted to say that **X** and **Y** both have the same value, we would have to write **X.EQ.Y** as indicated in the above chart. In algebra, of course, $x = y$ *does* mean that x and y have the same value.

In FORTRAN, calculations are performed in the following order: first all exponents, then all multiplications and divisions (in order from left to right) and, finally, all additions and subtractions. Parentheses are used as they are in algebra.

Example 2   Write each of the following algebraic expressions in FORTRAN.

a) 3×7 b) $5x^2$ c) $3x^2 + 7$

d) $a - b^2c$ e) $\dfrac{ab}{cd}$ f) b^{c+d}

g) $z(x - y)$ h) $(a + b) < d$ i) $x + y = 5$

Solution

Algebraic expression	FORTRAN expression
a) 3×7	3*7
b) $5x^2$	5*X**2
c) $3x^2 + 7$	3*X**2+7
d) $a - b^2c$	A−B**2*C
e) $\dfrac{ab}{cd}$	(A*B)/(C*D)
f) b^{c+d}	B**(C+D)
g) $z(x - y)$	Z*(X−Y)
h) $(a + b) < d$	(A+B).LT.D
i) $x + y = 5$	(X+Y).EQ.5

Example 3   Look at the FORTRAN statements in the solution of parts (e), (f), and (g) of Example 2. What happens if the parentheses are left out?

Solution   In part (e), if we leave out parentheses, the computer interprets the expression **A*B/C*D** as the algebraic statement $a\left(\dfrac{b}{c}\right)d$. This is not the same as $\dfrac{ab}{cd}$. This will happen because the machine does all multiplications and divisions in order from left to right unless told otherwise by means of parentheses.

Similarly, in part (f), the computer interprets **B**C+D** as the algebraic statement $b^c + d$.

In part (g), the computer would interpret **Z*X−Y** as $zx - y$.

Example 4   What is meant by the FORTRAN expression **X=X+1**?

Solution   In FORTRAN, the expression **X=X+1** means replace the value of **X** by the value **X+1**. So if **X** were 5, the statement **X=X+1** replaces the value of 5 with the "new" value of 6. From then on (unless told otherwise later), the computer would give **X** the value 6. Note that in algebra, the expression $x = x + 1$ can never be true.

Control Statements   A computer program is written as a sequence of steps to be followed in order. Each step is called a **statement**. It is sometimes necessary or helpful to number these statements, although not every statement has to be numbered.

There are four types of statements that are very important in writing any program. The first of these is the **GØ TØ statement**. This statement is of the form **GØ TØ** n, where n is any other statement number. Unless told otherwise, the computer will execute statements in the order in which they appear. The **GØ TØ** statement transfers control to a statement other than the next one in the sequence. Thus, if statement 2 says **GØ TØ 35**, then the computer will immediately go to statement 35 regardless of what the next statement is in the program. After doing what statement 35 says, it will then continue the program in sequence from statement 35 on.

The second type of control statement is the **computed GØ TØ statement**. This is of the form **GØ TØ** $(a, b, c, d, \ldots), n$, where a, b, c, d, etc., are statement numbers and n is a variable name. This tells the machine that if the variable named n is equal to 1, then it is to go to statement a, since a is the first statement number in the parentheses. If the variable named n is equal to 2, then it is to go to statement number b, since b is the second statement number in the parentheses. Similarly, if the variable named n is equal to 3, then the computer is to go to statement c. As an example, consider the computed **GØ TØ** statement **GØ TØ (3, 612, 49, 2, 78), IDIØT**. This tells the computer that

if **IDIØT** equals	go to statement number
1	3
2	612
3	49
4	2
5	78

The third type of control statement is the **arithmetic IF statement** which is of the form **IF** (n), a, b, c, where a, b, and c are any statement numbers and n is the name of a variable. This tells the computer that

if the variable named n is	then go to statement number
negative	a
zero	b
positive	c

Thus, the statement IF (X) 3, 28, 913 tells the computer that if X is negative it is to go to statement number 3. If X is 0, it is to go to statement number 28. If X is positive, it is to go to statement number 913.

The last type of control statement that we will consider is the **logical IF statement**, which is of the form IF (mXXn) S, where *m* and *n* are the names of variables, *S* is any statement number (with some exceptions), and *XX* stands for any one of the symbols .GT., .LT., .EQ., or .NE. (see p. 381). This statement tells the computer that if *mXXn* is true, then it is to go to statement *S*. Otherwise it is to continue to program in sequence.

Example 5   The statement IF (X.GT.Y) 17 tells the computer to go to statement number 17 if X is greater than Y. So if X is equal to 3 and Y is equal to 2, the computer would go to statement 17. If X were equal to 3 and Y were equal to 4, then the machine would just continue to the next statement.

Example 6   The statement IF((A−B).NE.3) 516 tells the computer to go to statement number 516 if A−B is *not* equal to 3. If A−B *is* equal to 3, the machine will just continue to the next statement.

Example 7   The statement GØ TØ (312, 41, 78, 7, 99), PICKLE gives the computer the following instructions.

if **PICKLE** is equal to	then go to statement number
1	312
2	41
3	78
4	7
5	99

Example 8   The statement IF (JØE) 17, 27, 38 tells the computer that

if the variable **JØE** is	then go to statement number
negative	17
zero	27
positive	38

EXERCISES

1. Which of the following are integer variables, real variables, or neither? Explain your answers.

   a) BLANK   b) 4 CRAZY   c) CRAZY 4

   d) KUMQUAT   e) 123456   f) GEE WIZ

   g) 6+12=18   h) MANIAC   i) LAKE

   j) SMOKEY*THE*BEAR k) /CHER   l) CHER/

2. Write each of the following FORTRAN statements in algebraic notation.

   a) X+Y.EQ.3   b) (X+Y)*Z   c) (X+Y)**Z

d) `5.NE.3`   e) `(A/B)**3`   f) `A/(B**3)`

g) `17.GT.(X-4)`   h) `X=X+5`

3. Write each of the following algebraic statements in FORTRAN notation.

a) $2 + 3 = 5$ b) xy c) $2^3 + 4$

d) $\dfrac{x}{3y}$ e) $(x + 3y^2)^3$ f) $\dfrac{x + 3y}{z}$

g) $x + \dfrac{3y}{z}$ h) $\dfrac{x}{3yz} + 5$

4. Which statement number does the computer go to after executing each of the following instructions (assume that X = 2 and Y = −3)?

a) `IF(X) 4,3,71`   b) `IF(Y) 4,3,71`

c) `IF((X-Y)**Z)1,2,3`   d) `GØ TØ 9`

e) `GØ TØ (3,6,25,1,102),X`   f) `IF (X.LT.Y) GØ TØ 9`

g) `IF (Y.LT.X) GØ TØ 200`   h) `GØ TØ (3,6,25,1,102),Y`

i) `IF (X.NE.Y) GØ TØ 111`   j) `IF (X*Y/3) 2,19,6`

5. In each of the following sequences of statements, give the final value of X. (Numbers on the left are statement numbers.)

a)
```
    X=5
    IF (X) 3, 2, 7
    X=6
 7  X=1
    GØ TØ 5
 4  X=3
 5  STØP
```

b)
```
    X=100
    GØ TØ 9
    IF (X.LE.200), 6, 1, 3
 6  GØ TØ 12
 9  X=X+1
 10 X=3
 12 STØP
```

c)
```
    X=6
    X=4
    STØP
```

d)
```
    X=14
    Y=3
    X=Y+1
    GØ TØ 8
    X=48
 8  STØP
```

e)
```
 X=10
 IF (3.LT.3**2) X=12
 STØP
```

f)
```
 X=2
 X=X+1
 STØP
```

g)
```
 4  X=9
    IF (X.EQ.5) GØ TØ 7
    IF (X.NE.5) GØ TØ 6
    GØ TØ 7
 7  GØ TØ 4
 6  STØP
```

h)
```
    X=2
    IF (9, 8, 31) X
    GØ TØ 12
 31 GØ TØ 40
 12 X=8
 40 STØP
```

*†5. SOME SAMPLE FORTRAN PROGRAMS

We are now ready to actually show how a program is written. Let us write a FORTRAN program to compute the area of a rectangle. Although we all know that the area is the length multiplied by the width, we will write a program that will automatically compute the area for *any* rectangle, regardless of its dimensions.

Suppose the rectangle that we are working with has length 2.399 and width 5.67838 centimeters. The first thing that we must do is feed these dimensions into the machine. This is done by means of a **READ** statement. The actual **READ** statements used in FORTRAN involve certain technical details that are beyond the scope of this text, so we will use instead a simplified version. Our form of the **READ** statement for this program is **READ (E, W)**. This tells the computer to read a "data card" on which the length and width of our rectangle have been punched. The letter **E** says that the name of the variable representing the length is *E*. The letter **W** says that the name of the variable representing the width is *W*. The data card that this statement tells the computer to read is shown in Fig. 5.

Figure 5

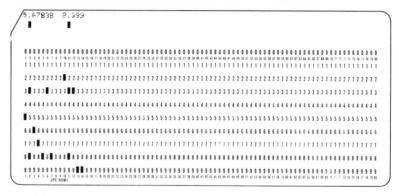

Now we can proceed to write the program, as shown below.

```
SAMPLE PRØGRAM—AREA ØF A RECTANGLE
READ (E, W)
AREA=E*W
```

This program will compute the area, but it is not quite finished. This is because we have not yet put anything in the program that tells the computer how to present the results. This can be accomplished by means of a **WRITE** statement. Again the actual **WRITE** statements used in FORTRAN are beyond the scope of this text, so we will use a simplified version. For this situation, it will be **WRITE (E, W, AREA)**. This statement tells the computer to print out in order the length, width, and area of the rectangle. After the computer has printed out the results, we want it to stop working on the program. We accomplish this by means of a **STØP** statement. Finally, we must inform the computer that the program is finished and there are no more instructions to follow. This can be accomplished by means of an **END** statement.

The complete program will look as follows.

```
SAMPLE PRØGRAM-AREA ØF A RECTANGLE
READ (E, W)
AREA=E*W
WRITE (E, W, AREA)
STØP
END
```

The computer will print out the results of this program as follows:

```
5.67838  2.399  13.62243362.
```

Comment   There is an important difference between a STØP statement and an END statement. The STØP statement tells the computer to stop its operation and await further instructions. On the other hand, an END statement tells the computer that the job is done and that no further instructions will be given for this job.

This same program can be used to compute the area of other rectangles. All we have to do is change the data card that gives the length and width. The actual program remains the same. In fact, this program can be modified to compute the area of many rectangles in one run of the program; the technique for doing this is illustrated in the program of Example 4 on p. 389.

Comment   In the above program, it would have been natural to denote the length by the variable *L*. We did not do this because *L* represents an integer variable and we were using a length that had a decimal part. Thus we used a real variable, *E*, instead.

Example 1   Write a FORTRAN program for finding the value of 100! [Remember that 100! means $(100) \cdot (99) \cdot (98) \cdot (97) \cdots (3) \cdot (2) \cdot (1)$.]

Solution   The following FORTRAN program will compute the value of 100!

```
    SAMPLE PRØGRAM-CØMPUTING 100!

 1  X=1.0
17  Y=1.0
25  X=X+1.0
31  Y=Y*X
10  IF (X.LT.100) GØ TØ 25
88  WRITE (Y)
12  STØP
11  END
```

Let us follow the first few steps of the program to convince ourselves that it will actually compute the value of 100! You should follow this with paper and pencil.

The first step sets **X** equal to 1.

The next step sets **Y** equal to 1.

The next step changes **X** to 2 (by adding 1 to the first value of **X**).

The next step multiplies the current values of **Y** and **X**, which are 1 and 2, to get 2. It makes this the new value of **Y**.

Next, statement 10 checks to see if **X** has reached 100 yet. If so, it will end the program. If not, it sends us back to statement 25. Since **X** is still only 2, we do go back to statement 25.

Now **X** is replaced by the new value of 3 (by adding 1 to the current value, which is 2).

Next, the current values of **Y** and **X**, which are 2 and 3, are multiplied to give 6. **Y** is given the value of 6.

Again, the next statement (number 10) checks to see if **X** is 100 yet. Since **X** is still only 3, this statement sends the computer back to statement 25 to continue the process.

Note, at this point, that **Y** is equal to 6 which was obtained by multiplying $(1) \cdot (2) \cdot (3)$. As we continue, **Y** will be multiplied by 4, then 5, then 6, then 7, etc., until finally it is multiplied by 100. So the final value of **Y** will be 100! The computer then prints this value out and stops.

Comment   A procedure such as the one used in the above program in which one process is repeated over and over again is called a **loop**.

Example 2   Write a FORTRAN program to compute the area of a circle.

Solution   As you probably know, the area of a circle is given by πr^2, where r is the radius of the circle and π is 3.14159 correct to 5 decimal places. We denote the radius by the variable name **RAD**. The FORTRAN program computing the area is

```
SAMPLE PRØGRAM-AREA ØF A CIRCLE
READ (RAD)
X=RAD**2
AREA=X*3.14159
WRITE (RAD, AREA)
STØP
END
```

This program prints out the value of the radius and the computed area. It can be used for any circle. It is only necessary to punch the radius of the circle on the data card to be read by the **READ** statement.

Example 3   Write a FORTRAN program to compute the take-home pay of workers, where take-home pay is computed as follows:

1. Multiply hours worked by hourly rate.

2. Multiply the number of exemptions claimed by $15.

3. Subtract the result of line 2 from line 1.

4. Multiply the result of line 3 by 0.05 to get the federal tax.

5. Subtract the result of line 4 from line 3 to get the take-home pay.

The program is to print the employee's social security number, gross pay, federal tax, and take-home pay.

Solution   Denote the employee's social security number by **SSN**.

Denote the hours worked by **H**.

Denote the hourly rate by **R**.

Denote the number of tax exemptions by **E**.

Denote the take-home pay by **PAY**.

We then have the following FORTRAN program:

```
SAMPLE PRØGRAM-TAKE-HØME PAY
READ (SSN, H, R, E)
X=H*R
Y=E*15
Z=X-Y
T=Z*0.05
PAY=Z-T
WRITE (SSN, X, T, PAY)
STØP
END
```

Example 4   A student has taken a multiple-choice test of 50 questions which will be graded by machine. The grading machine assigns each correct answer a value of 1 and each incorrect answer a value of -1. Answers that are left out are assigned a value of 0 by the grading machine. This machine prints a data card for each answer. The student's test score is computed by giving him or her two points for each correct answer and subtracting half a point for each incorrect answer (to discourage guessing). Nothing is added to or subtracted from the score for answers that are left out. Write a FORTRAN program to compute the student's score.

Solution   Let **X** represent the value of the student's answer to each question (that is, $+1$, -1, or 0). Then the program is as follows.

```
      SAMPLE PRØGRAM-TEST SCØRE
      SCØRE=0
    7 I=0
   16 READ (X)
      I=I+1
      IF (X)3, 101, 9
    9 SCØRE=SCØRE+2.0
```

```
100   IF (I.EQ.50) GØ TØ 20
      GØ TØ 16
  3   SCØRE=SCØRE-0.5
101   IF (I.EQ.50) GØ TØ 20
      GØ TØ 16
 20   WRITE (SCØRE)
      STØP
      END
```

In this program, the computer must read a data card for the first answer, **X**. Then it must add or subtract from the score according to whether the answer was correct or incorrect (that is, whether the **X** has the value +1 or −1). It has to repeat this process 50 times and then stop. The program must tell the computer when there are no more data cards to read. It does this by means of the "counter" **I**, which is introduced in statement number **7**. As each data card is read, the counter **I** is increased by 1. When **I** gets to 50, the last data card has been read and the computer is told to stop (by means of either statement number **100** or statement number **101**).

EXERCISES

1. Write a FORTRAN program to compute the area of a square.

2. Write a FORTRAN program to find the sum of the first 15 integers.

3. Write a FORTRAN program to find the product of the integers 10 through 20.

4. Write a FORTRAN program to compute the area of a triangle. (The area of a triangle is equal to one-half the base times the height.)

5. Write a FORTRAN program to compute the compound interest on a bank deposit of $1000, which earns annual interest of 6%, compounded annually for five years.

6. In a math course, three tests are given, plus a final exam. The final is counted as two tests. The grade in the course is the average of all the tests (with the final counted twice). For example, if a student's three test scores are 75, 81, and 82, and his or her final exam score is 84, then the grade is

$$\frac{75 + 81 + 82 + 2(84)}{5},$$

which equals 81.2. Write a FORTRAN program to compute a student's grade in this course.

---

6. HAND-HELD CALCULATORS

An important type of computer that is fast becoming more of a household item is the hand-held calculator. When these calculators first came on the market in the early 1950s, they were extremely expensive. Nowadays, as a result of improved technology, you can buy one for less than $10.

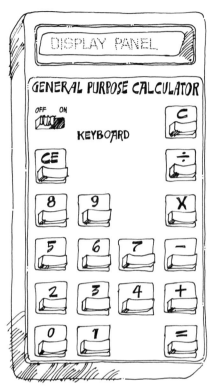

Figure 6

The major advantage in using these calculators is the speed and accuracy of their calculations. Since there are so many different models currently on the market, we will merely discuss some of the features common to all of them.

There are two types of calculators, the general-purpose and the more specialized scientific models. Although they all operate on the same principles, the scientific models have features that are especially useful to the engineer, mathematician, etc.

All the calculators consist of two parts, the keyboard and the display panel. The keyboard has separate buttons for each of the numbers from 0 to 9, and buttons for the four basic operations of +, −, ×, and ÷. In addition, there are C and CE buttons whose purpose we will explain shortly. A typical calculator is shown in Fig. 6. The more expensive calculators have additional buttons designed to perform special calculations.

Your calculator may use either *algebraic logic* or *reverse Polish notation*. This will determine how you enter your numbers. To illustrate these two systems, suppose we want to add 2 and 3. In the algebraic logic system, we push the button 2, then the + button, then the 3 button, and finally the = button. The display panel will then show 5. Notice that in this system, the numbers and symbols are entered in the normal order in which they are written.

In the reverse Polish notation, we would first enter the numbers 2 and 3. Then we would push the + button. Again the display will show 5. This is not the usual order of writing this calculation, so if your calculator uses this system, it takes time to get used to it. Consider this when you are buying a calculator. Since most calculators use algebraic logic, that is the system that we will use in the rest of this section.

The following examples show many calculations (using the algebraic logic system).

Example 1   Using a calculator, add 23 and 62.

Solution

What you do	What appears on display panel
1. Turn on/off button to on.	0.
2. Push 2 button.	2.
3. Push 3 button.	23.
4. Push + button.	23.
5. Push 6 button.	6.
6. Push 2 button.	62.
7. Push = button.	85.

Comment   If you make a mistake while entering a number, push the CE (clear error) button. This will remove *only* the last entries and leave everything else. To clear the calculator completely of *everything* in it, push the C (clear) button.

Example 2   Using a calculator, calculate 20,412 ÷ 28.

Solution

What you do	What appears on display panel
1. Turn on/off button to on.	0.
2. In order, push the 2, 0, 4, 1, and 2 buttons.	20412.
3. Push ÷ button.	20412.
4. Push 2 and 8 buttons in order.	28.
5. Push = button.	729.

Example 3   Using a calculator, calculate 20,413 ÷ 28.

Solution

What you do	What appears on display panel
1. Turn on/off button to on.	0.
2. In order, push the 2, 0, 4, 1, and 3 buttons.	20413.
3. Push ÷ button.	20413.
4. Push, in order, the 2 and 8 buttons.	28.
5. Push = button.	

Notice that in the last step of Example 3, we have not indicated what will appear on the display panel. That is because the display will depend on the particular calculator being used. Some calculators are programmed with a *fixed decimal point.* For example, a machine that has a fixed decimal point with two decimal places will give the answer to Example 3 as 729.04. The decimal point has been programmed to be in this "fixed" position and cannot be moved. All answers are rounded off to two decimal places. On the other hand, if the calculator has a *floating decimal point,* then the decimal point will "float" and appear where it belongs when calculations are performed. If Example 3 were done with a floating point calculator, the display panel would show 729.0357143 at the last step. (We are assuming here that the display panel has room for 10 digits.)

Example 4   Using a calculator, find $3 \times 4 + 5 \times 6$.

Solution   In mathematics, it is agreed that multiplications and divisions are always done before additions and subtractions. Thus this problem means: Multiply 3×4, getting 12. Then multiply 5×6, getting 30. Finally add the results 12 and 30 to get 42. The scientific calculators are programmed to always perform multiplication and division before addition and subtraction. Therefore, such calculators would treat this problem as if it were written as $(3 \times 4) + (5 \times 6)$.

Calculators that are not programmed this way (usually the general-purpose models) will perform the calculations exactly as they are entered. This is shown on the following page.

What you do	What appears on display panel when using scientific calculator programmed to do multiplication and division first	general-purpose calculator programmed to perform calculations as they are entered
Turn on/off button to on.	0	0.
Push 3 button.	3	3.
Push × button.	3.	3.
Push 4 button.	4	4.
Push + button.	12.	12.
Push 5 button.	5	5.
Push × button.	5.	17.
Push 6 button.	6	6.
Push = button.	42.	102.

Comment   In order to do the calculation above as $(3 \times 4) + (5 \times 6)$ on a general-purpose calculator, you would have to first compute 3×4 and record the result which is 12. Then you would compute 5×6, getting 30. Finally we add 12 to the 30 to get 42.

Before deciding upon which particular calculator to buy, you should consider exactly what you will be using it for. In many cases, the cheaper general-purpose models will be quite adequate. Purchase a scientific model only if your work requires it.

EXERCISES

1. Suppose you want to perform each of the following calculations. Explain how you can use the general-purpose hand calculator to obtain the correct answer.

   a) $(4 + 5)^3 - 3 \times 4^2$

   b) $\dfrac{700}{(100 + 10 + 30)}$

   c) $2^2 + 3^2$

   d) $7!$ (See the chapter on probability for the meaning of !.)

   e) $\dfrac{7!}{3!}$

   f) $\dfrac{5^2}{2^2 + 3^2}$

   g) $\dfrac{7 + 3^2}{4^2 - 2}$

   h) $5^3 + 7^2$

2. Using a calculator, perform all the calculations for each part of Exercise 1.

3. Try to perform the calculation $\dfrac{5}{0}$ on a calculator. What happens?

STUDY GUIDE   In this chapter, the following ideas were discussed.

Abacus (p. 369)
Counting board (p. 369)
First-generation computer (p. 373)

Second-generation computer (p. 373)
Third-generation computer (p. 373)
Time-sharing (p. 373)
Terminal (p. 373)
Applications of computer (p. 373)
Simulation (p. 375)
Cliometrics (p. 376)
Digital computers (p. 377)
Analog computers (p. 377)
Input unit (p. 378)
Compiler (p. 378)
Coder (p. 378)
Control unit (p. 378)
Arithmetic unit (p. 378)
Memory unit (p. 378)
Output unit (p. 379)
Hardware (p. 379)
Software (p. 379)

Computer program (p. 379)
FORTRAN (p. 380)
Variables (p. 380)
Integer variables (p. 380)
Real variables (p. 380)
FORTRAN symbols (p. 380)
Constants (p. 380)
Control statements (p. 383)
GØ TØ statements (p. 383)
Computed GØ TØ statement (p. 383)
Arithmetic IF statement (p. 383)
Logical IF statement (p. 384)
Loop (p. 388)
Hand-held calculators (p. 390)
Algebraic logic (p. 391)
Reverse Polish logic (p. 391)
Fixed decimal point (p. 392)
Floating decimal point (p. 392)

SUGGESTED FURTHER READINGS

Adler, I., *Thinking Machines*. New York: John Day, 1961. Discusses computers as thinking machines.

Berstein, J., *The Analytical Engine*. New York: Random House, 1964.

Computers and Computation (Readings from *Scientific American*). San Francisco: W. H. Freeman, 1971. Contains many interesting articles on the history, uses, and future of the computer. Articles 23–26 discuss the applications of the computer to technology, organization, education, and science.

Corliss, W. R., *Computers*. Oak Ridge, Tenn.: U.S. Atomic Energy Commission, 1967.

Eames, C., and R. Eames, *A Computer Perspective*. Cambridge, Mass.: Harvard University Press, 1973. Gives a pictorial history of the development of computers.

Halacy, D. S., Jr., *Computers, The Machines We Think With*. New York: Dell, 1964.

Hawkes, N., *The Computer Revolution*, World of Science Library. New York: Dutton, 1972. Contains detailed discussions on how the computer is applied. Specifically, Chapter 2 discusses computers in business, Chapter 3 computers in science, Chapter 4 the computer and the arts, and Chapter 8 the future.

Kemeny, J. G., *Man and the Computer*. New York: Scribner, 1972.

McCracken, Daniel D., *A Guide to FORTRAN IV Programming*. New York: Wiley, 1963. Contains a detailed discussion on FORTRAN programs.

Morrison, P., and E. Morrison, eds., *Charles Babbage and His Calculating Engines*. London: Dover, 1961.

Reichardt, J., *The Computer in Art*. New York: Van Nostrand Reinhold, 1971.

Rosen, S., "Electronic Computers: A Historical Survey." *Computing Surveys* 1(1) (March 1969).

Springer, C. H., et al., *The Mathematics for Management Series*. Homewood, Ill.: Richard D. Irwin, 1966. Volume 3 discusses several computer applications.

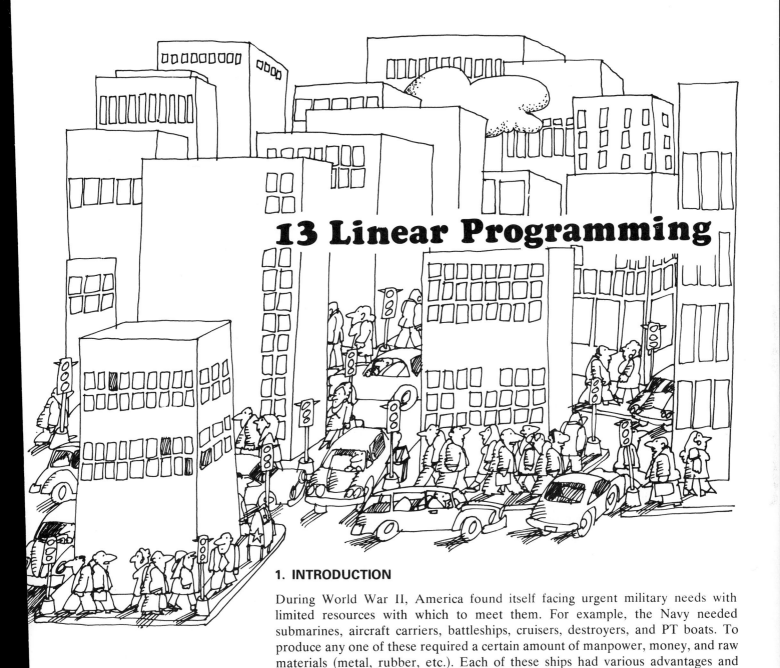

13 Linear Programming

1. INTRODUCTION

During World War II, America found itself facing urgent military needs with limited resources with which to meet them. For example, the Navy needed submarines, aircraft carriers, battleships, cruisers, destroyers, and PT boats. To produce any one of these required a certain amount of manpower, money, and raw materials (metal, rubber, etc.). Each of these ships had various advantages and limitations. Of course the cost varied greatly. Manufacturing a PT boat is obviously much cheaper than producing a battleship, but clearly a battleship has greater destructive power than a PT boat.

---

† This chapter requires some background in algebra.

(Wide World Photos, Inc.)

British and American transport planes carried supplies to West Berlin during the Berlin airlift in 1948. Planes landed at frequent and regular intervals, scheduled by linear programming.

Similarly, an aircraft carrier is very expensive to build. However, with its planes, it can inflict enormous damage on the enemy. On the other hand, the loss of an aircraft carrier may mean losing some planes also. The cost would then be very great. The government had to decide how to use its limited money and materials so as to maximize the destructive power and minimize losses.

The theory of linear programming was developed to answer this and similar questions. It turns out that in addition to military problems, linear programming can be applied to a wide variety of business and economic situations as well. One of the pioneers in this field was George Danzig. Working for the U.S. government, he developed the **simplex method**, which is of great use in solving such problems.

2. SOME LINEAR PROGRAMMING PROBLEMS

In this section, we will discuss some different types of problems that can be solved by linear programming methods. We will solve some of them. Others have solutions that are beyond the scope of this text. We include them merely to indicate the range of applicability.

Example 1   The Mark Toy Company has two best-selling toys, the *Clue* and the *Ringer*. To produce one case of Clues requires 3 hr on machine I and 2 hr on machine II and yields a profit of $14 per case. On the other hand, to produce one case of Ringers requires 2 hr on machine I and 4 hr on machine II. Ringers yield a profit of $12 a case. How many of each should the company produce in order to maximize its profit, if no machine can be used more than 8 hr per day?

Solution   Here we will merely translate the problem from words into mathematical expressions. The solution will be completed in a later section. The important expressions are marked with ● and are summarized at the end.

Let x = the number of cases of Clues to be produced.
Let y = the number of cases of Ringers to be produced.

We list all the information given, in the form of a box.

| | | Hours needed | | Profit from each |
		Machine I	Machine II	product (in dollars)
Products	Clue	3	2	14
	Ringer	2	4	12
Limit on machines		8	8	

Let us look at machine I. It gives 3 hr for each case of Clues and 2 hr for each case of Ringers. It cannot give more than 8 hr in one day. It could, of course, give less, since the machine does not have to operate all day long. The total time spent on Clues by machine I is

(time per case)　　·　　(number of cases)
$=$　　　3　　　times　　　x
$=$　　　$3x$.

Similarly, the total time spent on Ringers by machine I is

(time per case)　　·　　(number of cases)
$=$　　　2　　　times　　　y
$=$　　　$2y$.

In producing both toys, machine I has given (in one day) $3x + 2y$ hr. We know that machine I works at most 8 hr, and possibly less. Thus, the number of hours that machine I works is equal to or less than 8. We use the symbol $\leq$ to indicate "less than or equal to" (see p. 153). Therefore, for machine I we have

● 　$3x + 2y \leq 8$.　　(Machine I)

Similarly, the total time spent on Clues by machine II is

(time per case)　　·　　(number of cases)
$=$　　　2　　　times　　　x
$=$　　　$2x$.

Also, the total time spent by machine II on Ringers is

(time per case)　　·　　(number of cases)
$=$　　　4　　　times　　　y
$=$　　　$4y$.

In producing both toys, machine II works $2x + 4y$ hr. Since machine II also works 8 hr or less, this gives

● $2x + 4y \leq 8.$ (Machine II)

Now let's consider the profit. Mark Toy Company makes $14 on each case of Clues, and $12 on each case of Ringers. Its profit on Clues is

(profit per case)   ·   (number of cases)
=   14   times x
= $14y$.

Similarly, its profit on Ringers is

(profit per case)   ·   (number of cases)
=   12   times y
= $12y$.

The total profit from both products, then, is $14x + 12y$. Call this P. Obviously, the manufacturer wishes to maximize his profit. So we want to maximize $P = 14x + 12y$. We will abbreviate this as

● $\max P = 14x + 12y.$ (Profit)

It should be obvious that x and y can't be negative. They must be 0 or larger. Why?

We state this mathematically as

● $x \geq 0$ and $y \geq 0.$

The mathematical formulation of this problem can now be summarized as:

$$\begin{array}{l} 3x + 2y \leq 8, \\ 2x + 4y \leq 8, \\ x \geq 0 \quad \text{and} \quad y \geq 0, \\ \max P = 14x + 12y. \end{array}$$

Example 2   John has asked his instructor, Professor Hutton, if he can submit problems for extra credit. Professor Hutton says that he can submit up to 100 problems of three types, A, B, and C. Type A problems are worth 5 points each, type B problems are worth 4 points each, and type C problems are worth 6 points each. John finds that type A problems require 3 minutes, type B 2 minutes, and type C 4 minutes. John can spend at most $3\frac{1}{2}$ hr (210 minutes) on the assignment. Furthermore, problems of types A and B involve computations that give him a headache. So he cannot bear to spend more than $2\frac{1}{2}$ hr (150 minutes) on these. How many problems of each type should he do to maximize his credit?

Mathematical Formulation   Let x = the number of type A problems to be done.
Let y = the number of type B problems to be done.
Let z = the number of type C problems to be done.

Since John was asked to submit no more than 100 problems altogether, we have

• $x + y + z \leq 100.$

The time spent on type A problems is $3x$. The time spent on type B problems is $2y$. The time spent on type C problems is $4z$. The total time spent on all problems is $3x + 2y + 4z$.

Since he can devote, at most, 210 minutes to his mathematics assignment, we have

• $3x + 2y + 4z \leq 210.$

Furthermore, since he can spend at most 150 min on type A and B problems, we have

• $3x + 2y \leq 150.$

The credit John can receive for all the problems is

$P = 5x + 4y + 6z.$

Of course, John wants to maximize P.

Again notice that x, y, and z cannot be negative. Why? We denote this as

• $x \geq 0, \quad y \geq 0, \quad$ and $\quad z \geq 0.$

Finally, we summarize the formulation of the problem as

$$
\begin{array}{l}
x + y + z \leq 100, \\
3x + 2y + 4z \leq 210, \\
3x + 2y \leq 150, \\
x \geq 0, \ y \geq 0, \ \text{and} \ z \geq 0, \\
\max P = 5x + 4y + 6z.
\end{array}
$$

Example 3   Mr. Burns, the cook at Camp Boredom, has decided to serve foods A and B for dinner. He knows that each unit of food A contains 20 grams of protein and 25 grams of carbohydrates and costs 20¢. Each unit of food B contains 10 grams of protein and 20 grams of carbohydrates and costs 15¢. He wants the meal to contain at least 500 grams of protein and 800 grams of carbohydrates per person. The camp director has ordered him to save as much money as possible. What is the cheapest meal he can serve using these two foods only?

Mathematical Formulation   Let x = the number of units of food A to be prepared.
Let y = the number of units of food B to be prepared.

The following box summarizes the information given in the problem.

| | Nutritional content | | Cost (in cents) |
	Protein	Carbohydrates	
Food A	20	25	20
Food B	10	20	15
Minimum needed	500	800	

A camper can get 20 grams of protein from each unit of food A, and 10 grams of protein from each unit of food B. The total protein he gets from these foods is then $20x + 10y$. Since he must have *at least* 500 grams of protein, we then have

- $20x + 10y \geq 500$.

Similarly, he gets 25 grams of carbohydrates from each unit of food A, and 20 grams of carbohydrates from each unit of food B. Thus his total carbohydrate intake is

$25x + 20y$.

Again, he must have *at least* 800 grams of carbohydrates. Therefore,

- $25x + 20y \geq 800$.

The cost of food A is

(cost for one unit)   ·   (number of units)
=   20   times x
= $20x$.

The cost of food B is

(cost for one unit)   ·   (number of units)
=   15   times y
= $15y$.

The total cost, then, is $20x + 15y$. Of course, we would like to minimize this cost. We indicate this by

- min $P = 20x + 15y$.

Again, x and y can't be negative. Thus

- $x \geq 0$ and $y \geq 0$.

The problem can now be expressed as

$$20x + 10y \geq 500,$$
$$25x + 20y \geq 800,$$
$$x \geq 0 \quad \text{and} \quad y \geq 0,$$
$$\min P = 20x + 15y.$$

Example 4   The Helena Frankenstein Perfume Company markets two types of perfumes, My Error and Guess Who. To produce 1 gal of My Error requires 5 units of ingredient A and 3 units of ingredient B. It yields a profit of $300. In producing 1 gal of Guess Who, 2 units of ingredient A and 3 of ingredient B are used. The profit is $200. The Helena Frankenstein Company has available 180 units of ingredient A and 135 units of ingredient B. How many gallons of each perfume should they make in order to maximize profit?

Mathematical Formulation   Let x = the number of gallons of My Error to be made.
Let y = the number of gallons of Guess Who to be made.

We summarize the information given as follows:

	Ingredient A	Ingredient B	Profit (in dollars)
My Error	5	3	300
Guess Who	2	3	200
Amount available	180	135	

Amount used of

In producing My Error and Guess Who, we use $5x + 2y$ units of ingredient A. Since 180 units is the most that can be used, we have

● $5x + 2y \leq 180.$

Similarly, in producing the two perfumes, we use $3x + 3y$ units of ingredient B. Since, in this case, we have at most 135 units available, we get

● $3x + 3y \leq 135.$

The profit is

● $300x + 200y.$

Again, we have the obviously necessary restriction that x and y cannot be negative. Thus,

● $x \geq 0 \quad$ and $\quad y \geq 0.$

The problem can now be written as

$$
\begin{aligned}
5x + 2y &\leq 180, \\
3x + 3y &\leq 135, \\
x \geq 0 \quad &\text{and} \quad y \geq 0, \\
\max P &= 300x + 200y.
\end{aligned}
$$

In the following sections, we will complete the solutions of Examples 1, 2, and 4.

EXERCISES Formulate each of the following problems mathematically. Do not try to actually find the solution.

1. The Eastinghouse Appliance Company plans to produce washing machines and dryers. Washing machines require 15 min in shop I and 120 min in shop II. Dryers require 30 min in shop I and 180 min in shop II. The profit from washing machines is $70 and from dryers $102. The two shops can be kept open at most 8 hr (480 min) a day due to union regulations. Eastinghouse wants to know how many of each appliance should be produced so as to maximize profit.

2. The Rickety Furniture Company makes coffee tables, sofas, and bookcases. It has three workers, Moe, Larry, and Curly. Moe works at most 20 hr a week (he's the boss' brother-in-law). Larry and Curly each work at most 50 hr a week. The following chart shows each man's weekly production.

	Moe	Larry	Curly
Coffee tables	1	3	2
Sofas	3	5	4
Bookcases	5	4	8

If the profits are $10 for a coffee table, $25 for a sofa, and $35 for a bookcase, how many of each should be produced so as to maximize the profit?

3. Old MacDonald has a 200-acre farm on which he can plant two types of crops, alfalfa and spinach. Alfalfa requires two man-days of labor and $20 of capital for each acre planted. Spinach, on the other hand, requires 8 man-days of labor and $40 of capital for each acre planted. The farmer has only $2200 of capital and 320 man-days of labor available for the job. If alfalfa produces $80 of profit for each acre planted and spinach produces $32 of profit for each acre planted, how many acres of each crop should he plant to maximize his profit?

4. The I.M. Nuts Company has 180 lb (2880 oz) of Brazil nuts and 330 lb (5280 oz) of walnuts that it wants to mix and sell in 3-lb (48-oz) packages. It would like to have two types of mixtures. Mixture A will contain 18 oz of Brazil nuts and 30 oz of

walnuts. It will yield a profit of $2.04 a package. Mixture B will contain 15 oz of Brazil nuts and 33 oz of walnuts and will produce a profit of $2.10 per package. How many packages of each should the company make in order to maximize profit?

5. Two people are stranded on a desert island and the only foods available are whale blubber oil and castor oil. One unit of whale blubber oil contains 1 unit of carbohydrates, 3 units of vitamins, and 3 units of protein. One unit of castor oil contains 3 units of carbohydrates, 4 units of vitamins, and 3 units of proteins. This food is obtained by trading with the natives. One unit of whale blubber costs 50 seashells, and one unit of castor oil costs 25 seashells. The minimum daily nutritional requirements are 8 units of carbohydrates, 19 units of vitamins, and 7 units of protein. What menu will cost the least?

6. The Explosive Munitions Company makes jeeps and tanks in its two factories. Factory 1 performs the basic assembly operations. Factory 2 performs the finishing operations. For financial reasons, Factory 1 has only 185 man-days available per week. Factory 2 has 135 man-days available. Factory 1 needs 2 man-days on each jeep and 5 man-days on each tank. Factory 2 needs 3 man-days for each jeep or tank. How many of each should the company produce, if its profit is $200 on a jeep and $300 on a tank?

7. The Shmatta Clothing Company manufactures bikinis and tank suits. Manufacturing a bikini requires 2 hr of work by operator A and 4 hr of work by operator B. On the other hand, making a tank suit requires 4 hr by operator A and 2 hr by operator B. No operator can work more than 12 hr per day. If the profits are $3 from a bikini and $5 from a tank suit, how many of each should be produced to maximize profit?

8. Simon Shlock, an insurance salesman, is taking part in his company's sales contest. He sells policies on the north side and the south side of town. On the north side, it takes 3 hr to sell a health insurance policy and 4 hr to sell a life insurance policy. On the south side, it takes only 1 hr to sell a health insurance policy and 2 hr to sell a life insurance policy. His company requires him to spend at most 24 hr on the north side and at most 10 hr on the south side. If the company awards him 1 point for each policy that he sells, how many of each should he sell in order to maximize his points?

9. The Natural Foods Cereal Company has 6 bales each of rice and corn from which it can produce 2 foods, Special L and Krummy. Each dozen boxes of Special L requires 1 bale of rice and 2 bales of corn and yields a profit of $3. Each dozen boxes of Krummy requires 2 bales of rice and 1 bale of corn and yields a profit of $4. How many dozen boxes of each should the company produce so as to maximize profit?

10. A paper company makes two kinds of products, loose-leaf paper and notebooks, each of which must be processed by its two factories, A and B. In one day, Factory A can process at most 15 truckloads of paper, whereas Factory B can process at most 12 truckloads of paper. To make one truckload of loose-leaf paper requires 2 truckloads of paper processed by Factory A and 3 from Factory B. To produce one truckload of notebooks requires 3 truckloads of paper processed by Factory A and 1 from Factory B. If the paper company makes a profit of $25 from each truckload of loose-leaf paper and $35 from each truckload of notebooks, how many truckloads of each should be produced to maximize its profit?

*11. The Rickety Furniture Company is now making tables, chairs, and bookcases. It has available for the entire job 304 ft of lumber and 188 working hr. From past experience, the company knows that it should make at least 8 tables and 12 chairs, but at most 6 bookcases. To make a table requires 10 ft of lumber and 6 working hr. To make a chair requires 2 ft of lumber and 4 working hr. A bookcase needs 24 ft of lumber and 20 working hr. If the profits are $24 on a table, $10 on a chair, and $20 on a bookcase, how many of each should be produced to maximize profit?

---

3. THE GRAPHICAL METHOD

The first technique that we will discuss for solving linear programming problems is the **graphical method**. This requires a knowledge of how to graph a line. For a quick review of this technique, refer to Appendix C in the back of the book.

Inequalities and their Graphs

In linear programming, we are usually concerned with **inequalities**. These are expressions such as

$$2x - y < 4, \qquad x + y > 5, \qquad 4x \geq 2, \qquad y + 3 \leq 4.$$

Inequalities contain "less than" ($<$) or "greater than" ($>$) signs instead of only the equal sign. In this chapter we will deal only with inequalities where the sign in front of the y term is positive. If this is not the case, simple modifications are needed, which we will not discuss here. The interested reader can consult any standard algebra text for how to make the sign in front of the y term positive.

We will assume that all our inequalities are of the form in which the sign in front of the y-term is positive. (If this is not the case, the inequality can be put into that form. However, we will not discuss the technique here.)

The graph of the type of inequality to be discussed in this chapter can easily be drawn by the following procedure:

a) Change the inequality sign to an equal sign.
b) Graph the line given by step (a).
c) If the inequality is "less than," shade in the area below the line. If the inequality is "greater than," shade in the area above the line.
d) If the inequality is "$\leq$" or "$\geq$," include the line. Otherwise do not include the line in the shaded area.

Example 1   Graph the inequality $x + y > 5$.

Solution   We first change the "$>$" sign to an "$=$" sign, getting $x + y = 5$. We graph this line, using the table of values below.

x	y
0	5
2	3
5	0

Since the inequality is ">," we shade the area above the line. The line itself is not included. We indicate this by drawing it as a dotted line.

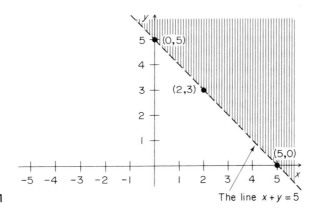

Figure 1

The line $x + y = 5$

Example 2 Graph the inequality $2x + 3y \leq 6$.

Solution We first change the "$\leq$" sign to an "$=$" sign, getting $2x + 3y = 6$. We graph this line as shown in Fig. 2. Since the inequality is "$\leq$," we shade in the area below the line and include the line as well.

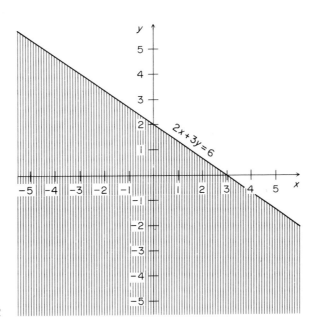

Figure 2

$2x + 3y = 6$

Solution of Linear Programming Problems by the Graphical Method

Now we will see how to solve some linear programming problems by using the graphical method. Let us look again at Example 1 of the last section (p. 397). The mathematical formulation was

$$3x + 2y \leq 8,$$
$$2x + 4y \leq 8,$$
$$x \geq 0 \quad \text{and} \quad y \geq 0,$$
$$\max P = 14x + 12y.$$

We first graph $3x + 2y \leq 8$ and shade it vertically (see Fig. 3). The third line of the mathematical formulation tells us that both x and y must be positive. Therefore, we shade nothing below the x axis (since y cannot be negative). Similarly, we shade nothing to the left of the y axis.

On the same set of axes we graph $2x + 4y \leq 8$ and shade it horizontally. This is also shown in Fig. 3.

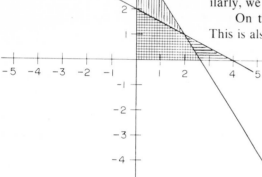

Figure 3

Notice that one part of the diagram has both kinds of shading. This region is the part that satisfies *both* inequalities. Since we are interested in this region only, we copy over just that part in Fig. 4. The region has four corners which we have labeled A, B, C, and D.

It can be proved in more advanced mathematics courses that a profit expression such as the P in our example will always have a maximum or minimum value at a *corner* of the region.

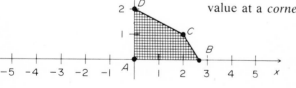

Figure 4

In our example, the corners are at $(0, 0)$, $(2\frac{2}{3}, 0)$, $(2, 1)$, and $(0, 2)$. These are the coordinates of the corners. They can be read from the graph or by using algebra. Let us now substitute each of these values into the profit expression, $P = 14x + 12y$.

At $x = 0$ and $y = 0$, the profit is

$$P = 14(0) + 12(0)$$
$$P = 0 + 0 = 0.$$

This profit is clearly a minimum.

At $x = 2\frac{2}{3}$ and $y = 0$, the profit is

$P = 14(2\frac{2}{3}) + 2 \cdot 0$

$P = 37\frac{1}{3} + 0 = 37\frac{1}{3}.$

At $x = 2$ and $y = 1$, the profit is

$P = 14(2) + 12(1)$

$P = 28 + 12 = 40.$

At $x = 0$ and $y = 2$, the profit is

$P = 14(0) + 12(2)$

$P = 0 + 24 = 24.$

Thus, the maximum profit occurs when $x = 2$ and $y = 1$, and it is 40. Referring back to Example 1 of the last section, this means that the company should produce 2 cases of Clues and 1 case of Ringers. This results in a maximum profit of $40.

We will do one more example to illustrate the technique.

Example 3   A manufacturer makes AM and FM radios. To produce an AM radio requires 2 hr in plant A and 3 hr in plant B. To produce an FM radio requires 3 hr in plant A and 1 hr in plant B. Plant A can operate for at most 15 hr a day, and plant B can operate for at most 12 hr a day. If the manufacturer makes a profit of $4 on an AM radio and $12 on an FM radio, how many of each should he produce in order to maximize his profit?

Solution   Let x = the number of AM radios to be produced.
Let y = the number of FM radios to be produced.

As in the previous section, we make a box which shows all the information given in the problem.

	Plant A	Plant B	Profit (in dollars)
AM radio	2	3	4
FM radio	3	1	12
Limitations	15	12	

The mathematical formulation, then, is

$2x + 3y \leq 15,$

$3x + y \leq 12,$

$x \geq 0 \quad \text{and} \quad y \geq 0,$

$\max P = 4x + 12y.$

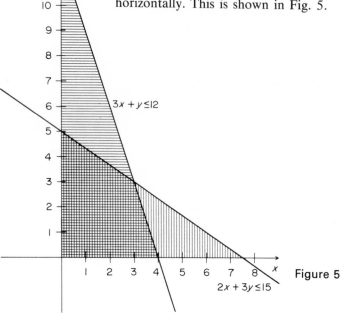

We first graph $2x + 3y \leq 15$ and shade it vertically. The third line of the mathematical formulation tells us that both x and y must be positive. Therefore we shade nothing below the x axis and nothing to the left of the y axis. We then draw, on the same set of axes, the graph of $3x + y \leq 12$ and shade it horizontally. This is shown in Fig. 5.

Figure 5

Again we redraw the doubly shaded region as shown in Fig. 6. There are four corners which we have labeled A, B, C, and D. These points are at $(0, 0)$, $(0, 5)$, $(3, 3)$, and $(4, 0)$, respectively.

We now substitute each of these values into the profit expression, $P = 4x + 12y$.

At $x = 0$ and $y = 0$, the profit is

$P = 4(0) + 12(0)$

$P = 0 + 0$

$P = 0.$

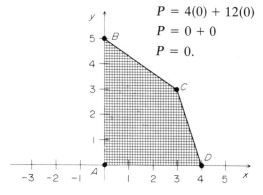

Figure 6

At $x = 0$ and $y = 5$, the profit is

$P = 4(0) + 12(5)$

$P = 0 + 60 = 60$

At $x = 3$ and $y = 3$, the profit is

$P = 4(3) + 12(3)$

$P = 12 + 36 = 48.$

At $x = 4$ and $y = 0$, the profit is

$P = 4(4) + 12(0)$

$P = 16 + 0 = 16.$

Thus the maximum profit occurs when $x = 0$ and $y = 5$, and it is $60. The manufacturer should produce no AM radios and 5 FM radios. This will result in a maximum profit of $60.

This technique for solving linear programming problems is very simple and convenient to use. However, it does not work in all cases. In particular, if there are more than three unknowns (letters x, y, and z), or if there are many inequalities, then this method becomes either very complicated or totally impossible. It is for this reason that we will study the simplex method.

EXERCISES

1. Using the graphical method, find the maximum and minimum values for the expression $P = 5x + 10y$ under the following conditions.

   a) $x \geq 0, y \geq 0$
   $2x + 3y \leq 15$
   $3x + y \leq 12$

   b) $x \geq 0, y \geq 0$
   $x + y \leq 5$
   $-2x + y \leq -1$

   c) $x \geq 0, y \geq 0$
   $2x + 3y \leq 12$
   $2x + y \leq 8$

   d) $x \geq 0, y \geq 0$
   $x + y \leq 7$
   $2x + y \leq 9$

2. Using the graphical method, solve Exercise 7 of Section 2 (p. 403).

3. Using the graphical method, solve Exercise 8 of Section 2 (p. 403).

4. Using the graphical method, solve Exercise 9, Section 2 (p. 403).

5. Using the graphical method, solve Exercise 10, Section 2 (p. 403).

6. Using the graphical method, solve Example 3 on p. 399.

7. Using the graphical method, solve Example 4 on p. 401.

8. Using the graphical method, solve Exercise 6, Section 2 (p. 403).

9. Using the graphical method, solve Exercise 1, Section 2 (p. 402).

10. Using the graphical method, solve Example 1 on p. 396.

4. SLACK AND SURPLUS VARIABLES

When we did the problems and exercises of Section 2, we found that the mathematical form always involved inequalities. In the next method that we are going to use to solve these problems (the simplex method), it is necessary to have *only* equalities (equations). Therefore, we convert the inequalities to equalities by adding on what are known as **slack** or **surplus** variables. This is done in the following way.

Let us look at the mathematical formulation of Example 1 given on p. 398.

$$3x + 2y \leq 8, \tag{1}$$

$$2x + 4y \leq 8, \tag{2}$$

$$x \geq 0 \quad \text{and} \quad y \geq 0, \tag{3}$$

$$\max P = 14x + 12y. \tag{4}$$

Statement (1) says that $3x + 2y$ is equal to or less than 8. We want to eliminate the "less than" part of the statement. We do this by adding on the *slack* something the *slack variable r*. Thus we have

● $3x + 2y + r = 8.$

Notice that r takes up the difference (if any) between $3x + 2y$ and 8. It "takes up the slack."

Statement (2) says that $2x + 4y$ is equal to or less than 8. Again we want to eliminate the "less than" part of the statement. We do this by adding on the *slack variable s*. This gives

● $2x + 4y + s = 8.$

You may be wondering why in the first case we use r as the slack variable and in the second case we use s. This is because each represents some unknown number and there is no reason for us to assume that they are the same. If we call them both r or both s, then we are assuming that they are the same.

Remember that we pointed out that x and y can't be negative. Similarly, each of the letters r and s can't be negative. (Why?) Thus, statement (3) becomes

● $x \geq 0, \quad y \geq 0, \quad r \geq 0, \quad \text{and} \quad s \geq 0.$

Let us now look at statement (4). It represents how much profit is to be made. (This is P which, of course, we want to maximize.) Our profit only comes from the Clues (x) and the Ringers (y). We can make no profit from r and s. Thus the profit associated with each is 0. Statement (4) now becomes

● $\max P = 14x + 12y + 0r + 0s.$

With the slack variables added, the problem can be written as follows:

$$3x + 2y + r = 8,$$
$$2x + 4y + s = 8,$$
$$x \geq 0, \quad y \geq 0, \quad r \geq 0, \quad \text{and} \quad s \geq 0,$$
$$\max P = 14x + 12y + 0r + 0s.$$

Comment   In the first two equations, we have left spaces. This was done on purpose and the reason will be explained in the next section.

Let us rewrite Example 2, p. 397, using slack variables. The mathematical formulation was

$$x + y + z \leq 100,$$
$$3x + 2y + 4z \leq 210,$$
$$3x + 2y \leq 150,$$
$$x \geq 0, \quad y \geq 0, \quad \text{and} \quad z \geq 0,$$
$$\max P = 5x + 4y + 6z.$$

With slack variables added, this becomes

$$x + y + z + r = 100,$$
$$3x + 2y + 4z + s = 210,$$
$$3x + 2y + t = 150,$$
$$x \geq 0, \quad y \geq 0, \quad z \geq 0, \quad r \geq 0, \quad s \geq 0, \quad \text{and} \quad t \geq 0,$$
$$\max P = 5x + 4y + 6z + 0r + 0s + 0t.$$

Finally, let us look at Example 3 of Section 2, p. 399. The mathematical statement of the problem was

$$20x + 10y \geq 500, \tag{5}$$

$$25x + 20y \geq 800, \tag{6}$$

$$x \geq 0 \quad \text{and} \quad y \geq 0, \tag{7}$$

$$\min P = 20x + 15y. \tag{8}$$

Statement (5) says that $20x + 10y$ is equal to or greater than 500. We want to eliminate the "greater than" part. This time we have no slack to take up. We may have *more than* 500. Thus to make it equal to 500, we must eliminate any *surplus. We do this by subtracting the surplus variable r.* Hence we have

• $20x + 10y - r = 500.$

In a similar manner, *in statement (6) we subtract the surplus variable s.* This gives

- $25x + 20y - s = 800.$

Statement (7) becomes

- $x \geq 0, \quad y \geq 0, \quad r \geq 0, \quad \text{and} \quad s \geq 0.$

Since the surplus variables r and s do not cost anything, statement (8) becomes

- $\min P = 20x + 15y + 0r + 0s.$

Summarizing, we have

$$
\begin{aligned}
20x + 10y - r \quad &= 500, \\
25x + 20y \quad - s &= 800, \\
x \geq 0, \quad y \geq 0, \quad r \geq 0, \quad \text{and} \quad s &\geq 0, \\
\min P = 20x + 15y + 0r + 0s. &
\end{aligned}
$$

Comment   In rewriting statement (8), we put $+0r + 0s$ rather than $-0r - 0s$. Since 0 times anything is 0, it makes no difference whether we use + or −. It is customary to write "$+0r + 0s$."

EXERCISES   1.–10. Rewrite Exercises 1–10 given at the end of Section 2 as equations by using slack and surplus variables.

11. Rewrite each of the following linear programming problems using slack and surplus variables.

a) $2x + \quad y \leq 16,$
   $x + 2y \leq 11,$
   $x + 3y \leq 15,$
   $x \geq 0 \quad \text{and} \quad y \geq 0,$
   $\max P = 30x + 50y.$

b) $5x + 2y \geq 180,$
   $x + \quad y \geq \quad 45,$
   $x \geq 0 \quad \text{and} \quad y \geq 0,$
   $\min P = 300x + 200y.$

c) $\quad x + 2y + 5z \geq 4,$
   $2x + \quad y + 2z \geq 3,$
   $3x + 5y + \quad z \leq 1,$
   $x \geq 0, \quad y \geq 0, \quad \text{and} \quad z \geq 0,$
   $\min P = 2x + 6y + 7z.$

d) $2x + \quad y \geq 7,$
   $x - 3y \leq 4,$
   $x \geq 0 \quad \text{and} \quad y \geq 0,$
   $\max P = 9x + 8y.$

---

5. THE SIMPLEX TABLEAU

Consider the following linear programming problem.

A farm equipment manufacturer produces reapers and tractors. He has available 900 tons of metal and he can afford to pay for 60,000 working hours. To make a reaper takes 2 tons of metal and 200 working hours. The profit is $500. To make a tractor requires 4 tons of metal and 150 working hours. The profit is $800. How many of each should he produce to maximize profit?

Before beginning the solution, we want to point out that the *explanation* is rather lengthy. This is because each and every step is discussed in detail. You should not be discouraged by the length. In actuality, the method is simple to learn and can be easily applied using only simple arithmetic. Just be patient!

Solution   We first rewrite the given problem as one which has inequalities. To help us do this, we make the following chart.

| | Per unit | | |
	Working hr	Metal	Profit (in dollars)
Reapers	200	2	500
Tractors	150	4	800
Limit	60,000	900	

Let x = the number of reapers to be made.
Let y = the number of tractors to be made.

By the techniques of Section 2, this problem can be expressed mathematically as

$$200x + 150y \leq 60{,}000,$$
$$2x + 4y \leq 900,$$
$$x \geq 0 \quad \text{and} \quad y \geq 0,$$
$$\max P = 500x + 800y.$$

After adding the slack variables r and s, as discussed in the last section, the problem now becomes

$$200x + 150y + r = 60{,}000,$$
$$2x + 4y + s = 900,$$
$$x \geq 0, \quad y \geq 0, \quad r \geq 0, \quad \text{and} \quad s \geq 0,$$
$$\max P = 500x + 800y + 0r + 0s.$$

We first rewrite the equations in matrix form, using only the numbers in front of the letters and disregarding the letters themselves. There are three matrices associated with this (and any similar) problem. These are called matrices A, b, and c, and they are obtained from the mathematical formulation of the problem.

Matrix A contains the numbers in front of the letters (excluding the P equation, and excluding the statement that all the letters are equal to or greater than 0). Thus, we have

$$A = \begin{matrix} a_1 & a_2 & a_3 & a_4 \\ \begin{pmatrix} 200 & 150 & 1 & 0 \\ 2 & 4 & 0 & 1 \end{pmatrix} \end{matrix}.$$

Notice that if there is no number in front of a letter, the number is understood to be 1. On the other hand, if a letter is missing, the number in front of it is understood to be 0. In this matrix, we have labeled the columns across with the letters a_1, a_2, a_3, and a_4. These correspond to x, y, r, and s. We use the a's instead of the x, y, r, and s, because it is standard notation in linear programming. The a's indicate that we are in the A matrix. Don't be frightened by the subscripts (little numbers) attached to the letter a. We could have used different letters but it is easier to use subscripts, as we shall see.

The b matrix contains the numbers to the right of the equal sign (not counting the P equation). Thus,

$$b = \begin{pmatrix} 60,000 \\ 900 \end{pmatrix}.$$

Finally, we have the c matrix, which we get from the P equation (the **profit** equation). It is

$$c = (500 \quad 800 \quad 0 \quad 0).$$

Do you remember what an identity matrix is? If not, refer back to p. 331. The 2×2 identity matrix is

$$\begin{pmatrix} 1 & 0 \\ 0 & 1 \end{pmatrix}.$$

We now examine matrix A, to see if it contains an identity matrix. In this case, it does. It is a 2×2 identity matrix made up of columns a_3 and a_4. (The reason we labeled the columns was so that we could refer to them in situations such as this.)

$$A = \begin{pmatrix} \overset{a_1}{200} & \overset{a_2}{150} & \overset{a_3}{1} & \overset{a_4}{0} \\ 2 & 4 & 0 & 1 \end{pmatrix}$$

The 2×2 identity matrix

Now that we know we have an identity matrix, we are ready to set up our first chart, which is known as Tableau 1. It is

Tableau 1

C_B[1]	Vectors in basis[1]	b	500 a_1	800 a_2	0 a_3	0 a_4
0	a_3	60,000	200	150	1	0
0	a_4	900	2	4	0	1
		0	-500	-800	0	0

---

1. The term "Vectors in basis" is a name that is commonly used for these entries. The same is true for the symbol C_B.

The entries are obtained as follows: The a_1, a_2, a_3, and a_4 columns are exactly the A matrix. Similarly the "b" column is the "b" matrix from before. Since we had an identity matrix consisting of columns a_3 and a_4, we write in the "Vectors in basis" column, a_3 and a_4, in that order. In the C_B column, we write in the profit to be derived from the "Vectors in basis." Since a_3 and a_4 correspond to the slack variables r and s, they do not yield any profit. Therefore, we write 0 for both.

Above a_1, a_2, a_3, and a_4 we write the profit or cost of x, y, r, and s. These are respectively 500, 800, 0, and 0.

The last line of the tableau is found by using the following two *formulas:*

$$Z = C_B \cdot b, \tag{1}$$

$$Z_J - C_J = C_B \cdot a_J - C_J. \tag{2}$$

These formulas look complicated, but they are really simple to use. They involve matrix multiplication, so if you have forgotten how to multiply matrices, you should review it.

Let us look at Formula (1). It says that Z is obtained by multiplying the two matrices C_B, which is written as $(0 \quad 0)$, and b, which is $\begin{pmatrix} 60,000 \\ 900 \end{pmatrix}$. Thus we have

$$Z = C_B \cdot b$$
$$= (0 \quad 0) \cdot \begin{pmatrix} 60,000 \\ 900 \end{pmatrix} = 0 \cdot 60,000 + 0 \cdot 900 = 0.^2$$

This gives us the number that belongs in the "b" column (third row).

Now let us look at Formula (2). The subscripts tell us which column we are working with. So for the first column, the formula would read

$$Z_1 - C_1 = C_B \cdot a_1 - C_1 \quad \text{(where } C_1 \text{ represents the number above } a_1)$$
$$= (0 \quad 0) \cdot \begin{pmatrix} 200 \\ 2 \end{pmatrix} - 500 = 0 \cdot 200 + 0 \cdot 2 - 500 = -500.$$

Similarly, for the second column we have

$$Z_2 - C_2 = C_B \cdot a_2 - C_2$$
$$= (0 \quad 0) \cdot \begin{pmatrix} 150 \\ 4 \end{pmatrix} - 800 = 0 \cdot 150 + 0 \cdot 4 - 800 = -800.$$

---

2. Notice that both matrices have only one column or one row. Such matrices are called vectors. When multiplying vectors, we write our answer without parentheses.

For the third column, we compute

$$Z_3 - C_3 = C_B \cdot a_3 - C_3$$

$$= (0 \quad 0) \cdot \binom{1}{0} - 0 = 0 \cdot 1 + 0 \cdot 0 - 0 = 0.$$

For the fourth column, we compute

$$Z_4 - C_4 = C_B \cdot a_4 - C_4$$

$$= (0 \quad 0) \cdot \binom{0}{1} - 0 = 0 \cdot 0 + 0 \cdot 1 - 0 = 0.$$

These numbers are now entered on the last line. This completes Tableau 1.

The technique will be further illustrated by several examples.

Example 1   Let us again look at the mathematical formulation of Example 1, Section 2, p. 398.

It was

$$3x + 2y \le 8,$$
$$2x + 4y \le 8,$$
$$x \ge 0 \quad \text{and} \quad y \ge 0,$$
$$\max P = 14x + 12y.$$

After adding the slack variables r and s, we get (see p. 410)

$$3x + 2y + r \qquad = 8,$$
$$2x + 4y \qquad + s = 8,$$
$$x \ge 0, \quad y \ge 0, \quad r \ge 0, \quad \text{and} \quad s \ge 0,$$
$$\max P = 14x + 12y + 0r + 0s.$$

In this case, matrix A is

$$
\begin{array}{cccc}
a_1 & a_2 & a_3 & a_4
\end{array}
$$
$$A = \begin{pmatrix} 3 & 2 & 1 & 0 \\ 2 & 4 & 0 & 1 \end{pmatrix}$$

Matrix b (which represents the numbers to the right of the "=" sign) is

$$b = \binom{8}{8}.$$

The profit matrix c is

$$c = (14 \quad 12 \quad 0 \quad 0).$$

We first examine the matrix A and see that columns a_3 and a_4 form a 2×2 identity matrix. We are now ready to write Tableau 1 for this problem.

Crewmen in the combat information center aboard an assault ship keep track of the location of other ships cruising with them during a naval operation. The theory of linear programming is a valuable tool in this process.

(Courtesy of U.S. Navy)

Tableau 1

C_B	Vectors in basis	b	14 a_1	12 a_2	0 a_3	0 a_4
0	a_3	8	3	2	1	0
0	a_4	8	2	4	0	1
		0	−14	−12	0	0

Again the numbers on the top are the profits from matrix c. The last line is obtained by using Formulas (1) and (2).

$$Z = C_B \cdot b$$

$$= (0 \quad 0) \cdot \begin{pmatrix} 8 \\ 8 \end{pmatrix}$$

$$= 0 \cdot 8 + 0 \cdot 8 = 0.$$

$$Z_1 - C_1 = C_B \cdot a_1 - C_1 = (0 \quad 0) \cdot \begin{pmatrix} 3 \\ 2 \end{pmatrix} - 14$$

$$= 0 \cdot 3 + 0 \cdot 2 - 14 = -14,$$

$$Z_2 - C_2 = C_B \cdot a_2 - C_2 = (0 \quad 0) \cdot \begin{pmatrix} 2 \\ 4 \end{pmatrix} - 12$$

$$= 0 \cdot 2 + 0 \cdot 4 - 12 = -12,$$

$$Z_3 - C_3 = C_B \cdot a_3 - C_3 = (0 \quad 0) \cdot \begin{pmatrix} 1 \\ 0 \end{pmatrix} - 0$$

$$= 0 \cdot 1 + 0 \cdot 0 - 0 = 0,$$

$$Z_4 - C_4 = C_B \cdot a_4 - C_4 = (0 \quad 0) \cdot \begin{pmatrix} 0 \\ 1 \end{pmatrix} - 0 = 0.$$

We enter these numbers on the last line. This completes Tableau 1.

Example 2   Let us look again at Example 4 of Section 2 (p. 401). The mathematical formulation, after adding the slack variables r and s, is

$$5x + 2y + r \qquad = 180,$$
$$3x + 3y \qquad + s = 135,$$
$$x \geq 0, \quad y \geq 0, \quad r \geq 0, \quad \text{and} \quad s \geq 0,$$
$$\max P = 300x + 200y + 0r + 0s.$$

The A matrix is

$$A = \begin{matrix} a_1 & a_2 & a_3 & a_4 \\ \begin{pmatrix} 5 & 2 & 1 & 0 \\ 3 & 3 & 0 & 1 \end{pmatrix} \end{matrix}.$$

The b matrix is

$$b = \begin{pmatrix} 180 \\ 135 \end{pmatrix}.$$

The profit matrix, c, is

$$c = (300 \quad 200 \quad 0 \quad 0).$$

We examine the matrix A and see that columns a_3 and a_4 again form a 2×2 identity matrix. Therefore, we are now ready for Tableau 1.

Tableau 1

C_B	Vectors in basis	b	300 a_1	200 a_2	0 a_3	0 a_4
0	a_3	180	5	2	1	0
0	a_4	135	3	3	0	1
		0	-300	-200	0	0

The numbers on top are the profits. We get the numbers for the last line by using Formulas (1) and (2).

$$Z = C_B \cdot b = (0 \quad 0) \cdot \begin{pmatrix} 180 \\ 135 \end{pmatrix} = 0 \cdot 180 + 0 \cdot 135 = 0, \leftarrow$$

$$Z_1 - C_1 = C_B \cdot a_1 - C_1 = (0 \quad 0) \cdot \begin{pmatrix} 5 \\ 3 \end{pmatrix} - 300$$

$$= 0 \cdot 5 + 0 \cdot 3 - 300 = -300,$$

$$Z_2 - C_2 = C_B \cdot a_2 - C_2 = (0 \quad 0) \cdot \begin{pmatrix} 2 \\ 3 \end{pmatrix} - 200$$

$$= 0 \cdot 2 + 0 \cdot 3 - 200 = -200,$$

$$Z_3 - C_3 = C_B \cdot a_3 - C_3 = (0 \quad 0) \cdot \begin{pmatrix} 1 \\ 0 \end{pmatrix} - 0$$

$$= 0 \cdot 1 + 0 \cdot 0 - 0 = 0,$$

$$Z_4 - C_4 = C_B \cdot a_4 - C_4 = (0 \quad 0) \cdot \begin{pmatrix} 0 \\ 1 \end{pmatrix} - 0$$

$$= 0 \cdot 0 + 0 \cdot 1 - 0 = 0.$$

We now enter these numbers on the last line in order to complete Tableau 1.

Example 3   As our final example, let us look at Example 2 of Section 2 (p. 398). After adding the slack variables r, s, and t, we get (p. 412):

$$\begin{aligned} x + y + z + r &= 100, \\ 3x + 2y + 4z + s &= 210, \\ 3x + 2y + t &= 150, \end{aligned}$$

$$x \geq 0, \quad y \geq 0, \quad z \geq 0, \quad r \geq 0, \quad s \geq 0, \quad \text{and} \quad t \geq 0,$$

$$\max P = 5x + 4y + 6z + 0r + 0s + 0t.$$

The A matrix here is

$$A = \begin{pmatrix} a_1 & a_2 & a_3 & a_4 & a_5 & a_6 \\ 1 & 1 & 1 & 1 & 0 & 0 \\ 3 & 2 & 4 & 0 & 1 & 0 \\ 3 & 2 & 0 & 0 & 0 & 1 \end{pmatrix}.$$

We first examine the A matrix. This time there are 3 equations and we must therefore try to find a 3×3 identity matrix. There is a 3×3 identity matrix consisting of columns a_4, a_5, and a_6.

The b matrix is

$$b = \begin{pmatrix} 100 \\ 210 \\ 150 \end{pmatrix}.$$

The profit matrix, c, is

$$c = (5 \quad 4 \quad 6 \quad 0 \quad 0 \quad 0).$$

We now enter these numbers into Tableau 1.

Tableau 1

C_B	Vectors in basis	b	5 a_1	4 a_2	6 a_3	0 a_4	0 a_5	0 a_6
0	a_4	100	1	1	1	1	0	0
0	a_5	210	3	2	4	0	1	0
0	a_6	150	3	2	0	0	0	1
		0	−5	−4	−6	0	0	0

We use Formulas (1) and (2) to compute the last line.

$$Z = C_B \cdot b = (0 \quad 0 \quad 0) \cdot \begin{pmatrix} 100 \\ 210 \\ 150 \end{pmatrix}$$

$$= 0 \cdot 100 + 0.210 + 0 \cdot 150 = 0.$$

$$Z_1 - C_1 = C_B \cdot a_1 - C_1 = (0 \quad 0 \quad 0) \cdot \begin{pmatrix} 1 \\ 3 \\ 3 \end{pmatrix} - 5 = -5.$$

Similarly, we calculate the other entries of the last line. This completes Tableau 1.

Comment   On p. 411, we pointed out that when you add slack and surplus variables, you must leave a space for any missing letter. This was done so that when you set up the A matrix, zeros would be put in for the missing letters. Otherwise the A matrix would be wrong.

EXERCISES   Make up Tableau 1 for Exercises 1, 2, 3, 4, 6, 7, 8, and 9 of Section 2.

---

6. THE SIMPLEX METHOD   Once we have completed Tableau 1, we examine the last line to see if any of the numbers are negative. If there are no negative numbers, then we will discuss later what happens. If there *are* negative numbers, then we will need to go on to a second tableau. Let us look again at Tableau 1 (on p. 414).

Tableau 1

C_B	Vectors in basis	b	500 a_1	800 a_2	0 a_3	0 a_4
0	a_3	60,000	200	150	1	0
0	a_4	900	2	4	0	1
		0	−500	−800	0	0

There are several negative numbers on the last line. In the simplex method we select the *smallest* one. In this case it is -800, which is in the a_2 column. (Remember that -800 is smaller than -500.) If there is a tie, you can select either one. We circle this column vertically as is shown.

In the second tableau, a_2, which corresponds to -800, will replace either a_3 or a_4 in the "Vectors in basis" column. To determine which one we remove, we must divide each number in the b column by the number on the same row of the circled column. We get

$$\frac{60{,}000}{150} \quad \text{and} \quad \frac{900}{4}.$$

We select the minimum of these. Since $60{,}000/150$ is 400 and $900/4$ is 225, we find that 225 is smaller. Thus a_4, which corresponds to $900/4$, will be removed and replaced by a_2. We circle the a_4 row horizontally in the above tableau. We will call the second row (which was a_4 and is now a_2) the **privileged row**. Our second tableau starts out like this.

Tableau 2

C_B	Vectors in basis	b	500 a_1	800 a_2	0 a_3	0 a_4
0	a_3					
800	a_2					

As before, the numbers in the C_B column are the profits. This tableau will have different numbers. We will now see how we get the numbers for this second tableau.

Notice that a number in Tableau 1 has been circled twice. It is 4.

The values for the second row (which is the privileged row) of Tableau 2 are computed in the following way. We divide each number of the second row of Tableau 1 by the doubly circled number, which is 4. Thus, we get

$$\frac{900}{4} = 225, \quad \frac{2}{4} = \frac{1}{2}, \quad \frac{4}{4} = 1, \quad \frac{0}{4} = 0, \quad \frac{1}{4} = \frac{1}{4}.$$

These we write in on Tableau 2, getting the following.

Tableau 2

C_B	Vectors in basis	b	500 a_1	800 a_2	0 a_3	0 a_4
0	a_3					
800	a_2	225	1/2	1	0	1/4

The numbers for the first and third rows (the nonprivileged ones) are found by a different technique called the "ring-around-the-rosy" method.

Ring around the rosy

1. Take a number in the row.
2. Go to the column that has been circled vertically (staying in the same row) and divide the number in that column by the doubly circled element.
3. Multiply the result of step 2 by the number in the column we started with *and* the circled row.
4. Subtract the result of step 3 from the number of step 1.
5. Enter the answer in the new tableau in the same *position* as the number of step 1.
6. Do the same thing for all the numbers in the nonprivileged rows.

This seems a bit complicated, but the following examples should clear it up.

For row 1 we have

Original number

$$60,000 \qquad -\frac{150\,(900)}{4}$$

Same column as the 60,000 and the same row that has been circled.

Doubly circled element

Thus we get

$$60,000 - \frac{150\,(\overset{225}{\cancel{900}})}{\cancel{4}} = 60,000 - 150(225) = 60,000 - 33,750 = 26,250.$$

So we enter 26,250 in the first position of the first row.

For the second position we get

$$200 - \frac{150}{\underset{2}{\cancel{4}}}(2) = 200 - \frac{150}{2} = 200 - 75 = 125.$$

Similarly, for the third position we have

$$150 - \frac{150}{4}(4) = 150 - 150 = 0.$$

And for the fourth position,

$$1 - \frac{150}{4}(0) = 1 - 0 = 1.$$

Finally, for the fifth position

$$0 - \frac{150}{4}(1) = \frac{-150}{4} = \frac{-75}{2}.$$

Entering these in Tableau 2, we get

Tableau 2

C_B	Vectors in basis	b	500 a_1	800 a_2	0 a_3	0 a_4
0	a_3	26,250	125	0	1	$-75/2$
800	a_2	225	1/2	1	0	1/4

We use the same procedure for the third line. We get

$$0 - \frac{-800}{4}(900) = 0 + 200(900) = 180,000,$$

$$-500 - \frac{-800}{4}(2) \quad = -500 + 400 = -100,$$

$$-800 - \frac{-800}{4}(4) \quad = -800 + 800 = 0,$$

$$0 - \frac{-800}{4}(0) \quad = 0 + 0 = 0,$$

$$0 - \frac{-800}{4}(1) \quad = 0 + \frac{800}{4} = 0 + 200 = 200.$$

We now enter these onto our chart. This completes Tableau 2.

Tableau 2

C_B	Vectors in basis	b	500 a_1	800 a_2	0 a_3	0 a_4
0	a_3	26,250	125	0	1	$-75/2$
800	a_2	225	1/2	1	0	1/4
		180,000	-100	0	0	200

Since the last line has a negative sign, the simplex method says that we are not done but will need to go on to another tableau. The next tableau will be found in exactly the same way as we found this one. First we determine which of the a's enters the "Vectors in basis" column. This is the minimum of all the negatives on the last line. Since in our case there is only one negative number, and this is in the a_1 column, we conclude that a_1 will enter the "Vectors in basis" column. This we circle vertically as we did on the last tableau.

Next we determine which vector leaves. The contest is between

$$\frac{26{,}250}{125} \quad \text{and} \quad \frac{225}{1/2},$$

or between

210   and   450.

Since 210 is less than 450 and 210 corresponds to the first row (the a_3), it will be thrown out. Circle this as shown in the complete Tableau 2 on p. 423. Our new tableau will look like this:

C_B	Vectors in basis	b	500 a_1	800 a_2	0 a_3	0 a_4
500	a_1					
800	a_2					

We now fill in all the remaining numbers which will (usually) be different from the numbers of Tableau 2. The procedure is exactly the same as the one we use in going from Tableau 1 to Tableau 2. In this case row 1 is the privileged row as shown in the complete Tableau 2 on p. 423.

We first find the values for the privileged row. These are found by dividing the original number by the doubly circled element which is 125. We then have

$$\frac{26{,}250}{125} = 210, \quad \frac{125}{125} = 1, \quad \frac{0}{125} = 0, \quad \frac{1}{125} = \frac{1}{125}, \text{ and}$$

$$\frac{-75/2}{125} = \frac{-75}{2} \div \frac{125}{1} = \frac{-75}{2} \cdot \frac{1}{125} = \frac{-75}{250} = \frac{-3}{10}.$$

We enter these on the chart, getting

Tableau 3

C_B	Vectors in basis	b	500 a_1	800 a_2	0 a_3	0 a_4
500	a_1	210	1	0	1/125	−3/10
800	a_2					

The second and third rows are found by using the ring-around-the-rosy method. We have for the second row:

$$225 - \frac{1/2}{125}(26{,}250) = 225 - \frac{1}{2}(210) = 225 - 105 = 120,$$

$$\frac{1}{2} - \frac{1/2}{125}(125) = \frac{1}{2} - \frac{1}{2} = 0,$$

$$1 - \frac{1/2}{125}(0) = 1 - 0 = 1,$$

$$0 - \frac{1/2}{125}(1) = -\frac{1}{2} \div \frac{125}{1} = -\frac{1}{2} \cdot \frac{1}{125} = -\frac{1}{250},$$

$$\frac{1}{4} - \frac{1/2}{125}\left(\frac{-75}{2}\right) = \frac{1}{4} - \left(\frac{1}{2}\right) \cdot \left(\frac{1}{125}\right) \cdot \left(\frac{-75}{2}\right) = \frac{1}{4} - \left(\frac{1}{250}\right)\left(\frac{-75}{2}\right)$$

$$= \frac{1}{4} + \frac{75}{500} = \frac{125}{500} + \frac{75}{500} = \frac{200}{500} = \frac{2}{5}.$$

For the third row we get

$$180{,}000 - \frac{-100}{125}(26{,}250) = 180{,}000 + 100(210) = 180{,}000 + 21{,}000$$
$$= 201{,}000,$$

$$-100 - \frac{-100}{125}(125) = -100 + 100 = 0,$$

$$0 - \frac{-100}{125}(0) = 0 + 0 = 0,$$

$$0 - \frac{-100}{125}(1) = 0 + \frac{100}{125} = \frac{100}{125} = \frac{4}{5},$$

$$200 - \frac{-100}{125}\left(\frac{-75}{2}\right) = 200 + \left(\frac{4}{5}\right)\left(\frac{-75}{2}\right) = 200 - 30 = 170.$$

We now enter all of these values on the chart, and thus we have

Tableau 3

C_B	Vectors in basis	b	500 a_1	800 a_2	0 a_3	0 a_4
500	a_1	210	1	0	1/125	−3/10
800	a_2	120	0	1	−1/250	2/5
		201,000	0	0	4/5	170

In a multilayered network of air routes, planes fly in layers 1000 ft apart and are protected on all sides by five miles of air space.

A computer printout schedules arriving and departing flights. Linear programming is used to efficiently assign airlines to scheduled routes.

This completes Tableau 3. Since there are no negative numbers in the last row, we are done. We can now read our answer from the chart. It is always to be found in the "b" column. Notice that the 210 in the b column corresponds to a_1 and 120 to a_2. This means that the manufacturer should produce 210 reapers and 120 tractors. The profit is on the third line. It is \$201,000. Any other combination will yield less profit or use up more material or working hours than is available.

We will illustrate the method further by solving Examples 1, 2, and 4 of Section 2.

Continuation of Example 1 of Section 2 (p. 396)   The mathematical formulation with the slack variables r and s added is

$3x + 2y + r \quad = 8,$

$2x + 4y \quad + s = 8,$

$x \geq 0, \quad y \geq 0, \quad r \geq 0, \quad \text{and} \quad s \geq 0,$

$\max P = 14x + 12y + 0r + 0s.$

The A matrix is easy to find:

$$A = \begin{pmatrix} \overset{a_1}{3} & \overset{a_2}{2} & \overset{a_3}{1} & \overset{a_4}{0} \\ 2 & 4 & 0 & 1 \end{pmatrix}.$$

2×2 identity matrix

The b matrix, representing the numbers to the right of the "=" sign, is

$$\begin{pmatrix} 8 \\ 8 \end{pmatrix}.$$

The c matrix, representing the profits, is the following:

$c = (14 \quad 12 \quad 0 \quad 0).$

Since the A matrix contains a 2×2 identity matrix consisting of columns a_3 and a_4, we may begin the simplex method. The a_3 and a_4 will go in the "Vectors in basis" column. We can now make up Tableau 1, as follows.

Tableau 1

C_B	Vectors in basis	b	14 a_1	12 a_2	0 a_3	0 a_4
0	a_3	8	3	2	1	0
0	a_4	8	2	4	0	1
		0	-14	-12	0	0

The b column is copied from the b matrix. Similarly, columns a_1, a_2, a_3, and a_4 are copied from the A matrix.

The last line is filled in by using Formulas (1) and (2) which we repeat here:

$$Z = C_B \cdot b \tag{1}$$

$$Z_J - C_J = C_B \cdot a_J - C_J. \tag{2}$$

Using these, we have

$$Z = (0 \quad 0) \cdot \binom{8}{8} = 0 \cdot 8 + 0 \cdot 8 = 0,$$

$$Z_1 - C_1 = C_B \cdot a_1 - C_1 = (0 \quad 0) \cdot \binom{3}{2} - 14$$

$$= 0 \cdot 3 + 0 \cdot 2 - 14 = -14,$$

$$Z_2 - C_2 = C_B \cdot a_2 - C_2 = (0 \quad 0) \cdot \binom{2}{4} - 12$$

$$= 0 \cdot 2 + 0 \cdot 4 - 12 = -12,$$

$$Z_3 - C_3 = C_B \cdot a_3 - C_3 = (0 \quad 0) \cdot \binom{1}{0} - 0$$

$$= 0 \cdot 1 + 0 \cdot 0 - 0 = 0,$$

$$Z_4 - C_4 = C_B \cdot a_4 - C_4 = (0 \quad 0) \cdot \binom{0}{1} - 0$$

$$= 0 \cdot 0 + 0 \cdot 1 - 0 = 0.$$

These we enter on the last line. We also write the profits for each above the "a" columns. In the C_B column we enter zeros, since these are the profits associated with a_3 and a_4.

This completes Tableau 1. Since there are negatives in the last line of Tableau 1 we will need to go on to Tableau 2. We first choose the smaller of the negatives. This is -14. It corresponds to a_1, so a_1 will be entering the "Vectors in basis" column in Tableau 2. We circle the a_1 column vertically. To determine which of the a's leaves the "Vectors in basis" column of Tableau 1, we divide each number in the b column of Tableau 1 by the number on the same horizontal line of the circled column. This gives

$$\frac{8}{3} \quad \text{and} \quad \frac{8}{2}.$$

We select the minimum of these which is 8/3. Since this corresponds to a_3, we will remove a_3 from the "Vectors in basis" column of Tableau 1 and replace it with a_1 in Tableau 2. We circle this row in Tableau 1 horizontally. The first row is now the privileged row.

Our second tableau with a_1 and a_4 (in that order) in the "Vectors in basis" column will look like the following.

Tableau 2

C_B	Vectors in basis	b	14 a_1	12 a_2	0 a_3	0 a_4
14	a_1	8/3	1	2/3	1/3	0
0	a_4	8/3	0	8/3	−2/3	1
		112/3	0	−8/3	14/3	0

We first compute the values for the privileged row of Tableau 2. They are obtained by dividing the original values of Tableau 1 by the doubly circled number 3. We get

$$\frac{8}{3}, \quad \frac{3}{3} = 1, \quad \frac{2}{3}, \quad \frac{1}{3}, \quad \frac{0}{3} = 0.$$

We enter these numbers in the first row of Tableau 2.

For the other rows of Tableau 2 we use the ring-around-the-rosy method (p. 422). Thus for row 2, we have

$$8 - \frac{2}{3}(8) = 8 - \frac{16}{3} = \frac{24}{3} - \frac{16}{3} = \frac{8}{3},$$

$$2 - \frac{2}{3}(3) = 2 - 2 = 0,$$

$$4 - \frac{2}{3}(2) = 4 - \frac{4}{3} = \frac{12}{3} - \frac{4}{3} = \frac{8}{3},$$

$$0 - \frac{2}{3}(1) = 0 - \frac{2}{3} = -\frac{2}{3},$$

$$1 - \frac{2}{3}(0) = 1 - 0 = 1.$$

We enter these in row 2 of Tableau 2. Again using the ring-around-the-rosy method for row 3, we get

$$0 - \frac{-14}{3}(8) = 0 + \frac{14}{3}(8) = \frac{112}{3},$$

$$-14 - \frac{-14}{3}(3) = -14 + 14 = 0,$$

$$-12 - \frac{-14}{3}(2) = -12 + \frac{14}{3}(2) = -12 + \frac{28}{3} = \frac{-36}{3} + \frac{28}{3} = \frac{-8}{3},$$

$$0 - \frac{-14}{3}(1) = 0 + \frac{14}{3} = \frac{14}{3},$$

$$0 - \frac{-14}{3}(0) = 0.$$

These we enter on row 3 of Tableau 2. Since there is a negative number in the last row, we will need to go on to another tableau.

There is *only one* negative number in the last row of Tableau 2 and this corresponds to the a_2 column. Thus we know that a_2 will enter the "Vectors in basis" column of Tableau 3. This we circle vertically in Tableau 2. To determine which one leaves, we compute

$$\frac{8/3}{2/3} = \frac{8}{3} \div \frac{2}{3} = \frac{8}{3} \cdot \frac{3}{2} = \frac{8}{2} = 4,$$

and

$$\frac{8/3}{8/3} = \frac{8}{3} \div \frac{8}{3} = \frac{8}{3} \cdot \frac{3}{8} = 1.$$

The minimum is 1, so that a_4 leaves. Circle the a_4 row horizontally in Tableau 2.

We first compute the values for the privileged row (row 2). We have

$$\frac{8/3}{8/3} = \frac{8}{3} \div \frac{8}{3} = \frac{8}{3} \cdot \frac{3}{8} = 1,$$

$$\frac{0}{8/3} = 0 \div \frac{8}{3} = 0 \cdot \frac{3}{8} = 0,$$

$$\frac{8/3}{8/3} = \frac{8}{3} \div \frac{8}{3} = \frac{8}{3} \cdot \frac{3}{8} = 1,$$

$$\frac{-2/3}{8/3} = \frac{-2}{3} \div \frac{8}{3} = \frac{-2}{3} \cdot \frac{3}{8} = -\frac{2}{8} = -\frac{1}{4},$$

$$\frac{1}{8/3} = 1 \div \frac{8}{3} = 1 \cdot \frac{3}{8} = \frac{3}{8}.$$

Now we compute the values for row 1 and row 3. For row 1 we have (using the ring-around-the-rosy method):

$$\frac{8}{3} - \frac{2/3}{8/3}\left(\frac{8}{3}\right) = \frac{8}{3} - \left(\frac{2}{3} \div \frac{8}{3}\right)\left(\frac{8}{3}\right) = \frac{8}{3} - \frac{2}{3} = \frac{6}{3} = 2$$

$$1 - \frac{2/3}{8/3}(0) = 1 - 0 = 1,$$

$$\frac{2}{3} - \frac{2/3}{8/3}\left(\frac{8}{3}\right) = \frac{2}{3} - \left(\frac{2}{3} \div \frac{8}{3}\right)\left(\frac{8}{3}\right) = \frac{2}{3} - \frac{2}{3} = 0$$

$$\frac{1}{3} - \frac{2/3}{8/3}\left(\frac{-2}{3}\right) = \frac{1}{3} - \left(\frac{2}{3} \div \frac{8}{3}\right)\left(\frac{-2}{3}\right) = \frac{1}{3} - \left(\frac{2}{8}\right)\left(\frac{-2}{3}\right) = \frac{1}{3} + \frac{4}{24} = \frac{1}{2},$$

$$0 - \frac{2/3}{8/3}(1) = 0 - \left(\frac{2}{3} \div \frac{8}{3}\right) = -\frac{2}{3} \cdot \frac{3}{8} = -\frac{1}{4}.$$

For the last line we have

$$\frac{112}{3} - \frac{-8/3}{8/3}\left(\frac{8}{3}\right) = \frac{112}{3} + \frac{8}{3} = \frac{120}{3} = 40,$$

$$0 - \frac{-8/3}{8/3}(0) = 0 + 0 = 0,$$

$$\frac{-8}{3} - \frac{-8/3}{8/3}\left(\frac{8}{3}\right) = \frac{-8}{3} + \frac{8}{3} = 0,$$

$$\frac{14}{3} - \frac{-8/3}{8/3}\left(\frac{-2}{3}\right) = \frac{14}{3} - \frac{2}{3} = \frac{12}{3} = 4,$$

$$0 - \frac{-8/3}{8/3}(1) = 0 + 1 = 1.$$

We now have Tableau 3.

Tableau 3

C_B	Vectors in basis	b	14 a_1	12 a_2	0 a_3	0 a_4
14	a_1	2	1	0	1/2	−1/4
12	a_2	1	0	1	−1/4	3/8
		40	0	0	4	1

Since there are no negative numbers in the last line we are finished. (Finally!). Once again our answer is obtained from the b column.

It is "produce 2 of the x and 1 of the y." Since x was the number of Clues and y was the number of Ringers, we have

> Produce 2 Clues
>   1 Ringer
> Profit $40

Continuation of Example 2 of Section 2 (p. 398)

With slack variables added, the mathematical formulation is

$$x + y + z + r \qquad = 100,$$
$$3x + 2y + 4z \quad + s \quad = 210,$$
$$3x + 2y \qquad\qquad + t = 150,$$
$$x \geq 0, \quad y \geq 0, \quad z \geq 0, \quad r \geq 0, \quad s \geq 0, \quad \text{and} \quad t \geq 0,$$
$$\max P = 5x + 4y + 6z + 0r + 0s + 0t.$$

We will merely give the tableaus needed for this problem. The reader should work through all the steps and verify that the entries are correct.

Tableau 1

C_B	Vectors in basis	b	5 a_1	4 a_2	6 a_3	0 a_4	0 a_5	0 a_6
0	a_4	100	1	1	1	1	0	0
0	a_5	210	3	2	4	0	1	0
0	a_6	150	3	2	0	0	0	1
			0	−5	−4	−6	0	0	0

Thus a_5 leaves and a_3 will replace it.

Comment The a_6 cannot leave the "Vectors in basis" column. As a matter of fact, it is not even a candidate to leave, since we would have 150/0, which means dividing by 0. Of course this is impossible. Therefore we can consider only a_4 and a_5 as possible candidates to leave the "Vectors in basis" column.

Tableau 2

C_B	Vectors in basis	b	5 a_1	4 a_2	6 a_3	0 a_4	0 a_5	0 a_6
0	a_4	95/2	1/4	1/2	0	1	−1/4	0
6	a_3	105/2	3/4	1/2	1	0	1/4	0
0	a_6	150	3	2	0	0	0	1
			315	−1/2	−1	0	0	+3/2	0

Thus a_2 enters and a_6 leaves.

Tableau 3

C_B	Vectors in basis	b	5 a_1	4 a_2	6 a_3	0 a_4	0 a_5	0 a_6
0	a_4	10	−1/2	0	0	1	−1/4	−1/4
6	a_3	15	0	0	1	0	1/4	−1/4
4	a_2	75	3/2	1	0	0	0	1/2
			390	1	0	0	0	3/2	1/2

Do 0 type A problems,
  75 type B problems,
  15 type C problems.
Point value is 390.

Since there are no negatives in the last row, we are finished. Our answer is that the student should do 15 of the a_3 and 75 of the a_2. Since a_1 does not appear in the "Vectors in basis" column, he should not do any of these problems at all. We are not interested in a_4 since it does not represent any problems but is merely a slack variable. Our answer is at the left.

Continuation of Example 4 of Section 2 (p. 401)

The mathematical formulation of the problem with slack variables added is

$$5x + 2y + r \quad\ = 180,$$
$$3x + 3y \quad\ + s = 135,$$
$$x \geq 0,\ \ y \geq 0,\ \ r \geq 0,\ \ \text{and}\ \ s \geq 0,$$
$$\max P = 300x + 200y + 0r + 0s.$$

Tableau 1

C_B	Vectors in basis	b	300 a_1	200 a_2	0 a_3	0 a_4
0	a_3	180	5	2	1	0
0	a_4	135	3	3	0	1
			0	-300	-200	0	0

Thus a_3 leaves and a_1 enters.

Tableau 2

C_B	Vectors in basis	b	300 a_1	200 a_2	0 a_3	0 a_4
300	a_1	36	1	2/5	1/5	0
0	a_4	27	0	9/5	$-3/5$	1
		10,800	0	-80	60	0

Here a_4 leaves and a_2 enters.

Tableau 3

C_B	Vectors in basis	b	300 a_1	200 a_2	0 a_3	0 a_4
300	a_1	30	1	0	1/3	$-2/9$
200	a_2	15	0	1	$-1/3$	5/9
		12,000	0	0	$\dfrac{100}{3}$	$\dfrac{+400}{9}$

The Helena Frankenstein Company should produce

> 30 gallons of My Error,
> 15 gallons of Guess Who.
> The profit is $12,000.

Comment   If, in Tableau 1, we have no negative numbers in the last row, then we can read off the solution immediately. No other tableaus are needed.

Comment   Suppose you have determined which of the a's is to enter the "Vectors in basis" column, and you are computing, by the method given, which of the a's must now leave. If the result is negative or involves a division by 0, then the problem has *no* solution.

Comment   The method used in finding the entries for the nonprivileged rows of the new tableaus is called "ring-around-the-rosy." This is because when we apply this technique we always go around in a "circle" of numbers this way:

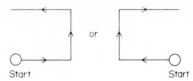

For convenience, we summarize the technique used in solving linear programming problems.

Method for Solving Linear Programming Problems
1. Express the problem in terms of inequalities.
2. Add slack or surplus variables as needed.
3. Make up Tableau 1, using Formulas (1) and (2).
4. Make up additional tableaus as needed[3] using the ring-around-the-rosy method and the privileged row formula.
5. When no negatives appear in the last row, read off your answer.

EXERCISES   Using the simplex method, complete the solutions to the following problems.

1. Exercise 1 of Section 2
2. Exercise 3 of Section 2
3. Exercise 9 of Section 2
4. Exercise 10 of Section 2
5. Exercise 7 of Section 2
6. Exercise 8 of Section 2
7. Exercise 4 of Section 2
8. Exercise 2 of Section 2
9. The Mickey Mouse Exterminating Company produces 2 kinds of mousetraps, the Normal trap and the Big trap. To produce these traps, each must be processed by two

---

3. In practice, many tableaus may be needed to solve a problem. The problems that you will be asked to solve can be done with only a few tableaus.

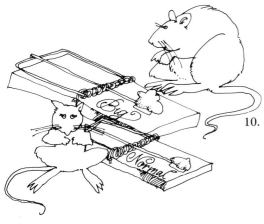

factories Y and Z. Factory Y assembles the traps and remains open at most 9 hr a day. Factory Z which paints and packages the traps, can stay open at most 8 hr a day. Each Normal trap requires 1 hr in factory Y and 2 hr in factory Z. It yields a profit of $1. Each Big trap requires 3 hr in factory Y and only 1 hr in factory Z. It yields a profit of $4. How many of each kind of trap should the company manufacture in order to maximize profit?

10. Disunited Airlines is planning to install pinball machines and hockey games in its terminal at Kennedy Airport. It can install these on either the upper or lower levels or on both levels. It has at most 9 sq yd of space on the upper level and at most 8 sq yd on the lower level available for the project. However, for security reasons, each pinball machine requires 1 sq yd of space on the upper level and 2 sq yd on the lower level. On the other hand, a hockey game requires 3 sq yd on the upper level and only 1 sq yd on the lower level. If the company expects a profit of $1 per game from each, how many pinball machines and hockey games should it install?

7. CONCLUDING REMARKS

The problems that we worked out in the last section all involved maximizing something (usually a profit). We have also given examples and exercises that involve minimizing something. These were not worked out. Such problems can be solved by the same techniques, with some simple modifications. The interested reader can consult the references given.

In all of the problems done, we have been fortunate in finding the identity matrix needed within the A matrix. It probably has occurred to you that one cannot always expect to find this. If an identity matrix cannot be found in A, then we can use techniques that exist for creating one artificially. The problem becomes complicated when this has to be done, so we have avoided it.

For further information on both the simplex and graphical methods, consult the references for this chapter. In the references, you can find the explanation as to why the simplex method works. (However, we do not suggest it unless you are well prepared in your math.) It is interesting to note that the simplex method has recently been applied to the study of game theory.

We wish to emphasize again the fact that not all linear programming problems can be solved. When a problem can be solved, it may involve large numbers of tableaus. Computers can be extremely helpful in doing the "dirty work," that is, all the lengthy computations involved.

One final word on the subject! An extremely important application of linear programming techniques is in the so-called "transportation" problem. In a typical situation of this kind, a manufacturer must decide how much of a product to send to which of his warehouses for distribution. For example, a large wholesale grocer, like the A & P, must be sure that each of its regional distribution centers has enough goods available to meet the local stores' demands. The problem is further complicated by the fact that there is limited storage space available, some goods are perishable, some goods are seasonal, and different neighborhoods vary in customer demands.

STUDY GUIDE   In this chapter, the following ideas about linear programming were discussed.

Graphical method

Graphs (p. 404)
Inequalities (p. 404)

Simplex method

Slack variable (p. 410)
Surplus variable (p. 410)

Simplex tableau (p. 412)
C_B matrix (p. 414)
Profit matrix (p. 415)
Ring-around-the-rosy method (p. 422)
Privileged row formula (p. 421)
Nonprivileged row formula (p. 422)

SUGGESTED FURTHER READINGS   Danzig, G., *Linear Programming and Extensions*. Princeton, N.J.: Princeton University Press, 1963. Contains a thorough discussion of the simplex method.

Kemeny, J., J. Snell, and J. Thompson, *Introduction to Finite Mathematics*, 3rd ed. Englewood Cliffs, N.J.: Prentice-Hall, 1974. Chapter 7 discusses graphical solutions of linear programming problems.

Kemeny, J., et al., *Finite Mathematics with Business Applications*. Englewood Cliffs, N.J.: Prentice-Hall, 1968. Chapter 7 discusses the graphical solution of business-type linear programming problems.

Lanchester, F. W., "Mathematics in Warfare." In James R. Newman, ed., *The World of Mathematics*, Vol. IV, Part 20. New York: Simon and Schuster, 1956. Article 1 discusses how mathematics is used in planning war strategies.

Mathematics in the Modern World (Readings from *Scientific American*). San Francisco: W. H. Freeman, 1968. Article 37 discusses mathematics in the social sciences, article 38 discusses the practice of quality control, article 42 discusses linear programming, and article 43 discusses operations research.

Morse, P., and G. E. Kimball, "How to Hunt a Submarine." In James R. Newman, *The World of Mathematics*, Vol. IV, Part 20. New York: Simon and Schuster, 1956. Article 2 discusses optimal strategies in hunting a submarine.

Wheeler, R., *Modern Mathematics for Business Students*. Belmont, Calif.: Brooks-Cole, 1968. Chapter 8 contains a discussion of the simplex method and the graphical solution of linear programming problems.

Appendixes

Appendix A: The Metric System

There are two standard systems of measurement in use today. One of these is known as the **metric system** and is used throughout most of the civilized world. The United States still uses the older **foot-pound** or **British system**, in which the foot is the standard unit of length. However, the Congress of the United States has passed legislation that would gradually convert our system of measures to the metric system.

In the metric system, the standard unit of length is the **meter**, a copy of which is kept in all the major capital cities of the world (see the discussion on p. 225). One major advantage of using this system is that all measurements are expressed in powers of 10 (that is, 10, 100, 1000, etc.). Thus the metric system is similar to our number system. The same is not true in our system. For example, in our system, there are 3 feet in a yard, 12 inches in a foot, 5280 feet in a mile, etc.

The lengths used in the metric system are shown in Table 1.

Table 1

Metric units	Commonly used abbreviation	How many meters?
Kilometer	km	1000
Hectometer		100
Dekameter		10
Meter	m	1
Decimeter		1/10 or 0.1
Centimeter	cm	1/100 or 0.01
Millimeter	mm	1/1000 or 0.001

In the above table, some abbreviations have been left out. Those that have been omitted are rarely used.

The following examples show how we convert from one unit to another in the metric system.

Example 1   How many centimeters are there in 325 meters?

Solution   Since 1 m equals 100 cm, we have 1 m = 100 cm. Therefore,

$$325 \times 1\,\text{m} = 325 \times 100\,\text{cm}$$
$$325\,\text{m} = 32,500\,\text{cm}.$$

Thus 325 m = 32,500 cm.

Example 2   How many kilometers are there in 7,342 millimeters?

Solution   Since 1 mm equals 0.001 m, we have 1 mm = 0.001 m.

Therefore, $7342 \times 1 \text{ mm} = 7342 \times 0.001 \text{ m}$

$$7342 \text{ mm} = 7.342 \text{ m}.$$

Now $1 \text{ km} = 1000 \text{ m}.$

This means that $1 \text{ m} = 1/1000 \text{ km or } 0.001 \text{ km}.$

Thus, $7.342 \times 1 \text{ m} = 7.342 \times 0.001 \text{ km or } 0.007342 \text{ km}.$

Therefore, $7342 \text{ mm} = .007342 \text{ km}.$

Since both the metric and British systems are widely used, we should know how to convert from one system to the other. Table 2 shows the relationships between the two systems.

Table 2

Converting from metric system to British system		Converting from British system to metric system	
1 millimeter	0.04 inch	1 inch	2.54 centimeters
1 centimeter	0.4 inch	1 foot	30.48 centimeters
1 meter	1.1 yards	1 yard	0.914 meter
1 kilometer	0.62 mile	1 mile	1.6 kilometers

Using the relationships given in Table 2, we can convert from metric system measure to the British system and vice versa. This is illustrated in the following examples.

Example 3   Convert 55 miles to kilometers.

Solution   From Table 2, we see that 1 mile is 1.6 km. Therefore, 55 miles = 55×1.6 km or 88 km.

Example 4   Convert 75 centimeters to inches.

Solution   From Table 2, we see that 1 centimeter is 0.4 inch. Therefore, 75 cm = 75×0.4 inch or 30 inches.

What about weights and liquid measure? In the metric system, the standard unit of weight is the *gram*, as opposed to our pound. Similarly, the standard unit of liquid measure in the metric system is the *liter*, as opposed to our quart. In Tables 3 and 4, we list the different measures, and the relationships between them.

In Table 5, we list the relationships between the metric system measurements and the British measurements (for weight and liquid measure). Using the results given in the table, we can convert from one system to the other, as the following examples will illustrate.

Table 3

Unit	Commonly used abbreviation	How many grams?
Kilogram	km	1000
Hectogram		100
Decagram		10
Gram	gm	1
Decigram		1/10 or 0.1
Centigram	cg	1/100 or 0.01
Milligram	mg	1/1000 or 0.001

Table 4

Unit	Commonly used abbreviation	How many liters?
Kiloliter	kl	1000
Hectoliter		100
Decaliter		10
Liter	ℓ	1
Deciliter		1/10 or 0.1
Centiliter		1/100 or 0.01
Milliliter	ml	1/1000 or 0.001

Table 5

Converting from metric system to British system		Converting from British system to metric system	
1 liter	1.05 liquid quarts	1 liquid quart	0.95 liter
1 liter	0.91 dry quart	1 dry quart	1.1 liters
1 kilogram	2.2 pounds	1 pound	0.45 kilogram

Example 5   Convert 763 centigrams to kilograms.

Solution   From Table 3 we find that 1 cg equals 0.01 gm.

Therefore,   1 cg = 0.01 gm,

so that 763×1 cg $= 763 \times 0.01$ gm,

or   763 cg = 7.63 gm.

Now we must convert 7.63 gm to kilograms. Since 1000 gm = 1 kg, we have 1 gm = 1/1000 kg or 0.001 kg. Thus 7.63 gm = 7.63×0.001 kg or 0.00763 kg.

Example 6   Convert 17.3 liters to liquid quarts.

Solution   From Table 5 we find that 1 ℓ = 1.05 liquid quarts, so that 17.3 ℓ = 17.3×1.05 liquid quarts or 17.3 ℓ = 18.165 liquid quarts.

Example 7   Convert 2133 pounds to kilograms.

Solution   From Table 5, we find that in 1 pound there are 0.45 kg. Therefore, 2133 pounds = 2133×0.45 kg, or 959.85 kg.

To make it easier to convert from one system to the other, we can use the results shown in Table 6.

Table 6

	To convert from	to	multiply by
Length	inches	millimeters	25
	inches	centimeters	2.54
	feet	centimeters	30.48
	feet	meters	0.3
	yards	meters	0.91
	miles	kilometers	1.6
	millimeters	inches	0.04
	centimeters	inches	0.4
	meters	feet	3.28
	meters	yards	1.1
	kilometers	miles	0.62
Weight	ounces	grams	28.3
	pounds	kilograms	0.45
	grams	ounces	0.035
	kilograms	pounds	2.2
Liquid Measure	ounces	milliliters	29.76
	pints	liters	0.476
	quarts	liters	0.95
	gallons	liters	3.81
	milliliters	ounces	0.034
	liters	pints	2.1
	liters	quarts	1.05
	liters	gallons	0.26

EXERCISES

1. Convert each of the following to the unit indicated.
   - a) 48 km to cm
   - b) 273 m to feet
   - c) 643 pounds to gm
   - d) 412 ℓ to liquid quarts
   - e) 623 mm to inches
   - f) 542 gm to pounds
   - g) 173 kg to cg
   - h) 423 pounds to cg
   - i) 816 cm to feet
   - j) 17 ℓ to dry quarts
   - k) 43 kg to pounds
   - l) 28 m to km
   - m) 323 mℓ to ℓ
   - n) 108 dry quarts to ℓ

2. A Miss America beauty contestant has the following measurements: 36–24–37. What are her measurements in the metric system?

3. Mary has just taken her baby to the doctor. The baby weighs 20 pounds, is 26 inches long, and just drank a 7-ounce bottle of milk. Express these measurements in the metric system.

4. An American motorist is traveling on a highway in Europe where there is a sign indicating that the maximum speed is 120 km. What is the maximum speed in miles?

Appendix B: Algebra Review

Some of the material in this book is preceded by a † indicating that a knowledge of algebra is needed. The algebra in this appendix, together with the material in Chapter 4, is sufficient to handle any of the material marked †.

This material is intended only as a *review* of the ideas in algebra needed for the text. It is by no means a complete discussion of all the ideas of algebra.

Variables and Constants

In algebra, two kinds of symbols are used to represent numbers, **variables** and **constants**.

A **constant** has a fixed value, such as 2, −4, $\frac{1}{2}$, π, etc. Sometimes we know that a number is fixed, but we do not know what its value is. Then we denote it by a letter at the beginning of the alphabet, such as a, b, c, etc.

A **variable** is a symbol that may be replaced by more than one number. We denote variables by letters at the end of the alphabet, such as x, y, z, etc. For example, in the expression "x is less than 4," x is not a fixed number, but can represent *any* number less than 4.

In algebra, multiplication is denoted in various ways. For example, "2 multiplied by x" can be denoted as $2 \cdot x$, $2(x)$, $(2)(x)$, or just $2x$. "2 multiplied by 4" can be denoted as $2 \cdot 4$, $2(4)$, or $(2)(4)$. It cannot be written as 24, because it would then be confused with the number twenty-four. Whenever such confusion is possible, the multiplication dot or parentheses *must* be used. They may be omitted, as in $2x$, if no other meaning is possible.

Example 1   In the expression $2x + 3y$, 2 and 3 are constants, while x and y are variables.

Example 2   In the equation $3s - 2.9t = 6$, the constants are 3, 2.9, and 6. The variables are s and t.

Example 3   In the equation $x - 2y = 4$, the constants are 2 and 4. (There is a constant 1 in front of the x, which is understood.) The variables are x and y, and they can represent many different values. For example, if x is 8, then y is 2; if x is 4, then y is 0; and if x is 10, then y is 3. There are actually infinitely many *pairs* of values that can replace x and y in this example. (Note that once we give a value to x, the value of y is specifically determined.)

Example 4   In the expression $1/t$, 1 is a constant, and t is a variable that can represent any value except 0. (See p. 126.)

Example 5   Let us evaluate the expression $2x - y + 3z$ for several different values of x, y, and z.

If x is 1, y is 0, and z is 3, we have $2x - y + 3z = 2(1) - 0 + 3(3)$ which equals 11.

If x is -2, y is 4, and z is $\frac{1}{3}$, we have $2x - y + 3z = 2(-2) - 4 + 3(\frac{1}{3})$ which equals -7.

Operations on Algebraic Expressions   An **algebraic expression** is a combination of letters and numbers joined together by the operations $+$, $-$, $\cdot$, or $\div$. The **terms** of the expression are the parts preceded directly by a $+$ or a $-$.

Example 6 $2x - 3yz + 4$ is an expression. The terms are $2x$, $-3yz$, and 4. (Notice that a term includes the sign that precedes it.)

Example 7 $5xy^2 - 6\sqrt{x} + 4y - 2$ is an expression. The terms are $5xy^2$, $-6\sqrt{x}$, $4y$, and -2.

Example 8 $6a^2bc$ is an expression. It has only one term, namely, $6a^2bc$.

Like terms are terms with identical letter parts.

Example 9 $4a$ and $3a$ are like terms.

Example 10 $4a$ and $3a^2$ are *not* like terms, because in $4a$ the letter part is a, and in $3a^2$ the letter part is a^2.

Example 11 $5abc$ and $2abc$ are like terms.

To *add expressions*, we add like terms. In the following examples, the combined terms are linked by curved lines.

Example 12 $(2x - 3y - 2) + (-4x + 3y + z) = -2x - 2 + z.$

Example 13 $(-x + x^2 - 3) + (3x^2 - 2x + 1) = 4x^2 - 3x - 2.$

Example 14 $(-5a + 2b - c) + (3a^2 - 4b + 6c) = 3a^2 - 5a - 2b + 5c.$

To *subtract expressions*, we change *all* the signs of the second expression and add.

Example 15 $(2x - 3y + 6) - (x + 2y - 1) = (2x - 3y + 6) + (-x - 2y + 1)$
$$= x - 5y + 7.$$

Example 16 $(a^2 - 2a + 3) - (-4a^2 + 7a - 2) = (a^2 - 2a + 3) + (+4a^2 - 7a + 2)$
$$= 5a^2 - 9a + 5.$$

To *multiply expressions,* we multiply each term of the first by each term of the second. Then we simplify. In the following examples, terms multiplied together are linked by curved lines.

Example 17 $(4a)(3x + 5y) = 12ax + 20ay$

Example 18 $(2x - 3y)(x + 2y) = 2x^2 + 4xy - 3xy - 6y^2$
$$= 2x^2 + xy - 6y^2.$$

Example 19 $(s + 3t)(s - 3t) = s^2 + 3st - 3st - 9t^2$
$$= s^2 - 9t^2.$$

Equations An **equation** consists of two expressions joined by an equal sign.

Example 20 $2x = 4$

Example 21 $3x + 5 = 6x - 2$

Example 22 $4x^2 - 2x + 3 = 0$

Example 23 $5x - y = 7$

An equation containing x raised only to the exponent 1 (and to no higher or lower exponent) is called a **linear equation** in x.

Example 24 $2x - 2 = 6$ is a linear equation in x.

Example 25 $4x^2 - 2x + 3 = 0$ is not a linear equation in x because it contains an x^2 term.

Example 26 $3x - 2z + y^2$ is a linear equation in x. It is also a linear equation in z. It is not a linear equation in y.

To *solve* a linear equation in x means to find a value of x that makes the equation true.

Example 27 $2x = 6$ has the solution $x = 3$, because $2(3) = 6$ is true.

Example 28 $y^2 - 4 = 0$ has the solutions $y = 2$ and $y = -2$, because $(2)^2 - 4 = 0$ is true, and $(-2)^2 - 4 = 0$ is true.

Example 29 $5 + z = z$ has no solution because there is no value of z that will make the equation true.

To solve an equation for x, we must get the equation into the form "$x =$ some expression." To do this, we can use the following rules.

1. Any number can be added to or subtracted from *both* sides of the equation.
2. *Both* sides of the equation may be multiplied or divided by the same number (except that we cannot divide by 0).

Example 30 Solve the equation $2x = 4x - 3$.

Solution

$$2x = 4x - 3$$

Subtract $4x$ from both sides.

$$\underline{-4x \qquad -4x}$$
$$-2x = -3$$

Divide both sides by -2.

$$\frac{\cancel{-2}x}{\cancel{-2}} = \frac{-3}{-2}$$

So the solution is

$$x = \frac{3}{2}$$

Example 31 Solve the equation $3y - 6 = 2y + 1$ for y.

Solution

$$3y - 6 = 2y + 1$$

Add 6 to both sides.

$$\underline{\quad + 6 \qquad\qquad + 6\quad}$$
$$3y = 2y + 7$$

Subtract $2y$ from both sides.
So the solution is

$$\underline{-2y \qquad\qquad -2y\quad}$$
$$y = 7$$

Example 32 Solve the equation $ax - 2 + y = 3y$ for x.

Solution

$$ax - 2 + y = 3y$$

Add $2 - y$ to both sides.

$$\underline{\quad 2 - y \qquad 2 - y\quad}$$
$$ax = 2 + 2y$$

Divide both sides by a.

$$\frac{\cancel{a}x}{\cancel{a}} = \frac{2 + 2y}{a}$$

So the solution is

$$x = \frac{2 + 2y}{a}$$

Example 33 Solve the equation $ax - 2 + y = 3y$ for y.

Solution

$$ax - 2 + y = 3y$$

Subtract ax from both sides.

$$\underline{-ax \qquad\qquad\qquad - ax\quad}$$
$$- 2 + y = 3y - ax$$

Add 2 to both sides.

$$\underline{\quad + 2 \qquad\qquad\qquad + 2\quad}$$
$$y = 3y - ax + 2$$

Subtract $3y$ from both sides.

$$\underline{-3y = -3y\qquad\qquad}$$
$$-2y = -ax + 2$$

Divide both sides by -2.

$$\frac{\cancel{-2}y}{\cancel{-2}} = \frac{-ax + 2}{-2}$$

So the solution is

$$y = \frac{-ax + 2}{-2}.$$

Appendix C: Functions and Graphs

This appendix is intended as a brief elementary introduction to the concepts of functions and graphs. Other approaches are possible. Consult any standard algebra or precalculus text for a complete treatment.

Functions   A **function** is a relationship between two variables x and y (any letters can be used) such that for each value substituted for x, there is obtained a *unique* value of y.

Example 1 $y = x^2$ is a function.

If we replace x by 1, we get $y = 1$. If we replace x by -2, we get $y = 4$, etc. For each value of x, we get a *unique* value of y.

Example 2 $y = 3x + 2$ is a function.

Example 3 $y = 1/x$ is a function.

Example 4 $y = \pm\sqrt{x}$ (which means y equals either the positive or negative square root of x) is not a function. The reason for this is that if we replace x by any number, we do *not* get a *unique* value for y. Thus, if x is 4, we get $y = 2$ *and* $y = -2$.

Example 5 $y = -\sqrt{x}$ *is* a function, because now there *is* a unique value of y for each value of x. Thus, if x is 4, y can only have the value -2.

When we have a relationship like this, we say "y is a function of x," which we abbreviate as $y = f(x)$ (read as "$y = f$ of x").

A function is like a coin-operated machine: each time you put in a coin, you get an item out. We put x-values into the "function machine." For each x-value we put in, we get a unique y-value out of the machine.

Like any coin-operated machine, the function will "accept" some values of x, but may reject others. For example, in the function $y = 1/x$, we cannot put in the value $x = 0$. (Why not?)

The values of x that we can put in the function are called the **domain** of the function. (In the elementary study of functions, we use only real numbers; we do not use complex numbers.)

Example 6   In the function $y = 1/x$, we can put in any value for x except 0. So the domain of this function is all real numbers except 0.

Example 7   In the function $y = -\sqrt{x}$, we can put in any positive number or 0 for x. We cannot put in negative values for x, because these would give us complex numbers. Thus the domain of this function is $x \geq 0$.

Example 8   In the function $y = 2x - 3$, we can put in *any* value for x. So the domain is all real numbers.

The **range** of a function is the set of y-values that come out of the function machine.

Example 9   In the function $y = -\sqrt{x}$, only negative values of y or 0 come out, so the range is $y \leq 0$.

Example 10   In the function $y = x^2$, only positive values of y or 0 come out, so the range is $y \geq 0$.

Example 11   In the function $y = |x|$ (y is the absolute value of x), only positive values of y or 0 come out, so the range is $y \geq 0$.

If we have a function $y = f(x)$, then the symbol $f(a)$ means the value of the function when $x = a$.

Example 12   For the function $y = 2x - 3$, $f(1) = 2(1) - 3$ or -1.
For the function $y = 2x - 3$, $f(0) = 2(0) - 3$ or -3.
For the function $y = 2x - 3$, $f(-1) = 2(-1) - 3$ or -5.
For the function $y = 2x - 3$, $f(a) = 2a - 3$.

Example 13   For the function $y = x^2$, $f(3) = 3^2$ or 9.
For the function $y = x^2$, $f(1) = 1^2$ or 1.
For the function $y = x^2$, $f(t) = t^2$.

Example 14   For the function $\dfrac{3x - 4}{5}$, $f(2) = \dfrac{3(2) - 4}{5}$ or $\dfrac{2}{5}$.

For the function $\dfrac{3x - 4}{5}$, $f(-6) = \dfrac{3(-6) - 4}{5}$ or $\dfrac{-22}{5}$.

For the function $\dfrac{3x - 4}{5}$, $f(0) = \dfrac{3(0) - 4}{5}$ or $\dfrac{-4}{5}$.

Functions are usually given by equations, as in the examples we have given. It is often convenient to be able to draw a "picture" of an equation, called a **graph**. To do this, we need a **rectangular coordinate system**.

Rectangular Coordinate System

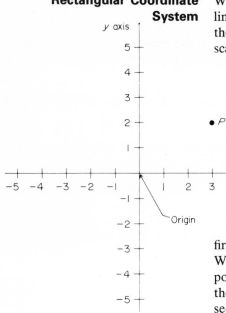

We draw two straight lines perpendicular to each other as shown. The horizontal line is called the *x* **axis** and the vertical line is called the *y* **axis**. The point where the *x* axis and *y* axis meet is called the **origin**. The axes are marked with a number scale, as shown.

If we want to get to the point *P* in the diagram, starting from the origin, we first move 3 units to the right on the *x* axis and from that point move up 2 units. We call 3 the *x* **coordinate** of point *P* and 2 the *y* **coordinate** of point *P*. We label point *P* as $(3, 2)$ and we call these numbers the **coordinates** of point *P*. In writing the coordinates of a point, we write the *x* coordinate first, the *y* coordinate second, and enclose them within parentheses.

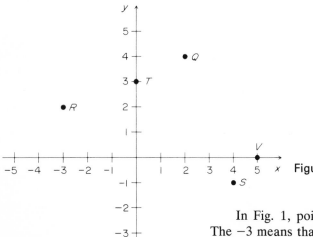

Figure 1

In Fig. 1, point *Q* has coordinates $(2, 4)$. Point *R* has coordinates $(-3, 2)$. The -3 means that we start at the origin and go 3 units to the left on the *x* axis. The 2 means that we go 2 up from that point.

Point *S* has coordinates $(4, -1)$. This means that we start at the origin and go 4 units to the right and then 1 down.

Point *T* has coordinates $(0, 3)$. This means that we start at the origin and move nothing to the right or left. We simply move 3 units up.

Point *V* has coordinates $(5, 0)$. This means that we start at the origin and go 5 units to the right and no units up or down.

A system such as the one described is called a **rectangular coordinate system**.

Graphs Consider the equation $y = 2x - 4$. One solution of this equation is $x = 3$ and $y = 2$. We write this in abbreviated form as $(3, 2)$. Another solution is $x = 0$ and $y = -4$, which we write as $(0, -4)$. Similarly, other solutions are $(4, 4)$, $(1, -2)$, etc. Let us put these values into a table as shown at the left. We have found only four solutions to the equation. Actually there are infinitely many such pairs of values. To find any solution, just let x be any convenient number, and then find y. The **graph** of the equation consists of all the points whose coordinates (x, y) are solutions of the equation.

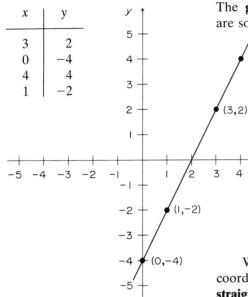

x	y
3	2
0	-4
4	4
1	-2

Figure 2

We can plot the points $(3, 2)$, $(0, -4)$, $(1, -2)$, and $(4, 4)$ in a rectangular coordinate system as shown in Fig. 2. If we join these points, we see that we get a **straight line**. This line is called the **graph** of the equation.

Example 15 Draw the graph of $y = x + 1$.

Solution Three pairs of values that satisfy this equation are shown in the table. The graph of this equation is shown on the right. It is also a straight line.

x	y
0	1
3	4
4	5

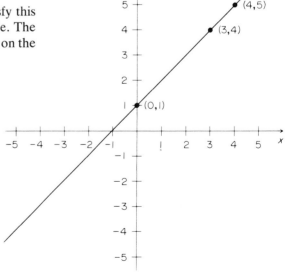

Example 16   Draw the graph of $2x + 3y = 6$.

Solution   Three pairs of values that satisfy this equation are shown in the table.

x	y
0	2
3	0
6	−2

The graph of this equation is shown here.

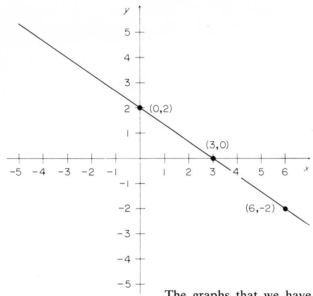

The graphs that we have drawn so far are all straight lines. Actually any equation of the form $ax + by + c = 0$ (where a, b, and c are constants, and a and b are not both 0) will have a straight-line graph. An equation of the form $ax + by + c = 0$ (where a and b are not both 0) is called a **linear equation** for this reason.

Example 17 $2x − y − 4 = 0$ is a linear equation. a is 2, b is −1, and c is −4.

Example 18 $3x + 5y = 6$ can be rewritten as $3x + 5y − 6 = 0$. It is a linear equation, with $a = 3$, $b = 5$, and $c = −6$.

Example 19 $x − 4 = 0$ is a linear equation, with $a = 1$, $b = 0$, and $c = −4$.

Example 20 $y = 9 − x$ can be rewritten as $x + y − 9 = 0$. It is a linear equation, with $a = 1$, $b = 1$, and $c = −9$.

Example 21 $y = −7$ can be rewritten as $y + 7 = 0$. It is a linear equation, with $a = 0$, $b = 1$, and $c = 7$.

Of course, graphs need not always be straight lines.

Example 22 Let us draw the graph of $y = x^2$. First we find several solutions of this equation. Some solutions are $(0, 0)$, $(1, 1)$, $(2, 4)$, $(-1, +1)$, $(-2, +4)$. We put these into a table as shown. Then we plot these points. The resulting graph is called a **parabola**.

x	y
0	0
1	1
2	4
−1	+1
−2	−4

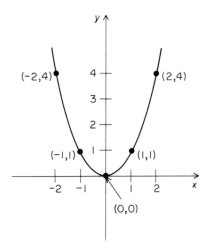

We drew this graph using only 5 points. The number of points needed to draw a straight-line graph is 2. (Why?) The number of points needed to draw any graph depends on the particular equation.

Example 23 Let us draw the graph of $y = |x|$.

First we find several solutions of this equation. Some solutions are $(0, 0)$, $(1, 1)$, $(2, 2)$, $(3, 3)$, $(-1, 1)$, $(-2, 2)$, and $(-3, 3)$. We put these into a table as shown and join them. The resulting graph is a "v-shape."

x	y
0	0
1	1
2	2
3	3
−1	1
−2	2
−3	3

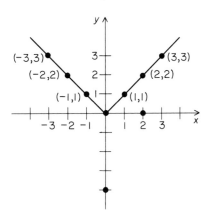

Answers

Chapter 1, Section 2, Page 6

2. (a) Inductive;   (b) inductive;
(c) deductive;   (d) deductive.

5. Probably but not necessarily.

6. The job with the semiannual raise.

7. (a) 25;   (b) 9;   (c) 8;   (d) 17;   (e) 21;
(f) $\frac{1}{16}$;   (g) 5;   (h) $\frac{4}{3}$;   (i) 8;   (j) 3.

8.

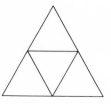

9. (a) Yes;   (b) no;   (c) no.

Chapter 1, Section 3, Page 10

2. (a)

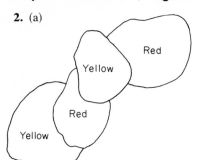

(b)

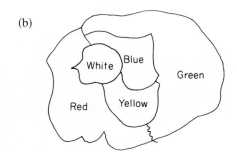

(c)

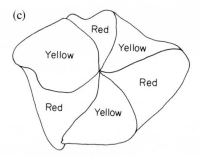

Chapter 1, Section 4, Page 15

1. Valid.   **2.** Invalid.   **3.** Valid.   **4.** Invalid.

5. (a) Invalid;   (b) invalid;   (c) invalid.

6. Valid.   **7.** Valid.   **8.** Valid.

9. (a) Invalid;   (b) invalid;   (c) invalid.

10. (a) Invalid;   (b) invalid;   (c) invalid.

11. (a) Valid;   (b) valid;   (c) invalid;
(d) invalid;   (e) invalid

12. (a) Invalid;   (b) invalid;   (c) invalid;
(d) invalid;   (e) invalid.

13. (a) Invalid;   (b) invalid;   (c) invalid.

14. (a) Valid;   (b) invalid;   (e) valid;
(f) invalid;   (g) invalid.

15. (a) Invalid;   (b) invalid;   (c) valid;   (d) invalid.

16. (a) Invalid;   (b) invalid;   (d) invalid.

17. (a) Invalid;   (c) invalid.

18. (a) Valid;   (b) valid;   (c) valid;
(d) invalid;   (e) invalid.

19. Valid.

Chapter 1, Section 5, Page 21

2. Valid.   **3.** Invalid.   **4.** Invalid.

Chapter 1, Section 6, Page 26

1. (a) Proposition;   (b) not a proposition;
(c) not a proposition;   (d) proposition;   (e) proposition;
(f) not a proposition;   (g) not a proposition.

2. (a) I am a monkey's uncle and I eat bananas.

(b) I am not a monkey's uncle.

(c) Either I am a monkey's uncle or I eat bananas.

(d) I am not a monkey's uncle and I eat bananas.

(e) I am not a monkey's uncle and I do not eat bananas.

(f) I am a monkey's uncle and I do not eat bananas.

(g) I do not eat bananas.

(h) Either I am not a monkey's uncle or I eat bananas.

(i) Either I am not a monkey's uncle or I do not eat bananas.

(j) It is not true that I am a monkey's uncle and I eat bananas.

3. (a) $p \wedge q$;   (b) $\sim p$;   (c) $(\sim q) \wedge p$;

(d) $\sim(p \wedge q)$;   (e) $p \vee q$;   (f) $p \vee (\sim q)$.

Chapter 1, Section 7, Page 30

1. (a) If they raise the toll on the Verrazano bridge, then I will swim to school.

(b) If I swim to school, they will raise the toll on the Verrazano bridge.

(c) If they raise the toll on the Verrazano bridge, then I will not swim to school.

(d) If they do not raise the toll on the Verrazano bridge, then I will swim to school.

(e) If they do not raise the toll on the Verrazano bridge, then I will not swim to school.

(f) If I do not swim to school, then they will not raise the toll on the Verrazano bridge.

2. (a) $r \to s$;   (b) $(\sim s) \to (\sim r)$;
(c) $(\sim r) \to (\sim s)$;   (d) $s \to r$.

3. *Converse:*   If I am strong, I have eaten my spinach.
Inverse: If I do not eat my spinach, then I will not be strong.

Contrapositive: If I am not strong, then I have not eaten my spinach.

(b) *Converse:*   If I am a genius, then I can understand mathematics.
Inverse: If I cannot understand mathematics, then I am not a genius.

Contrapositive: If I am not a genius, then I cannot understand mathematics.

4. (b) and (d) must be true.

Chapter 1, Section 8, Page 35

1. (a)

p	q	$p \to q$	$\sim(p \to q)$
T	T	T	F
T	F	F	T
F	T	T	F
F	F	T	F

(c)

p	q	$\sim q$	$p \vee (\sim q)$	$\sim[p \vee (\sim q)]$
T	T	F	T	F
T	F	T	T	F
F	T	F	F	T
F	F	T	T	F

(e)

p	q	$\sim q$	$p \wedge (\sim q)$	$[p \wedge (\sim q)] \to p$
T	T	F	F	T
T	F	T	T	T
F	T	F	F	T
F	F	T	F	T

(g)

p	q	$p \to q$	$(p \to q) \to p$
T	T	T	T
T	F	F	T
F	T	T	F
F	F	T	F

(i)

p	q	$p \vee q$	$q \vee p$	$(p \vee q) \to (q \vee p)$	$\sim[(p \vee q) \to (q \vee p)]$
T	T	T	T	T	F
T	F	T	T	T	F
F	T	T	T	T	F
F	F	F	F	T	F

(k)

p	r	s	~p	r ∨ s	(~p) → (r ∨ s)
T	T	T	F	T	T
T	T	F	F	T	T
T	F	T	F	T	T
T	F	F	F	F	T
F	T	T	T	T	T
F	T	F	T	T	T
F	F	T	T	T	T
F	F	F	T	F	F

(m)

p	q	r	~r	p ∧ q	(p ∧ q) ∧ (~r)	[(p ∧ q) ∧ (~r)] → r
T	T	T	F	T	F	T
T	T	F	T	T	T	F
T	F	T	F	F	F	T
T	F	F	T	F	F	T
F	T	T	F	F	F	T
F	T	F	T	F	F	T
F	F	T	F	F	F	T
F	F	F	T	F	F	T

(o)

p	q	~p	~q	p ∧ q	~(p ∧ q)	(~p) ∨ (~q)	[~(p ∧ q)] → [(~p) ∨ (~q)]
T	T	F	F	T	F	F	T
T	F	F	T	F	T	T	T
F	T	T	F	F	T	T	T
F	F	T	T	F	T	T	T

(q)

p	q	~p	~q	p ∨ q	~(p ∨ q)	(~p) ∧ (~q)	[~(p ∨ q)] → [(~p) ∧ (~q)]
T	T	F	F	T	F	F	T
T	F	F	T	T	F	F	T
F	T	T	F	T	F	F	T
F	F	T	T	F	T	T	T

2. (a) Neither; (b) neither; (c) neither; (d) tautology;
(e) tautology; (g) neither; (i) self-contradiction; (j) neither;
(k) neither; (l) tautology; (m) neither; (o) tautology;
(p) tautology; (q) tautology; (r) tautology.

5. (a) 4 letters require 16 rows; (b) 5 letters require 32 rows;
(c) n letters require 2^n rows.

6.

p	q	p ⊻ q
T	T	F
T	F	T
F	T	T
F	F	F

7.

p	q	p ↓ q
T	T	F
T	F	F
F	T	F
F	F	T

Chapter 1, Section 9, Page 39

1. Invalid **2.** Valid **3.** Valid **4.** Valid **5.** Valid
6. Invalid **7.** Valid **8.** Invalid **9.** Invalid **10.** Invalid
11. Invalid **12.** Valid **13.** Invalid **14.** Invalid **15.** Invalid

Chapter 1, Section 10, Page 43

1. (a) The formula for the circuit is $P \wedge (\sim Q)$.

P	Q	$\sim Q$	$P \wedge (\sim Q)$
T	T	F	F
T	F	T	T
F	T	F	F
F	F	T	F

(b) The formula for the circuit is $[P \wedge (\sim Q)] \vee Q$.

P	Q	$\sim Q$	$P \wedge (\sim Q)$	$[P \wedge (\sim Q)] \vee Q$
T	T	F	F	T
T	F	T	T	T
F	T	F	F	T
F	F	T	F	F

(c) The formula for the circuit is $[P \vee Q] \vee [P \wedge (\sim Q)]$.

P	Q	$\sim Q$	$P \vee Q$	$P \wedge (\sim Q)$	$[P \vee Q] \vee [P \wedge (\sim Q)]$
T	T	F	T	F	T
T	F	T	T	T	T
F	T	F	T	F	T
F	F	T	F	F	F

(d) The formula for the circuit is $\{[P \wedge (\sim Q)] \vee [(\sim P) \wedge Q]\} \vee [P \wedge Q]$.

P	Q	$\sim P$	$\sim Q$	$P \wedge (\sim Q)$	$(\sim P) \wedge Q$	$[P \wedge (\sim Q)] \vee [(\sim P) \wedge Q]$	$P \wedge Q$	$\{[P \wedge (\sim Q)] \vee [(\sim P) \wedge Q]\} \vee (P \wedge Q)$
T	T	F	F	F	F	F	T	T
T	F	F	T	T	F	T	F	T
F	T	T	F	F	T	T	F	T
F	F	T	T	F	F	F	F	F

(e) The formula for the circuit is $(P \wedge Q) \vee \{P \wedge [(\sim Q) \vee (\sim P)]\}$.

P	Q	$\sim P$	$\sim Q$	$P \wedge Q$	$(\sim Q) \vee (\sim P)$	$P \wedge [(\sim Q) \vee (\sim P)]$	$(P \wedge Q) \vee \{P \wedge [(\sim Q) \vee (\sim P)]\}$
T	T	F	F	T	F	F	T
T	F	F	T	F	T	T	T
F	T	T	F	F	T	F	F
F	F	T	T	F	T	F	F

(f) The formula for the circuit is $[P \vee (\sim Q)] \vee [Q \vee (\sim P)]$.

P	Q	$\sim P$	$\sim Q$	$P \vee (\sim Q)$	$Q \vee (\sim P)$	$[P \vee (\sim Q)] \vee [Q \vee (\sim P)]$
T	T	F	F	T	T	T
T	F	F	T	T	F	T
F	T	T	F	F	T	T
F	F	T	T	T	T	T

(g) The formula for the circuit is $(P \lor Q) \lor \{[(\sim P) \lor (\sim Q)] \land Q\}$.

P	Q	$\sim P$	$\sim Q$	$P \lor Q$	$(\sim P) \lor (\sim Q)$	$[(\sim P) \lor (\sim Q)] \land Q$	$(P \lor Q) \lor \{[(\sim P) \lor (\sim Q)] \land Q\}$
T	T	F	F	T	F	F	T
T	F	F	T	T	T	F	T
F	T	T	F	T	T	T	T
F	F	T	T	F	T	F	F

2. a) $P \lor Q$

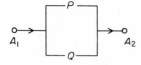

b) Q

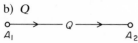

(c) P

d) $P \lor Q$

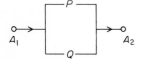

3. All can be simplified to $P \lor Q$.

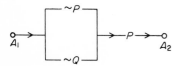

4. a)

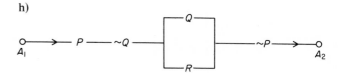

b)

c)

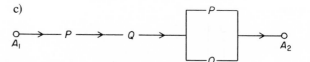

d)

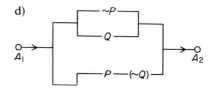

e)

f)

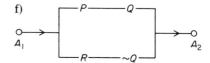

g)

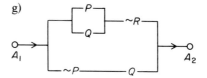

h)

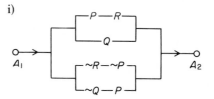

i)

Chapter 1, Section 11, Page 46

1.

p	q	$p \wedge q$
T	T	T
T	F	F
T	M	M
M	T	M
M	F	F
M	M	M
F	T	F
F	F	F
F	M	F

2.

p	q	$p \vee q$
T	T	T
T	F	T
T	M	T
M	T	T
M	F	M
M	M	M
F	T	T
F	F	F
F	M	M

3.

p	q	$\sim q$	$p \wedge (\sim q)$
T	T	F	F
T	F	T	T
T	M	M	M
F	T	F	F
F	F	T	F
F	M	M	F
M	T	F	F
M	F	T	M
M	M	M	M

5. Mr Chrysler.

6. Bill Holland is going to England; Pat Canada is going to Holland; Debbie English is going to Canada.

7. The blind prisoner wore a white hat.

9. Red

10. Would a member of the other tribe tell me to take this road?

11. One.

13. Mr. Gusher did it.

Chapter 2, Section 4, Page 55

1. (a) The set of odd numbers between 1 and 19 inclusive; (b) the set of months with 30 days; (d) the set of different letters in the word bellybutton; (e) the set of all parts of a car; (f) the set of all positive numbers that are divisible by 3.

2. (a) {s, u, p, e, r, c, a, l, y, f, g, i, t, x, d, o, h}; (b) { }, (c) {Eisenhower, Kennedy, Johnson, Nixon, Ford, Carter}; (d) {2, 4, 6, 8, . . .}; (e) { }.

3. (a) No; (b) yes; (c) no; (d) no.

4. (a) Finite; (b) finite; (c) infinite; (d) null, finite; (e) unit, finite; (f) finite

5. (a) True; (b) false; (c) false; (e) false.

6. (a) Yes; (b) not necessarily.

7. $X = \{e, f, 1, 5\}$ $X = \{e, f, 1, 5\}$
$Y = \{k, e, 3, m\}$ $Y = \{k, e, 3, m\}$
(Many others are possible.)

8. (a) Equivalent; (b) equal and equivalent; (c) equivalent; (d) neither; (e) equal and equivalent; (f) neither.

9. (a) 5; (b) 1; (c) 12; (d) 0; (f) 0.

10. One example of each is:
(a) {1, 2, 3, 4, 5, 6, 7}
(b) {all radios over 200 years old}
(c) {1, 2} and {⚉, □}
(d) {M}
(e) {1, 4, 7}
(f) {1, $\frac{1}{2}$, $\frac{1}{3}$, $\frac{1}{4}$ . . .}

Chapter 2, Section 5, Page 59

1. (a) {Chinese, Japanese} and {Chinese}
(b) {1, 2, ⚉} and ∅
(c) {piano, guitar, violin} and {drums}
(d) {5} and ∅
(e) {cucumber, pepper} and {lettuce}
(f) {the tragedies} and {the comedies}

2. One suitable universal set is given. Other answers may be possible.
(a) {all mammals}; (b) {all planets};
(c) {all college courses}; (d) {all colors};
(e) {all cars}; (f) {all multiples of 5};
(g) {all prime numbers};
(h) {all geographic locations}.

3. (a) Proper; (b) improper; (c) improper; (d) proper; (e) neither; (f) proper.

4. (a) True; (b) true; (c) true; (d) false; (e) false; (f) true; (g) false; (h) true; (i) true; (j) false; (k) true.

5. (a) { }; (b) {x}, $\varnothing$; (c) {15, y}, {15}, {y}, $\varnothing$
(d) {20, $\frac{1}{2}$, $\triangle$}, {20, $\frac{1}{2}$}, {20, $\triangle$}, {$\frac{1}{2}$, $\triangle$}, {20}, {$\frac{1}{2}$}, {$\triangle$}, $\varnothing$
(e) {eeny, meeny, miny mo}, {eeny, meeny, miny}, {eeny, meeny, mo}, {eeny, miny, mo}, {meeny, miny, mo}, {eeny, meeny}, {eeny, miny}, {eeny, mo}, {meeny, miny}, {meeny, mo}, {miny, mo}, {eeny}, {meeny}, {miny}, {mo}, $\varnothing$

6. (a) 1, 2, 4, 8, 16, 8, and 4
(b) Yes; 2^n where n is the number of elements in the set.

7. (a) 3; (b) 1; (c) 7; (d) 15.

Chapter 2, Section 6, Page 63

1. (a) {1, 2, 4, 6, 8, 9, 10}; (b) {4, 8, 9}; (c) {2, 3, 5, 7, 10}; (d) {1, 3, 5, 6, 7}; (e) {1, 6}; (f) {2, 10}; (g) {1, 4, 6, 8, 9}.

2. (a) {all members of congress}   (b) the null set
(c) {all male Republican members of congress}
(d) {all female Republican members of congress}
(e) {all non-Republican members of congress}
(f) {all non-Republican male members of congress}
(g) {all non-Democrat members of congress}
(h) {members of congress with hair}
(i) {all non-Republican members of Congress who have hair}
(j) {all female Democrats in Congress}

3. (a) {all consonants}; (b) {all vowels};
(c) {all letters of the alphabet}; (d) {all vowels}; (e) $\varnothing$;
(f) {all vowels}; (g) {all letters of the alphabet}; (h) $\varnothing$;
(i) {all vowels}; (j) $\varnothing$; (k) {all consonants}; (l) $\varnothing$.

4. (a) $\bigcup$; (b) A; (c) A; (d) $\varnothing$.

5. {$\triangle$, 2, 4}; (b) {$\triangle$, 2, 4, 6}; (c) {$\triangle$, 2, 4}; (d) {$\triangle$, 2, 4};
(e) $\varnothing$;   (f) {6};   (g) $\varnothing$

6. (a) A is a subset of B; (b) B is a subset of A.

7. (a) B is a subset of A; (b) A is a subset of B.

8. $A = \varnothing$.

9. (a) set Z
(b) {all points between A and B inclusive and all points between C and D inclusive}
(c) $\varnothing$; (d) {point B}; (e) set Y
(f) {all points between A and D inclusive}

10. (a) 12; (b) 18; (c) 90; (d) 110; (e) 69; (f) 43; (g) 1;
(h) 30; (i) 62; (j) 50; (k) 96.

11. (a) 37; (b) 21; (c) 56; (d) 19;   (e) 86; (f) 75; (g) 19;
(h) 56; (i) 64; (j) 88.

12. (a) 133; (b) 101; (c) 21; (d) 159; (e) 27; (f) 103.

13. (a) 394; (b) 61; (c) 310; (d) 73;   (e) 690; (f) 525; (g) 99;
(h) 182; (i) 244; (j) 320; (k) 887.

Chapter 2, Section 8, Page 70

1. a)

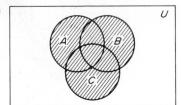

b)

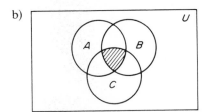

c)

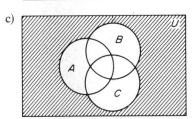

d)

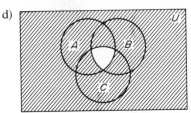

e)

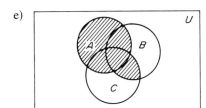

c)

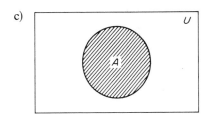

f)

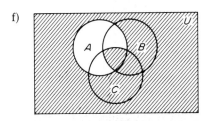

d)

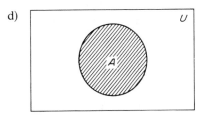

g)

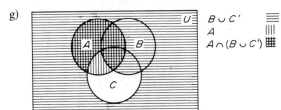

$B \cup C'$ ≡≡≡
A ‖‖‖
$A \cap (B \cup C')$ ▦

e)

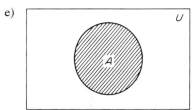

2. (2) $A \cap (B \cup C)'$; (3) $A \cap B \cap C'$; (4) $B \cap (A \cup C)'$;
(5) $B \cap C \cap A'$; (6) $A \cap B \cap C$; (7) $A \cap C \cap B'$;
(8) $C \cap (A \cup B)'$. (*Note:* Other answers are possible.)

f)

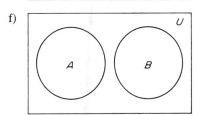

3. a)

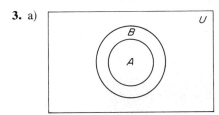

g)

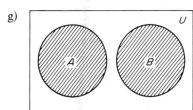

b)

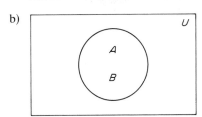

h)

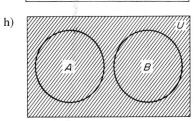

i)

4. (a) 12; (b) 2;   (c) 24; (d) 10.
5. (a) 60; (b) 3;   (c) 73; (d) 57; (e) 16; (f) 14.
6. (a) 15; (b) 50; (c) 5;   (d) 25; (e) 15.
7. (a) 80; (b) 20; (c) 60; (d) 60; (e) 50.
8. (a) 195; (b) 140; (c) 50; (d) 30; (e) 15.
9. (a) 100; (b) 70; (c) 85; (d) 60; (e) 0.
10. (a) 85; (b) 115; (c) 112; (d) 68; (e) 0.

11. The numbers are inconsistent.

12. No.

13. If analyzed by means of Venn diagrams, we conclude that there were 823 customers interviewed. The agency claimed that it interviewed 800 customers.

14. (a) 1160; (b) 760.

Chapter 2, Section 9, Page 78

1. a) {1, 3, 5, 7, ..., n, ...}
⇕ ⇕ ⇕ ⇕   ⇕
{4, 12, 20, 28, ..., $4n$, ...}

b) {3, 6, 9, 12, ..., $3n$, ...}
⇕ ⇕ ⇕ ⇕   ⇕
{4, 8, 12, 16, ..., $4n$, ...}

c) {1, 2, 3, 4, ..., n, ...}
⇕ ⇕ ⇕ ⇕   ⇕
$\left\{1, \frac{1}{2}, \frac{1}{3}, \frac{1}{4}, ..., \frac{1}{n}, ...\right\}$

d) {1, 2, 3, 4, ..., n, ...}
⇕ ⇕ ⇕ ⇕   ⇕
{1, 4, 9, 16, ..., n^2, ...}

e) {2, 4, 8, 16, ..., 2^n, ...}
⇕ ⇕ ⇕ ⇕   ⇕
{2, 4, 6, 8, ..., $2n$, ...}

2. Draw lines from the center of the circle to the square.

3. Draw lines from the center of the inner figure to the outside figure.

Chapter 3, Section 2, Page 85

1. (a) 112; (b) 37; (c) 319; (d) 1017; (e) 418; (f) 116; (g) 322; (h) 1029.

2. a)

b)

c)

d)

e)

f)

g)

h)

3. a)

b)

4. a)

b)

Chapter 3, Section 3, Page 89

1. (a) LXXIV; (b) LXXXVI; (c) XLIII; (d) DCXXXII.

2. (a) 111; (b) 71; (c) 304; (d) 62.

3. (a) CLXV; (b) CXVII.

5. b) (i)

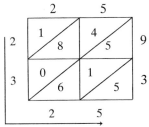

Answer: 2325

(ii)

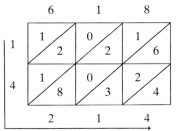

Answer: 14,214

(iii)

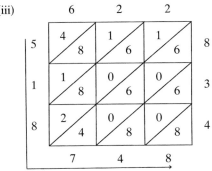

Answer: 518,748

(iv)

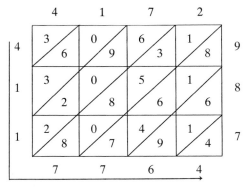

Answer: 4,117,764

(v)

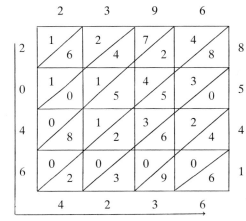

Answer: 20, 464, 236

(vi)

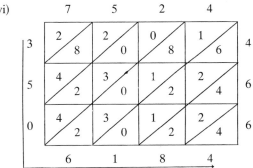

Answer: 3,506,184

6. (a)

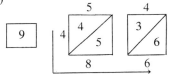

Answer: 486

(b)

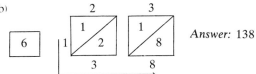

Answer: 138

(c) 13,104;   (d) 43,725;   (e) 64,410;

(f) 35,568   (g) 157,115.

Chapter 3, Section 4, Page 96

1. (a) $100111_{(2)}$; (b) $22201_{(3)}$; (c) $1838_{(9)}$; (d) $41231_{(2)}$;
(e) $e08t_{(12)}$; (f) $3320_{(4)}$;   (g) $21532_{(6)}$; (h) $1061_{(7)}$;
(i) $1617_{(8)}$;   (j) $4256t8_{(12)}$; (k) $111011100_{(2)}$;
(l) $2210_{(3)}$;   (m) $114_{(5)}$;   (n) $63631_{(9)}$; (o) $1677_{(8)}$.

2. (a) 8; (b) 9; (c) 3; (d) 20.

4.

+	0	1
0	0	1
1	1	10

×	0	1
0	0	0
1	0	1

Base 2

+	0	1	2	3
0	0	1	2	3
1	1	2	3	10
2	2	3	10	11
3	3	10	11	12

×	0	1	2	3
0	0	0	0	0
1	0	1	2	3
2	0	2	10	12
3	0	3	12	21

Base 4

9.

+	L	O	V	E
L	L	O	V	E
O	O	V	E	OL
V	V	E	OL	OO
E	E	OL	OO	OV

10. b) (i)

63	55
31	110
15	220
7	440
3	880
1	1760
	3465

(iii)

68	~~325~~
34	~~650~~
17	1300
8	~~2600~~
4	~~5200~~
2	~~10,400~~
1	20,800
	22,100

(v)

78	~~49~~
39	98
19	196
9	392
4	~~784~~
2	~~1568~~
1	3136
	3822

(vii)

69	223
34	~~446~~
17	892
8	~~1784~~
4	~~3568~~
2	~~7136~~
1	14,272
	15,387

(ix)

45	216
22	~~432~~
11	864
5	1728
2	~~3456~~
1	6912
	9720

(x)

17	219
8	~~438~~
4	~~876~~
2	~~1752~~
1	3504
	3723

Chapter 3, Section 5, Page 100

1. (a) $2320_{(5)}$; (b) $31021_{(4)}$; (c) $11111101_{(2)}$; (d) $525_{(12)}$;
(e) $22211_{(3)}$; (f) $133_{(9)}$;   (g) $470_{(8)}$;   (h) $1451_{(6)}$;
(i) $1556_{(7)}$;   (j) $26t_{(12)}$; (k) $10010011_{(2)}$;
(l) $11010_{(3)}$; (m) $505_{(8)}$;   (n) $149_{(16)}$.

Chapter 3, Section 6, Page 103

1. (a) $155_{(10)}$; (b) $54_{(10)}$; (c) $89_{(10)}$; (d) $117_{(10)}$; (e) $59_{(10)}$;
(f) $500_{(10)}$; (g) $70_{(10)}$; (h) $223_{(10)}$; (i) $476_{(10)}$; (j) $31_{(10)}$;
(k) $1225_{(10)}$; (l) $294_{(10)}$; (m) $1727_{(10)}$; (n) $239_{(10)}$;
(o) $304_{(10)}$; (p) $142_{(10)}$; (q) $34_{(10)}$; (r) $207_{(10)}$.

2. Mark is 32 years old and Sharon is 26 years old.

Chapter 3, Section 8, Page 111

1. (a) Good; (b) good; (c) bad; (d) good.

2. (a) Take 16 from the 19 pile.
(b) Take 2 from the 7 pile.
(c) Take 8 from the 15 pile.

Chapter 4, Section 2, Page 123

1. (a) Commutative law of addition; (b) associative law of multiplication; (c) commutative law of multiplication; (d) law of closure for addition; (e) associative law of addition; (f) commutative law of addition; (g) commutative law of multiplication; (h) commutative law of multiplication; (i) law closure for addition; (j) law closure for multiplication.

2. (a) Not commutative; (b) not commutative; (d) not commutative; (e) not commutative; (f) not commutative; (g) commutative; (h) not commutative.

3. (a) No; (b) yes; (c) yes; (d) yes; (e) yes; (f) no; (g) no.

4. (a) Yes; (b) yes; (c) yes.

5. (a) Yes; (b) no; (c) no.

6. (a) Yes; (b) yes.

Chapter 4, Section 3, Page 127

1. The identity for multiplication is the number 1.

2. No.

6. Both as a place holder and as a quantity.

Chapter 4, Section 4, Page 130

1. (a) +23; (b) +4;   (c) −4; (d) −14; (e) +3;   (f) −11;
(g) −1;   (h) +42; (i) +2; (j) +3;   (k) −42; (l) +48;
(m) −24; (n) +7;   (o) 0;   (p) 0;   (q) 0.

3. Associative law for
addition: $a + (b + c) = (a + b) + c.$
Associative law for
multiplication: $a(b \cdot c) = (ab)c.$
Law of closure for
addition: $a + b$ is an integer.
Law of closure for
multiplication: $a \cdot b$ is an integer.
Commutative law: $a \cdot b = b \cdot a$
for multiplication

4. 0   **5.** Yes   **6.** No

8. (a) Commutative law of addition; (b) distributive law;
(c) commutative law of addition; (d) associative law of
addition; (e) commutative law of multiplication; (f) law of
closure for multiplication; (g) closure for addition, or additive inverse; (h) associative law of multiplication;
(i) multiplication property of 0, or closure for multiplication.

9. No.

10. (a) 5; (b) 5; (c) 4; (d) 0; (e) 4; (f) 2; (g) 0.

12. (a) Positive; (b) negative.

Chapter 4, Section 5, Page 143

1. (a) 9/10; (b) −3/5; (c) 1/4; (d) −6/5; (e) −1; (f) 4.

2. (a) Equal; (b) not equal; (c) equal; (d) equal;
(e) not equal; (f) not equal.

4. (a) 2; (b) 2/9; (c) 4/5; (d) $\dfrac{92}{63}$; (e) $\dfrac{-5}{6}$; (f) $\dfrac{20}{99}$; (g) 1;
(h) $\dfrac{-5}{24}$; (i) 1; (j) $\dfrac{6}{7}$; (k) $\dfrac{-9}{25}$; (l) 64; (m) 3; (n) $\dfrac{5}{56}$.

7. (a) True; (b) true; (c) false; (d) false; (e) true; (f) true.

11. (a) Terminating; (b) repeating; (c) terminating;
(d) repeating; (e) nonterminating and nonrepeating;
(f) terminating; (g) nonterminating and nonrepeating;
(h) nonterminating and nonrepeating.

12. Yes

13. Not necessarily

14. (a) 0.111...; (b) 0.5714285...; (c) 0.28;
(d) 0.363636....

15. (a) $\dfrac{86}{100}$ or $\dfrac{43}{50}$; (b) $\dfrac{3}{10}$; (c) $\dfrac{58}{10}$ or $\dfrac{29}{5}$; (d) $\dfrac{7}{1000}$.

16. (a) $\dfrac{31}{99}$;   (b) $\dfrac{97}{99}$; (c) $\dfrac{8}{99}$;   (d) $\dfrac{567}{999}$.

Chapter 4, Section 7, Page 153

1. (a) I; (b) R; (c) I; (d) $\varnothing$; (e) R; (f) $\varnothing$; (g) R; (h) W;
(i) Q; (j) I.

2. (a) True; (b) false; (c) true; (d) false; (e) true; (f) true.

5. (a) $4.7 < 4.9$; (b) $(−5) < (+4)$; (c) $(−4) > (−5)$;
(d) $\dfrac{1}{2} > \dfrac{1}{3}$; (e) $\dfrac{1}{2} = 0.5$;   (g) $\sqrt{2} < 2$.

6. a)

c)

e)

8. (b) $2 \cdot 0 = 3 \cdot 0$. However, 2 is not equal to 3.

Chapter 5, Section 2, Page 162

1. (a)

⊕	0	1	2	3	4	5
0	0	1	2	3	4	5
1	1	2	3	4	5	0
2	2	3	4	5	0	1
3	3	4	5	0	1	2
4	4	5	0	1	2	3
5	5	0	1	2	3	4

(c)

⊕	0	1	2
0	0	1	2
1	1	2	0
2	2	1	0

2. (a) 1; (b) 2; (c) 4; (d) 10; (e) 4; (f) 0; (g) 0; (h) 8; (i) 2; (j) 0; (k) 6; (l) 0; (m) 6; (n) 4.

3. (a) 4; (b) 9; (c) 3; (d) 0; (e) 0; (f) 1; (g) 3; (h) 0.

4. (a) mod 5; (b) mod 5; (c) mod 3; (d) mod 8; (e) mod 9; (f) mod 6; (g) mod 2; (h) mod 7. Other answers are also possible.

6. Inverse of 0 is 0
Inverse of 1 is 5
Inverse of 2 is 4
Inverse of 3 is 3
Inverse of 4 is 2
Inverse of 5 is 1

Chapter 5, Section 3, Page 166

1. (a) 0; (b) 1; (c) 6; (d) 2; (e) 1; (f) 2; (g) 3; (h) 4; (i) 2; (j) 0.

2. (a) True; (b) true; (c) true; (d) true; (e) true; (f) false; (g) true; (h) false; (i) false; (j) true.

3. Two possible answers are given. Others are possible.
(a) 3, 8; (b) 7, 17; (c) 3, 11; (d) 0, 4; (e) 4, 13; (f) 1, 6; (g) 5, 11; (h) 0, 6; (i) 5, 12; (j) 0, 2.

4. (a) 4; (b) 3; (c) 11; (d) any number; (e) 10; (f) 7; (g) 4; (h) 7.

5. Monday **6.** Thursday **7.** Tuesday **8.** Wednesday

Chapter 5, Section 4, Page 170

1. (a) 1358; (b) 11,306; (c) 198; (d) 47,969; (e) 744; (f) 67,510; (g) 4128; (h) 735,696; (i) 15,924; (j) 4444; (k) 9,948,015; (l) 1,522,756.

Chapter 5, Section 6, Page 170

1. (a) Yes; (b) no; (c) no; (d) no; (e) no; (f) yes; (g) no; (h) no; (i) yes; (j) no.

2. (a) Yes; (b) no; (c) yes; (d) no.

3. In Exercise 2, only (a) and (c) are abelian.

5. (a) x; (b) The inverse of x is x; the inverse of y is y; the inverse of z is z; (c) no.

6. Yes

7. No

8. (a) Yes; (b) 2; (c) no; (d) yes, 6; (e) yes; (f) no.

9. Yes

10. (a) Yes; (b) yes; (c) yes;
(d) the inverse of 1 is 1;
the inverse of 5 is 5;
the inverse of 3 is 3;
the inverse of 7 is 7.
(e) Yes

Chapter 5, Section 7, Page 180

2. $r \circ (s \circ t) = (r \circ s) \circ t$ $q \circ (u \circ v) = (q \circ u) \circ v$
$r \circ v = q \circ t$ $q \circ s = v \circ v$
$w = w$ $p = p$
$v \circ (w \circ r) = (v \circ w) \circ r$
$v \circ v = r \circ r$
$p = p$

3. The inverse of X is X; the inverse of Y is Z; the inverse of Z is Y; the inverse of P is P; the inverse of Q is Q; the inverse of R is R.

7. (a)

∘	r	l	a	s
r	a	s	l	r
l	s	a	r	l
a	l	r	s	a
s	r	l	a	s

(b) Yes; (c) yes; (d) yes, s; (e) the inverse of r is l; the inverse of l is r; the inverse of a is a; the inverse of s is s; (f) r for both; (g) yes.

8. $Z \circ P$, which is Q is not equal to $P \circ Z$, which is R.

Chapter 6, Section 2, Page 187

1. (a) 49; (b) 2500; (c) 2464.

2. (a) 60; (b) 56; (c) 991; (d) 2; (e) $10n$.

3. 19683

4. (i)   11   (ii)   88   (iii)   9
       111   888   98
      1111   8888   987

   (iv) 111 and $1 + 1 + 1 = 3$ (v)   1
       222 and $2 + 2 + 2 = 6$ 121
      333 and $3 + 3 + 3 = 9$ 12321

   (vi)   49   (vii) 111111
      4489   222222
   444889   333333

7. Definitely not

Chapter 6, Section 3, Page 192

1. 377, 610, and 987

2. a) The sum of the first 4 numbers is one less than the sixth number, which is 8.
   b) The sum of the first 5 numbers is one less than the seventh number, which is 13.
   c) 20

3. 1, 1, 2, 3, 1, 0, 1, 1, 2, 3, 1, 0, 1, 1, 2, 3, 1, 0, 1, 1; yes

4. There are 13 possible ways by which the bee can get to cell 5. These are

$0 \rightarrow 1 \rightarrow 3 \rightarrow 4 \rightarrow 5$
$0 \rightarrow 1 \rightarrow 3 \rightarrow 5$
$0 \rightarrow 2 \rightarrow 3 \rightarrow 4 \rightarrow 5$
$0 \rightarrow 2 \rightarrow 3 \rightarrow 5$
$0 \rightarrow 2 \rightarrow 4 \rightarrow 5$
$0 \rightarrow 1 \rightarrow 2 \rightarrow 3 \rightarrow 4 \rightarrow 5$
$0 \rightarrow 1 \rightarrow 2 \rightarrow 3 \rightarrow 5$
$0 \rightarrow 1 \rightarrow 3 \rightarrow 5$
$1 \rightarrow 2 \rightarrow 3 \rightarrow 4 \rightarrow 5$
$1 \rightarrow 2 \rightarrow 3 \rightarrow 5$
$1 \rightarrow 2 \rightarrow 4 \rightarrow 5$
$1 \rightarrow 3 \rightarrow 4 \rightarrow 5$
$1 \rightarrow 3 \rightarrow 5$

5. The numbers that divide 10 and are less than 10 are 1, 2, and 5. Their sum, $1 + 2 + 5 = 8$, is not equal to 10.

6. $6_{(10)} = 110_{(2)}$
     $28_{(10)} = 11100_{(2)}$
  $496_{(10)} = 111110000_{(2)}$
$8128_{(10)} = 1111111000000_{(2)}$

Chapter 6, Section 4, Page 195

1. 1, 2, 3, 4, 6, 9, 12, 18, 36

2. 1, 2, 3, 4, 5, 6, 9, 10, 12, 15, 18, 20, 30, 36, 45, 60, 90, 180

3. 1, 3, 43, and 129

4. Every number divides 0.

6. 1210

7. (a) 10; (b) 165; (c) 1540; (d) the product of any three consecutive integers is always divisible by 6.

8. (a) Divisible by 2; (b) none; (c) none; (d) none; (e) none; (f) divisible by 2, 4; (g) divisible by 3; (h) divisible by 2, 3, and 5.

9. (b) Only 5430 is divisible by 6.

10. None are

11. One possible answer is 92520.

12. One possible answer is 901,260.

13. One possible answer is 100,150.

15. (a) No; (b) no; (c) yes; (d) yes; (e) no; (f) yes.

16. One possible answer is 144.

Chapter 6, Section 5, Page 201

1. 2, 3, 5, 7, 11, 13, 17, 19, 23, 29, 31, 37, 41, 43, 47

2. 2, 3

3. 14, 15, and 16 are one possible answer. Others are also possible.

4. 24, 25, 26, and 27 are one possible answer.

6. (a) $2 \times 2 \times 2 \times 2 \times 3$; (b) $3 \times 3 \times 3 \times 3$; (c) $2 \times 2 \times 2 \times 2 \times 5 \times 5 \times 5$; (d) $2 \times 2 \times 2 \times 2 \times 3 \times 11$; (e) 5×61; (f) $2 \times 2 \times 2 \times 3 \times 3 \times 5$; (g) $2 \times 2 \times 3 \times 5 \times 5$; (h) $2 \times 3 \times 5 \times 29$; (i) $2 \times 3 \times 3 \times 5 \times 7$; (j) $2 \times 2 \times 2 \times 2 \times 3 \times 5 \times 5$.

7. No, yes.

8. $17 = 4 \cdot 4 + 1$
$\quad 5 = 4 \cdot 1 + 1$ are three possible answers.
$\quad 29 = 4 \cdot 7 + 1$

9. $11 = 4 \cdot 2 + 3$
$\quad 19 = 4 \cdot 4 + 3$ are three possible answers.
$\quad 23 = 4 \cdot 5 + 3$

10. 3, 31, and 127 are three possible answers.

11. (a) 2 and 3 are two possible answers; (b) 4 and 5 are two possible answers.

12. 11 and 13
17 and 19 are three possible answers.
29 and 31

13. 3, 5, and 7

14. No

15. (a) 15; (b) 25.

16. (a) 17; (b) 257; (c) 1279.

Chapter 7, Section 2, Page 209

1. 3

2. 6

3. 6

4. (a) Point B; (b) $\overline{AC}$; (c) $\overline{CD}$; (d) the whole line; (e) point D; (f) $\overline{CD}$; (g) $\overrightarrow{EA}$; (h) $\varnothing$; (i) all points between B and C inclusive and all points between D and E inclusive; (j) $\overleftrightarrow{BC}$.

5. (a) All points on the line segments $\overline{BC}$ and $\overline{CD}$; (b) all points on the sides of the triangle BCD; (c) point C; (d) point E; (e) all points on the sides of the quadrilateral (four-sided figure) $ABCD$; (f) point C; (g) point B; (h) all points on the line segments $\overline{AD}$ and $\overline{DC}$.

6. No

7. Not necessarily

8. 4

Chapter 7, Section 3, Page 215

1. (a)

(b)

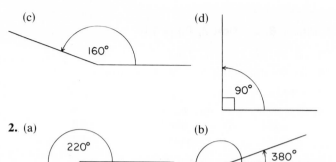

(c) 160°   (d) 90°

2. (a) 220°   (b) 380°

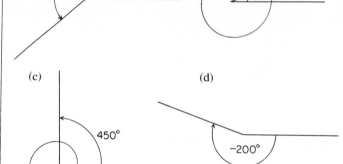

(c) 450°   (d) −200°

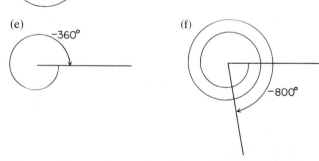

(e) −360°   (f) −800°

3. (a) e and f are adjacent; f and h are adjacent; h and g are adjacent; g and e are adjacent; a and b are adjacent; b and d are adjacent; d and c are adjacent; c and a are adjacent; f and g are vertical; e and h are vertical; a and d are vertical; b and c are vertical.

(b) i and j are adjacent; j and l are adjacent; l and k are adjacent; k and i are adjacent; m and n are adjacent; n and p are adjacent; p and o are adjacent; o and m are adjacent; i and l are vertical; k and j are vertical; m and p are vertical; n and o are vertical.

(c) q and s are adjacent; no vertical angles.

(d) s and t are adjacent and t and u are adjacent; there are no vertical angles.

4. (a) 135°; (b) −157.5°; (c) 1800°; (d) 330°; (e) 0°; (f) −288°; (g) −90°; (h) 240°; (i) −300°; (j) 60°; (k) −1080°; (l) 540°/π or 171.97°.

5. (a) $\frac{\pi}{36}$; (b) $\frac{2\pi}{3}$; (c) $\frac{-3\pi}{4}$; (d) $\frac{52\pi}{9}$; (e) $\frac{7\pi}{18}$; (f) $\frac{-25\pi}{36}$; (g) $\frac{43}{36}\pi$; (h) $\frac{-5\pi}{12}$; (i) $\frac{\pi}{180}$; (j) $\frac{5\pi}{3}$; (k) $\frac{-\pi}{4}$; (l) $\frac{(\pi)^2}{180}$.

Chapter 7, Section 4, Page 219

1. (a) Yes; (b) no; (c) yes; (d) yes; (e) no; (f) no.

2. (a) Yes; (b) no; (c) yes; (d) no.

3.

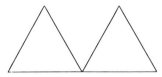

4.

5.

6. The sum of the angles is always 360°.

7. (a)

(b)

(c)

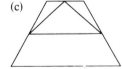

9. No

Chapter 7, Section 5, Page 223

1. (a) $m(\overline{DE}) = 52$. $m(\overline{EF}) = 48$;
(b) $m(\angle C) = 70°$, $m(\angle D) = 58°$, $m(\angle E) = 52°$;
(c) $m(\overline{BC}) = 8$, $m(\overline{ED}) = 15$;
(d) $m(\angle C) = 65°$, $m(\angle D) = 45°$, $m(\angle E) = 70°$.

2. (a) No; (b) yes. **3.** (a) No; (b) no.

4. (a) No; (b) yes; (c) no; (d) no; (e) yes; (f) yes; (g) yes; (h) yes.

5. 27 inches **6.** 5.71 feet

Chapter 7, Section 6, Page 227

1. (a) 18 square units; (b) 210 square units; (c) 25 square units; (d) 103.5 square units; (e) 108 square units; (f) 58 square units; (g) 70 square units; (h) 16π square units.

2. The second room is wider. **3.** Area $= b \cdot h$.

4. Area $= \frac{1}{2}bh + ah + \frac{1}{2}dh$ or Area $= h(a + \frac{1}{2}b + \frac{1}{2}d)$.

Chapter 8, Section 2, Page 246

1. 32 **2.** $6 \times 6 \times 6 = 216$ **3.** $7 \times 4 \times 3 = 84$

4. (a) (i) 27; (ii) 256; (iii) 3125; (b) (i) 6; (ii) 24; (iii) 120.

5. $3 \times 5 = 15$ **6.** 120 **7.** 1872 **8.** 456,976 **9.** 12

10. 240 **11.** (a) 132; (b) 12; (c) 32. **12.** 12 **13.** 18

Chapter 8, Section 3, Page 252

1. (a) 6; (b) 42; (c) 1; (d) 1; (e) 1; (f) 56; (g) 56; (h) 70; (i) 360; (j) 60; (k) 2; (l) 1; (m) 720; (n) 362,880; (o) 6; (p) 1.

2. (a) 60; (b) 90,720; (c) 120; (d) 45,360; (e) 302,400; (f) 129,729,600.

3. $6! = 720$ **4.** $_{10}P_4 = 5040$

5. (a) $9! = 362,880$; (b) $8! = 40,320$. **6.** $_6P_4 = 360$

7. $_7P_7 = 7! = 5040$

8. (a) $7! = 5040$; (b) $(3!)(4!) + (4!)(3!) = 288$.

9. $5! = 120$ **10.** $_4P_4 = 4! = 24$ **11.** $_6P_6 = 6! = 720$

12. $_5P_5 = 5! = 120$ **13.** (a) 9,000,000; (b) 604,800; (c) 10,000,000; (d) 1,000,000.

14. (a) $8! = 40,320$; (b) 5040; (c) 720.

15. (a) 135,200; (b) 121,680; (c) 93,600.

16. $_9P_4 = 3024$ **17.** $_{50}P_{10} = \dfrac{50!}{(50-10)!} = \dfrac{50!}{40!}$

Chapter 8, Section 4, Page 258

1. (a) 21; (b) 28; (c) 4; (d) 10; (e) 1; (f) 7; (g) 15; (h) 70.
3. $_8C_4 = 70$ **4.** $_{16}C_{12} = 1820$ **5.** $_9C_2 = 36$
6. $_7C_4 \cdot _9C_5 = 4410$ **7.** $_{52}C_5 = \dfrac{52!}{47!\,5!}$ **8.** $_5C_3 = 10$ **9.** 24
10. $_{10}C_3 \cdot _7C_6 \cdot _4C_2 = 5040$ **11.** $_{15}C_4 = 1365$
12. $_7C_3 \cdot _4C_4 + _7C_4 \cdot _4C_3 + _7C_5 \cdot _4C_2 + _7C_6 \cdot _4C_1 + _7C_7 \cdot _4C_0 = 330$
13. $_8C_5 = 56$
14. (a) $_{10}C_3 = 120$; (b) $_{10}C_7 = 120$; (c) they are the same.
15. $_{30}C_{15} \cdot _{20}C_{10} = \dfrac{30!}{15!\,15!} \cdot \dfrac{20!}{10!\,10!}$
16. (a) 1; (b) 2; (c) 3; (d) 5.

Chapter 8, Section 5, Page 265

1. (a) $\frac{1}{2}$; (b) $\frac{1}{12}$; (c) $\frac{1}{2}$; (d) 0; (e) 1.
2. (a) $\frac{1}{4}$; (b) $\frac{1}{13}$; (c) $\frac{1}{2}$; (d) $\frac{5}{13}$. **3.** (a) $\frac{1}{36}$; (b) $\frac{8}{9}$; (c) 0; (d) $\frac{7}{12}$.
4. (a) $\frac{3}{5}$; (b) $\frac{2}{5}$; (c) 0. **5.** (a) 0; (b) $\frac{1}{365}$. **6.** $\frac{3}{8}$
7. (a) $\frac{1}{15}$; (b) $\frac{2}{3}$; (c) $\frac{1}{3}$. **8.** $\frac{1}{24}$ **9.** $\frac{1}{4}$ **10.** $\frac{1}{2}$ **11.** $\frac{1}{2^{48}}$ **12.** No
13. $-\frac{3}{5}, \frac{5}{3}$ **14.** $\frac{1}{4}$ **15.** (a) $\frac{5}{8}$; (b) $\frac{11}{24}$; (c) $\frac{3}{8}$
16. $\frac{1}{10^7} = \frac{1}{10,000,000}$ **17.** $\frac{1}{63}$ **18.** (a) 0.97206; (b) 0.73429
19. (a) 0.143; (b) greater than.

Chapter 8, Section 6, Page 273

1. (a) Not mutually exclusive; (b) not mutually exclusive;
(c) mutually exclusive; (d) mutually exclusive;
(e) not mutually exclusive; (f) mutually exclusive;
(g) not mutually exclusive.
2. $\frac{9}{16}$ **3.** 0.40 **4.** $\frac{11}{40}$ **5.** 0.4 **6.** 0.979 **7.** 0.1 **8.** 1
9. 0.46 **10.** 0.35 **11.** 0.51 **12.** (a) $\frac{9}{29}$; (b) $\frac{3}{29}$. **13.** $\frac{1}{2}$ **14.** $\frac{61}{168}$

Chapter 8, Section 7, Page 278

1. 19 to 1 **2.** 3 to 247 **3.** 3 to 10 **4.** 903 to 97 **5.** 5 to 7
6. 0 **7.** 2.61 people **8.** $107.45
9. $\frac{82}{19}$ or $4.32 **10.** $-7.86 **11.** Location B

Chapter 8, Section 8, Page 282

1. All car owners whose license plate numbers are listed in
columns 1 and 2. (These are *all* the numbers.)
2. Those volunteers whose numbers are 150, 69, 143, 127,
42, 47, 185, 75, 3, 104, 15, 62, 110, 54, and 55.
5. Those restaurants whose numbers are 141, 248, 187, 58,
176, 298, 136, 47, 263, 287, 153, 147, 222, 86, 205, 254,
258, 253, 81, and 300.

Chapter 8, Section 9, Page 284

1. $\frac{1}{4}$
2. (a) 1; (b) 0; (c) cannot be determined from the given information; (d) cannot be determined from the given information.

Chapter 9, Section 3, Page 293

1. Mean = 17, median = 17.5, and mode = 14 and 18.
2. $761.36 **3.** Mean = 4.7, median = 3.5, and mode = 3.
4. Mean = 77, median = 78, and mode = none; probably the
mean is most important to her. **5.** $319.87
6. Mean = 90, median = 90, mode = 100.
8. The second student. **9.** Mode
10. No; there is no reason to believe that both classes have the
same number of students.
11. (a) Mean = 4.17, median = 4, and mode = 4; (b) the mean
would be 3.17, the mode would be 3, and the median would
be 3. Each would be decreased by 1.
12. (b) Chain A's prices vary considerably.

Chapter 9, Section 4, Page 297

1. (a) 7.5th percentile; (b) 85th percentile; (c) 45th percentile.
2. (a) 64th percentile; (b) not affected at all.

Chapter 9, Section 5, Page 300

1. Range = 13, standard deviation = $\sqrt{15.67}$ or 3.96, and
variance = 15.67
2. Range = 60, standard deviation = $\sqrt{263.33}$ or 16.23, and
variance = 263.33

3. Range = 45, standard deviation = $\sqrt{172.83}$ or 13.15, and variance = 172.83

4. Range = 16, standard deviation = $\sqrt{25.33}$ or 5.03, and variance = 25.33

5. Range = 28, standard deviation = $\sqrt{73.11}$ or 8.55, and variance = 73.11

6. (a) Mean = \$427, and median = \$432; there is no mode; (b) each is increased by 10; (c) Range = 86, standard deviation = $\sqrt{910.67}$ or 30.18, and variance = 910.67; (d) there is no change in the range, standard deviation, or variance.

7. The mean remains the same. The standard deviation is four times as great.

Chapter 9, Section 6, Page 305

1.

Number of rapes	Tally	Frequency
0	l	1
1		0
2	l	1
3	ШТ	5
4	ШТ l	6
5	ll	2
6		0
7	l	1
8	ll	2
9	llll	4
10	l	1
11	llll	4
12	l	1
13		0
14		0
15		0
16		0
17	l	1

2.

Height	Tally	Frequency
61	l	1
62	ll	2
63	lll	3
64	ll	2
65	ШТ l	6
66	ll	2
67	lll	3
68	lll	3
69	ll	2
70	l	1
71	ll	2
72	ll	2
73	l	1

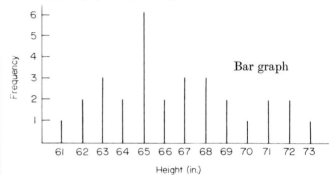

Bar graph

Height (in.)

3.

Number of people returning merchandise	Tally	Frequency
1		0
2	l	1
3		0
4		0
5	l	1
6		0
7	l	1
8	l	1
9	l	1
10	l	1
11		0
12	ll	2
13	l	1
14	l	1
15	lll	3
16	l	1
17	ШТ	5
18	l	1

Number of rapes reported

Bar graph for the preceding distribution

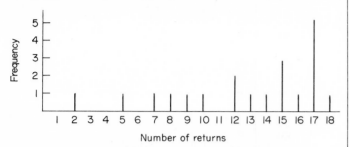

4.

Number of performers interviewed	Tally	Frequency
0	l	1
1	lll	3
2	lll	3
3	lll	3
4	lll	3
5	ll	2
6	l	1
7	llll	5
8	ll	2
9	ll	2

Bar graph for this distribution

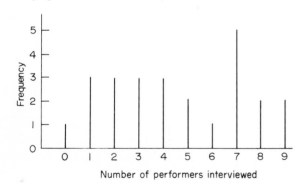

5.

Number of smugglers arrested	Tally	Frequency
0	l	1
1	l	1
2	l	1
3	ll	2
4	lll	3
5	l	1
6	lll	3
7	ll	2
8	lll	3
9	llll	4
10	ll	2
11	ll	2
12	llll	4
13	l	1
14	l	1

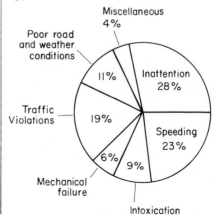

Chapter 9, Section 7, Page 310

1. (a) 1974 and 1976; (b) 1969 and 1973; (c) between 1973 and 1974.

2. (a) January; (b) July; (c) between April and May.

5.

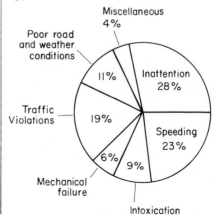

Chapter 9, Section 8, Page 313

2. (a) 68%; (b) 84%; (c) 16%; (d) 9600.

3. (a) 96%; (b) 14%; (c) 50%.

4. (a) 68%; (b) 16%; (c) 98%.

5. (a) 98%; (b) 84%; (c) 34%; (d) 68%; (e) one is twice as much as the other.

Chapter 10, Section 3, Page 326

1. a) $\begin{pmatrix} -5 & 8 \\ 6 & 8 \end{pmatrix}$ b) $\begin{pmatrix} -13 & -1 \\ +6 & 1 \end{pmatrix}$ c) $\begin{pmatrix} -2 & -2 \\ 5 & -7 \\ 3 & -1 \\ 5 & 4 \end{pmatrix}$

d) $(1 \quad -1)$ e) cannot be done f) $\begin{pmatrix} 7 & 1 & -7 \\ -4 & -10 & -10 \\ 4 & -6 & -10 \end{pmatrix}$

g) cannot be done h) $\begin{pmatrix} 3 \\ 1 \\ -5 \\ 8 \\ 9 \end{pmatrix}$ i) $\begin{pmatrix} 1 & -3 \\ 9 & -5 \\ -3 & -7 \\ 2 & 2 \\ -4 & 4 \end{pmatrix}$

j) cannot be done k) $\begin{pmatrix} -4 \\ 8 \end{pmatrix}$ l) $\begin{pmatrix} 0 & 0 \\ 0 & 0 \end{pmatrix}$ m) $\begin{pmatrix} -5 & -7 \\ -3 & -9 \end{pmatrix}$

3. a) $\begin{pmatrix} -1 & 6 \\ 10 & 9 \end{pmatrix}$ b) $\begin{pmatrix} -1 & 6 \\ 10 & 9 \end{pmatrix}$

Chapter 10, Section 4, Page 332

1. a) $\begin{pmatrix} 4 & -13 \\ 12 & 24 \end{pmatrix}$ b) $\begin{pmatrix} 11 \\ 18 \\ -13 \end{pmatrix}$ c) $\begin{pmatrix} 24 & 18 & 42 & -6 \\ 0 & 0 & 0 & 0 \\ 0 & 0 & 0 & 0 \\ 20 & 15 & 35 & -5 \end{pmatrix}$

d) (19) e) cannot be done f) (10) g) (-21)

h) $\begin{pmatrix} -16 & -28 \\ 2 & 4 \\ -7 & -17 \end{pmatrix}$ i) $\begin{pmatrix} 5 & -3 & 7 \\ 2 & -21 & 17 \\ 4 & 3 & -4 \end{pmatrix}$ j) (56)

2. $\begin{pmatrix} 1 & 0 & 0 & 0 \\ 0 & 1 & 0 & 0 \\ 0 & 0 & 1 & 0 \\ 0 & 0 & 0 & 1 \end{pmatrix}$

3. One possible answer is $M = \begin{pmatrix} 0 & 0 \\ 2 & -2 \end{pmatrix}$ and $N = \begin{pmatrix} 2 & 0 \\ 2 & 0 \end{pmatrix}$. Other answers are possible.

4. a) $\begin{pmatrix} -2 & 8 \\ 6 & 0 \end{pmatrix}$ b) $\begin{pmatrix} -3 & 15 & 18 \\ 6 & 0 & 6 \\ 15 & 0 & 21 \end{pmatrix}$ c) $\begin{pmatrix} -16 & 12 & -10 \\ 112 & 14 & 60 \\ -26 & -66 & -14 \end{pmatrix}$

d) $\begin{pmatrix} 9 & -7 \\ -17 & 55 \end{pmatrix}$ e) $\begin{pmatrix} -1 & -53 & -46 \\ 11 & -30 & -27 \\ 8 & -38 & 6 \end{pmatrix}$

5. $\begin{pmatrix} 0 & 0 \\ 0 & 0 \end{pmatrix}$ 6. $\begin{pmatrix} 1 & 0 \\ 0 & 1 \end{pmatrix}$

Chapter 10, Section 5, Page 338

1. (a) 141; (b) 145; (c) 318.

2. (a) 8; (b) 19; (c) nuts and orange juice.

3. (a) 105,000; (b) 33,000; (c) 45,000; (d) 60,000; (e) 37,000; (f) 5000

(g) $\begin{pmatrix} 80 \\ 110 \\ 120 \\ 150 \end{pmatrix}$ h) $(10 \quad 28 \quad 7 \quad 12) \cdot \begin{pmatrix} 80 \\ 110 \\ 120 \\ 150 \end{pmatrix} = \$6520$

4. a) $\begin{pmatrix} 10 & 3 & 18 & 7 & 9 \\ 2 & 15 & 19 & 10 & 0 \\ 7 & 10 & 15 & 14 & 0 \\ 0 & 12 & 10 & 14 & 8 \end{pmatrix}$ b) 183; (c) 19; (d) 17;

(e) 44; (f) 687

5. a)

	Frank	Coke	Salad	Veal cutlet
Bill	2	1	1	0
Phil	0	2	0	1
Will	1	1	1	0
Gil	1	1	1	$\frac{1}{2}$
Jill	3	2	0	$\frac{1}{2}$

b) $\begin{pmatrix} 50 \\ 35 \\ 55 \\ 90 \end{pmatrix}$ c) $1.40   d) 7

e) $\begin{pmatrix} 2 & 1 & 1 & 0 \\ 0 & 2 & 0 & 1 \\ 1 & 1 & 1 & 0 \\ 1 & 1 & 1 & \frac{1}{2} \\ 3 & 2 & 0 & \frac{1}{2} \end{pmatrix} \cdot \begin{pmatrix} 50 \\ 35 \\ 55 \\ 90 \end{pmatrix} = \begin{pmatrix} 190 \\ 160 \\ 140 \\ 185 \\ 265 \end{pmatrix}$

f) total spent is $9.40

6. a) $\begin{pmatrix} 23 & 27 & 26 \\ 6 & 5 & 12 \\ 32 & 32 & 31 \\ 10 & 15 & 12 \\ 51 & 58 & 46 \end{pmatrix}$ b) $\begin{matrix} \text{Oct} \\ \text{Nov} \\ \text{Dec} \end{matrix} \begin{pmatrix} 30 & 20 & 50 & 65 & 100 \\ 30 & 20 & 50 & 65 & 100 \\ 30 & 20 & 50 & 65 & 100 \end{pmatrix}$

c) $\begin{pmatrix} 23 & 27 & 26 \\ 6 & 5 & 12 \\ 32 & 32 & 31 \\ 10 & 15 & 12 \\ 51 & 58 & 46 \end{pmatrix} \cdot \begin{pmatrix} 30 & 20 & 50 & 65 & 100 \\ 30 & 20 & 50 & 65 & 100 \\ 30 & 20 & 50 & 65 & 100 \end{pmatrix}$

The total cost is $25,395.

Chapter 10, Section 6, Page 343

1. Because no one can communicate with himself.

2. (a) No;   (b) no.

3. (a) Lolly's, Fred's, and Jed's;
(b) Write to Jed to send her Fred's address.

4.

	Patient	Wife	Son	Girl-friend	Wife's mother	Wife's father
Patient	0	1	1	0	0	0
Wife	0	0	0	0	1	1
Son	0	0	0	1	0	0
Girlfriend	0	0	0	0	0	0
Wife's mother	0	0	0	0	0	0
Wife's father	0	0	0	0	0	0

5.

	Bill	Sam	Fred	Sidney
Bill	0	1	1	0
Sam	0	0	0	1
Fred	0	1	0	1
Sidney	1	0	0	0

6. (a) Only the president;   (b) 3;   (c) everyone.

Chapter 10, Section 7, Page 349

2. (a) $x = 1$, $y = 2$;   (b) $x = 1$, $y = -2$;
(c) $x = 13$, $y = -6$; (d) $x = 3$, $y = -1$; (e) $x = 1$, $y = 1$;
(f) $x = 2$, $y = 4$;   (g) $x = 7$, $y = 3$;   (h) $x = 3$, $y = 0$;
(i) this system of equations has no solution.

Chapter 11, Section 2, Page 359

1. Sue should show two fingers; Pete should show one finger.

3. Since the value of the game is $+1$.   **4.** No

5. Since we can find a number that is the minimum of its row and also the maximum of its column.

6. a) Not strictly determined
b) Strictly determined; value is 4; game is not fair. R plays row 3, C plays column 3.
c) Not strictly determined
d) Strictly determined; value is 5; game is not fair. R plays row 1, C plays column 2.
e) Strictly determined; value is 0; game is fair. R plays either row 1 or row 2, C plays column 2.
f) Strictly determined; value is 0; game is fair. R plays row 1, C plays column 1.
g) Not strictly determined
h) Strictly determined; value is 1; game is not fair. R plays row 1, C plays column 3.

7.

Ms. Casino

	Bk 6	Rd 4	Bk 8
Bk 3	9	−7	11
Rd 8	−14	12	−16
Rd 2	−8	6	−10

Mr. Rummy

The game is not strictly determined.

Dave

	1	2
1	2	−2
2	−2	4

8. Jean The game is not strictly determined.

9. (a)

Speedy

Shlep		Stonehill	Benton
	Stonehill	0	−6
	Benton	8	0

(b) Yes; (c) value is 0; (d) yes; (e) both companies should open in Benton.

Chapter 11, Section 3, Page 365

1. (a) $a = 2$, $b = 5$, $c = 4$, $d = 1$, $R_1 = \frac{1}{2}$, $r_2 = \frac{1}{2}$, $C_1 = \frac{2}{3}$, $C_2 = \frac{1}{3}$, $v = +3$
(b) $a = 5$, $b = -3$, $c = -4$, $d = -2$, $R_1 = \frac{1}{5}$, $R_2 = \frac{4}{5}$, $C_1 = \frac{1}{10}$, $C_2 = \frac{9}{10}$, $v = -\frac{11}{5}$
(c) $a = -3$, $b = 0$, $c = 0$, $d = -4$, $R_1 = \frac{4}{7}$, $R_2 = \frac{3}{7}$, $C_1 = \frac{4}{7}$, $C_2 = \frac{3}{7}$, $v = -\frac{12}{7}$
(d) $a = 5$, $b = 6$, $c = 7$, $d = 4$, $R_1 = \frac{3}{4}$, $R_2 = \frac{1}{4}$, $C_1 = \frac{1}{2}$, $C_2 = \frac{1}{2}$, $v = \frac{11}{2}$
(e) $a = 0$, $b = 7$, $c = 1$, $d = -4$, $R_1 = \frac{5}{12}$, $R_2 = \frac{7}{12}$, $C_1 = \frac{11}{12}$, $C_2 = \frac{1}{12}$, $v = \frac{7}{12}$
(f) $a = 9$, $b = -3$, $c = -2$, $d = -1$, $R_1 = \frac{1}{13}$, $R_2 = \frac{12}{13}$, $C_1 = \frac{2}{13}$, $C_2 = \frac{11}{13}$, $v = -\frac{15}{13}$

3. No

4. (a)

Chester guesses

Virginia hides		15	20
	15	−15	20
	20	15	−20

(c) 0; (d) yes; (e) Virginia: show 15 cigarettes $\frac{1}{2}$ of the time, and show 20 cigarettes $\frac{1}{2}$ of the time; Chester: guess 15 cigarettes $\frac{4}{7}$ of the time, and guess 20 cigarettes $\frac{3}{7}$ of the time; (f) Virginia should use a coin.

5. (a)

Jim

Sam		1	2
	1	1	−3
	2	−3	4

(b) $v = -\frac{5}{11}$; (c) no, Jim's; (d) Sam: show one finger $\frac{7}{11}$ of the time, and show two fingers $\frac{4}{11}$ of the time; Jim: show one finger $\frac{7}{11}$ of the time, and show two fingers $\frac{4}{11}$ of the time.

6. (a)

Sam

Jim		1	2
	1	2	−3
	2	−3	4

(b) $v = -\frac{1}{12}$; (c) no, Sam's; (d) Jim: show one finger $\frac{7}{12}$ of the time, and show two fingers $\frac{5}{12}$ of the time; Sam: show one finger $\frac{7}{12}$ of the time, and show two fingers $\frac{5}{12}$ of the time.

7. (a)

Enemy guns

Navy ship		Short range	Long range
	Near	−9 + 5	8 + 5
	Far	8	−9

Which can be simplified as

Enemy guns

Navy ship		Short range	Long range
	Near	−4	13
	Far	8	−9

(b) Navy ship: go near shore half the time, and stay far out half the time; enemy guns: short range $\frac{11}{17}$ of the time, and long range $\frac{6}{17}$ of the time.

8. Croke: Beach 1, $\frac{19}{31}$ of the time, and Beach 2, $\frac{12}{31}$ of the time; Seven-Down: Beach 1, $\frac{17}{31}$ of the time, and Beach 2, $\frac{14}{31}$ of the time.

Chapter 12, Section 4, Page 384

1. (a) Real variable; (b) neither; (c) real variable;
   (d) neither; (e) neither; (f) real variable; (g) neither;
   (h) integer variable; (i) integer variable; (j) neither;
   (k) neither; (l) real variable.

2. (a) $x + y = 3$; (b) $(x + y)z$; (c) $(x + y)^z$; (d) $5 \neq 3$;
   (e) $\left(\dfrac{a}{b}\right)^3$; (f) $\dfrac{a}{b^3}$; (g) $17 > x - 4$; (h) replace X by $X + 5$.

3. (a) `2 + 3. EQ. 5`; (b) `X * Y`; (c) `2 ** 3 + 4`;
   (d) `X / (3 * Y)`; (e) `(X + 3 * Y ** 2) ** 3`;
   (f) `(X + 3 * Y) / Z`; (g) `X + 3 * Y / Z`;
   (h) `X / (3 * Y * Z) + 5`.

4. (a) Statement number 71; (b) statement number 4;
   (c) statement number 3; (d) statement number 9;
   (e) statement number 6; (f) next statement number in the
   program; (g) statement number 200; (h) next statement
   number in the program; (i) statement number 111;
   (j) statement number 2.

5. (a) $X = 1$; (b) $X = 3$; (c) $X = 4$; (d) $X = 4$;
   (e) $X = 12$; (f) $X = 3$; (g) $X = 9$; (h) $X = 2$.

Chapter 12, Section 6, Page 393

2. (a) 681; (b) 5; (c) 13; (d) 5040; (e) 840; (f) $\frac{25}{13}$ or 1.92 (to
   two decimal places); (g) $\frac{16}{14}$ or 1.14 (to two decimal places);
   (h) 174.

Chapter 13, Section 2, Page 402

1. Let x = number of washing machines to be produced.
   Let y = number of dryers to be produced.
   $$15x + 30y \leq 480,$$
   $$120x + 180y \leq 480,$$
   $$x \geq 0 \quad \text{and} \quad y \geq 0,$$
   $$\max P = 70x + 102y.$$

2. Let x = number of coffee tables to be produced.
   Let y = number of sofas to be produced.
   Let z = number of bookcases to be produced.
   $$x + 3y + 5z \leq 20,$$
   $$3x + 5y + 4z \leq 50,$$
   $$2x + 4y + 8z \leq 50,$$
   $$x \geq 0, \quad y \geq 0, z \geq 0,$$
   $$\max P = 10x + 25y + 35z.$$

3. Let x = number of acres of alfalfa to be planted.
   Let y = number of acres of spinach to be planted.
   $$x + y \leq 200,$$
   $$2x + 8y \leq 320,$$
   $$20x + 40y \leq 2200,$$
   $$x \geq 0 \quad \text{and} \quad y \geq 0,$$
   $$\max P = 80x + 32y.$$

4. Let x = number of packages of mixture A to be made.
   Let y = number of packages of mixture B to be made.
   All weights are expressed in ounces.
   All profits are expressed in cents.
   $$18x + 15y \leq 2880,$$
   $$30x + 33y \leq 5280,$$
   $$x \geq 0 \quad \text{and} \quad y \geq 0,$$
   $$\max P = 204x + 210y.$$

5. Let x = number of units of whale blubber oil to be
   included in menu.
   Let y = number of units of castor oil to be included in
   menu.
   $$x + 3y \geq 8,$$
   $$3x + 4y \geq 19,$$
   $$3x + 3y \geq 7,$$
   $$x \geq 0 \quad \text{and} \quad y \geq 0,$$
   $$\min P = 50x + 25y.$$

6. Let x = number of jeeps to be produced.
   Let y = number of tanks to be produced.
   $$2x + 5y \leq 185,$$
   $$3x + 3y \leq 135,$$
   $$x \geq 0 \quad \text{and} \quad y \geq 0,$$
   $$\max P = 200x + 300y.$$

7. Let x = number of bikinis to be made.
   Let y = number of tanksuits to be made.
   $$2x + 4y \leq 12,$$
   $$4x + 2y \leq 12,$$
   $$x \geq 0 \quad \text{and} \quad y \geq 0,$$
   $$\max P = 3x + 5y.$$

8. Let x = number of health insurance policies to be sold.
   Let y = number of life insurance policies to be sold.
   $$3x + 4y \leq 24,$$
   $$x + 2y \leq 10,$$
   $$x \geq 0 \quad \text{and} \quad y \geq 0,$$
   $$\max P = x + y.$$

9. Let x = number of dozens of Special L to be made.
Let y = number of dozens of Krummy to be made.

$x + 2y \le 6$,
$2x + y \le 6$,
$x \ge 0$ and $y \ge 0$,
max $P = 3x + 4y$.

10. Let x = number of truckloads of loose-leaf paper to be made.
Let y = number of truckloads of notebooks to be made.

$2x + 3y \le 15$,
$3x + y \le 12$,
$x \ge 0$ and $y \ge 0$,
max $P = 25x + 35y$.

*11. Let x = number of tables to be made.
Let y = number of chairs to be made.
Let z = number of bookcases to be made.

$x \ge 8$,
$y \ge 12$,
$z \le 6$,
$10x + 2y + 24z \le 304$,
 $6x + 4y + 20z \le 188$,
$x \ge 0$, $y \ge 0$,   and $z \ge 0$,
max $P = 24x + 10y + 20z$.

Chapter 13, Section 3, Page 409

1. (a)

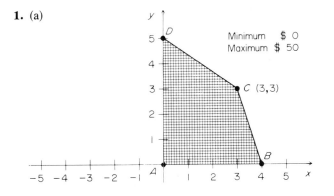

(b)

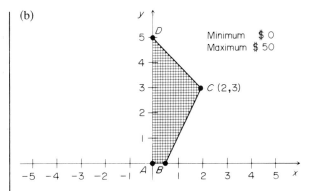

(c)

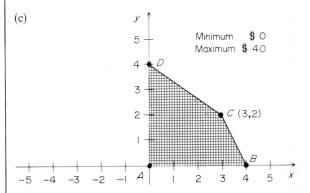

(d)

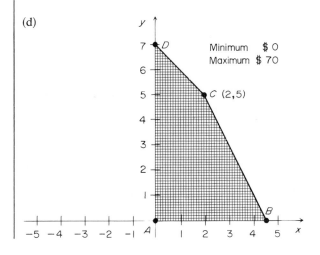

2.

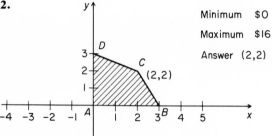

Minimum $0
Maximum $16
Answer (2,2)

4.

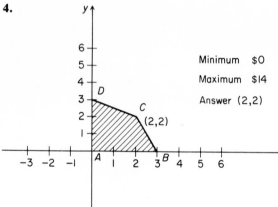

Minimum $0
Maximum $14
Answer (2,2)

5.

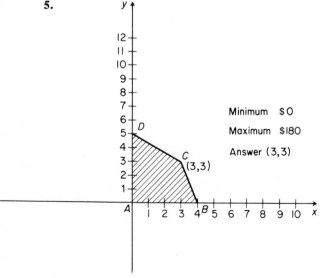

Minimum $0
Maximum $180
Answer (3,3)

10.

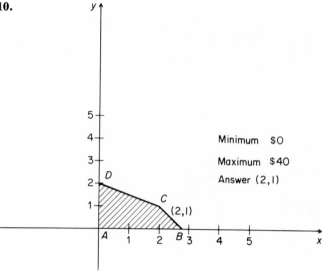

Minimum $0
Maximum $40
Answer (2,1)

Chapter 13, Section 4, Page 412

1. $15x + 30y + r = 480,$
$120x + 180y + s = 480,$
$x \geq 0, \quad y \geq 0, \quad r \geq 0, \quad \text{and} \quad s \geq 0,$
$\max P = 70x + 102y + 0r + 0s.$

2. $x + 3y + 5z + r = 20,$
$3x + 5y + 4z + s = 50,$
$2x + 4y + 8z + t = 50,$
$x \geq 0, y \geq 0, z \geq 0, r \geq 0, s \geq 0, \text{and } t \geq 0,$
$\max P = 10x + 25y + 35z + 0r + 0s + 0t.$

3. $x + y + r = 200,$
$2x + 8y + s = 320,$
$20x + 40y + t = 2200,$
$x \geq 0, \quad y \geq 0, \quad r \geq 0, \quad s \geq 0, \quad \text{and} \quad t \geq 0,$
$\max P = 80x + 32y + 0r + 0s + 0t.$

4. $18x + 15y + r = 2880,$
$30x + 33y + s = 5280,$
$x \geq 0, \quad y \geq 0, \quad r \geq 0, \quad \text{and} \quad s \geq 0,$
$\max P = 204x + 210y + 0r + 0s.$

5. $x + 3y - r = 8,$
$3x + 4y - s = 19,$
$3x + 3y - t = 7,$
$x \geq 0, \quad y \geq 0, \quad r \geq 0, \quad s \geq 0, \quad \text{and} \quad t \geq 0,$
$\min P = 50x + 25y + 0r + 0s + 0t.$

6. $2x + 5y + r \quad = 185,$
$3x + 3y \quad + s = 135,$
$x \geqslant 0, \quad y \geqslant 0, \quad r \geqslant 0, \quad \text{and} \quad s \geqslant 0,$
$\max P = 200x + 300y + 0r + 0s.$

7. $2x + 4y + r \quad = 12,$
$4x + 2y \quad + s = 12,$
$x \geqslant 0, \quad y \geqslant 0, \quad r \geqslant 0, \quad \text{and} \quad s \geqslant 0,$
$\max P = 3x + 5y + 0r + 0s.$

8. $3x + 4y + r \quad = 24,$
$x + 2y \quad + s = 10,$
$x \geqslant 0, \quad y \geqslant 0, \quad r \geqslant 0, \quad \text{and} \quad s \geqslant 0,$
$\max P = x + y + 0r + 0s.$

9. $x + 2y + r \quad = 6,$
$2x + y \quad + s = 6,$
$x \geqslant 0, \quad y \geqslant 0, \quad r \geqslant 0, \quad \text{and} \quad s \geqslant 0,$
$\max P = 3x + 4y + 0r + 0s.$

10. $2x + 3y + r \quad = 15,$
$3x + y \quad + s = 12,$
$x \geqslant 0, \quad y \geqslant 0, \quad r \geqslant 0, \quad \text{and} \quad s \geqslant 0,$
$\max P = 25x + 35y + 0r + 0s.$

11. a) $2x + y + r \quad = 16,$
$x + 2y \quad + s \quad = 11,$
$x + 3y \quad + t = 15,$
$x \geqslant 0, \quad y \geqslant 0, \quad r \geqslant 0, \quad s \geqslant 0, \quad \text{and} \quad t \geqslant 0,$
$\max P = 30x + 50y + 0r + 0s + 0t.$

b) $5x + 2y - r \quad = 180,$
$x + y \quad - s = 45,$
$x \geqslant 0, \quad y \geqslant 0, \quad r \geqslant 0, \quad \text{and} \quad s \geqslant 0,$
$\min P = 300x + 200y + 0r + 0s.$

c) $x + 2y + 5z - r \quad = 4,$
$2x + y + 2z \quad - s \quad = 3,$
$3x + 5y + z \quad + t = 1,$
$x \geqslant 0, \quad y \geqslant 0, \quad z \geqslant 0, \quad r \geqslant 0, \quad s \geqslant 0, \quad \text{and} \quad t \geqslant 0.$
$\min P = 2x + 6y + 7z + 0r + 0s + 0t.$

d) $2x + y - r \quad = 7,$
$x - 3y \quad + s = 4,$
$x \geqslant 0, \quad y \geqslant 0, \quad r \geqslant 0 \quad \text{and} \quad s \geqslant 0,$
$\max P = 9x + 8y + 0r + 0s.$

Chapter 13, Section 5, Page 420

1.

C_B	Vectors in basis	b	70 a_1	102 a_2	0 a_3	0 a_4
0	a_3	480	15	30	1	0
0	a_4	480	120	180	0	1
		0	-70	-102	0	0

2.

C_B	Vectors in basis	b	10 a_1	25 a_2	35 a_3	0 a_4	0 a_5	0 a_6
0	a_4	20	1	3	5	1	0	0
0	a_5	50	3	5	4	0	1	0
0	a_6	50	2	4	8	0	0	1
		0	-10	-25	-35	0	0	0

3.

C_B	Vectors in basis	b	80 a_1	32 a_2	0 a_3	0 a_4	0 a_5
0	a_3	200	1	1	1	0	0
0	a_4	320	2	8	0	1	0
0	a_5	2200	20	40	0	0	1
		0	-80	-32	0	0	0

4.

C_B	Vectors in basis	b	204 a_1	210 a_2	0 a_3	0 a_4
0	a_3	2880	18	15	1	0
0	a_4	5280	30	33	0	1
		0	-204	-210	0	0

6.

C_B	Vectors in basis	b	200 a_1	300 a_2	0 a_3	0 a_4
0	a_3	185	2	5	1	0
0	a_4	135	3	3	0	1
		0	-200	-300	0	0

7.

C_B	Vectors in basis	b	3 a_1	5 a_2	0 a_3	0 a_4
0	a_3	12	2	4	1	0
0	a_4	12	4	2	0	1
			0	-3	-5	0	0

8.

C_B	Vectors in basis	b	1 a_1	1 a_2	0 a_3	0 a_4
0	a_3	24	3	4	1	0
0	a_4	10	1	2	0	1
			0	-1	-1	0	1

9.

C_B	Vectors in basis	b	3 a_1	4 a_2	0 a_3	0 a_4
0	a_3	6	1	2	1	0
0	a_4	6	2	1	0	1
			0	-3	-4	0	0

Chapter 13, Section 6, Page 434

1. Tableau 1

C_B	Vectors in basis	b	70 a_1	102 a_2	0 a_3	0 a_4
0	a_3	480	15	30	1	0
0	a_4	480	120	180	0	1
			0	-70	-102	0	0

Tableau 2

C_B	Vectors in basis	b	70 a_1	102 a_2	0 a_3	0 a_4
0	a_3	400	-5	0	1	$-1/6$
102	a_2	8/3	2/3	1	0	1/180
			272	-2	0	0	17/30

Tableau 3

C_B	Vectors in basis	b	70 a_1	102 a_2	0 a_3	0 a_4
0	a_3	420	0	15/2	1	$-1/8$
70	a_1	4	1	3/2	0	1/120
			280	0	3	0	7/12

Answer: Produce 4 washing machines, 0 dryers.
Profit $280.

2. Tableau 1

C_B	Vectors in basis	b	80 a_1	32 a_2	0 a_3	0 a_4	0 a_5
0	a_3	200	1	1	1	0	0
0	a_4	320	2	8	0	1	0
0	a_5	2200	20	40	0	0	1
			0	-80	-32	0	0	0

Tableau 2

C_B	Vectors in basis	b	80 a_1	32 a_2	0 a_3	0 a_4	0 a_5
0	a_3	90	0	-1	1	0	$-1/20$
0	a_4	100	0	6	0	1	$-1/10$
80	a_1	110	1	2	0	0	1/20
			8800	0	128	0	0	4

Answer: Plant   110 acres of alfalfa,
0 acres of spinach.
Profit $8,000.

3. Tableau 1

C_B	Vectors in basis	b	3 a_1	4 a_2	0 a_3	0 a_4
0	a_3	6	1	2	1	0
0	a_4	6	2	1	0	1
			0	-3	-4	0	0

Tableau 2

C_B	Vectors in basis	b	3 a_1	4 a_2	0 a_3	0 a_4
4	a_2	3	1/2	1	1/2	0
0	a_4	3	3/2	0	-1/2	1
		12	-1	0	2	0

Tableau 3

C_B	Vectors in basis	b	3 a_1	4 a_2	0 a_3	0 a_4
4	a_2	2	0	1	2/3	-1/3
3	a_1	2	1	0	-1/3	2/3
		14	0	0	5/3	2/3

Answer: Produce 2 dozen of Special L
2 dozen of Krummy.
Profit $14.

4. Tableau 1

C_B	Vectors in basis	b	25 a_1	35 a_2	0 a_3	0 a_4
0	a_3	15	2	3	1	0
0	a_4	12	3	1	0	1
		0	-25	-35	0	0

Tableau 2

C_B	Vectors in basis	b	25 a_1	35 a_2	0 a_3	0 a_4
35	a_2	5	2/3	1	1/3	0
0	a_4	7	7/3	0	-1/3	1
		175	-5/3	0	35/3	0

Tableau 3

C_B	Vectors in basis	b	25 a_1	35 a_2	0 a_3	0 a_4
35	a_2	3	0	1	3/7	-2/7
25	a_1	3	1	0	-1/7	3/7
		180	0	0	80/7	5/7

Answer: Produce 3 truckloads of loose-leaf paper,
3 truckloads of notebooks.
Profit $180.

5. Tableau 1

C_B	Vectors in basis	b	3 a_1	5 a_2	0 a_3	0 a_4
0	a_3	12	2	4	1	0
0	a_4	12	4	2	0	1
		0	-3	-5	0	0

Tableau 2

C_B	Vectors in basis	b	3 a_1	5 a_2	0 a_3	0 a_4
5	a_2	3	1/2	1	1/4	0
0	a_4	6	3	0	-1/2	1
		15	-1/2	0	5/4	0

Tableau 3

C_B	Vectors in basis	b	3 a_1	5 a_2	0 a_3	0 a_4
5	a_2	2	0	1	1/3	-1/6
3	a_1	2	1	0	-1/6	1/3
		16	0	0	7/6	1/6

Answer: Produce 2 bikinis,
2 tank suits.
Profit $16.

6. Tableau 1

C_B	Vectors in basis	b	1 a_1	1 a_2	0 a_3	0 a_4
0	a_3	24	3	4	1	0
0	a_4	10	1	2	0	1
		0	−1	−1	0	0

Tableau 2

C_B	Vectors in basis	b	1 a_1	1 a_2	0 a_3	0 a_4
1	a_1	8	1	4/3	1/3	0
0	a_4	2	0	2/3	−1/3	1
		8	0	1/3	1/3	0

Answer: Sell 8 health insurance policies,
0 life insurance policies.
Points 8.

Note If we chose to insert a_2 instead of a_1 in getting Tableau 2, we would eventually get the same answers. However, four tableaus would then be needed.

7. Tableau 1

C_B	Vectors in basis	b	204 a_1	210 a_2	0 a_3	0 a_4
0	a_3	2880	18	15	1	0
0	a_4	5280	30	33	0	1
		0	−204	−210	0	0

Tableau 2

C_B	Vectors in basis	b	204 a_1	210 a_2	0 a_3	0 a_4
0	a_3	480	48/11	0	1	−15/33
210	a_2	160	10/11	1	0	1/33
		33600	−144/11	0	0	70/11

Tableau 3

C_B	Vectors in basis	b	204 a_1	210 a_2	0 a_3	0 a_4
204	a_1	110	1	0	11/48	−5/48
210	a_2	60	0	1	−5/24	1/8
		35040	0	0	3	55/11

Answer: Produce 110 packages of mixture A,
60 packages of mixture B.
Profit $350.40

8. Tableau 1

C_B	Vectors in basis	b	10 a_1	25 a_2	35 a_3	0 a_4	0 a_5	0 a_6
0	a_4	20	1	3	5	1	0	0
0	a_5	50	3	5	4	0	1	0
0	a_6	50	2	4	8	0	0	1
		0	−10	−25	−35	0	0	0

Tableau 2

C_B	Vectors in basis	b	10 a_1	25 a_2	35 a_3	0 a_4	0 a_5	0 a_6
35	a_3	4	1/5	3/5	1	1/5	0	0
0	a_5	34	11/5	13/5	0	−4/5	1	0
0	a_6	18	2/5	−4/5	0	−8/5	0	1
		140	−3	−4	0	7	0	0

Tableau 3

C_B	Vectors in basis	b	10 a_1	25 a_2	35 a_3	0 a_4	0 a_5	0 a_6
25	a_2	20/3	1/3	1	5/3	1/3	0	0
0	a_5	50/3	4/3	0	−13/3	−5/3	1	0
0	a_6	70/3	2/3	0	4/3	−4/3	0	1
		500/3	−5/3	0	20/3	25/3	0	0

Tableau 4

C_B	Vectors in basis	b	10 a_1	25 a_2	35 a_3	0 a_4	0 a_5	0 a_6
25	a_2	5/2	0	1	7/12	3/4	$-1/4$	0
35	a_1	25/2	1	0	$-13/4$	$-5/4$	3/4	0
0	a_6	15	0	0	7/2	$-1/2$	$-1/2$	1
		375/2	0	0	5/4	25/4	5/4	0

Answer: Produce $\dfrac{25}{2}$ coffee tables,

$\dfrac{5}{2}$ sofas,

0 bookcases.

Profit $\$\dfrac{375}{2}$.

9. Let x = the number of *Normal* mousetraps to be produced.

Let y = the number of *Big* traps to be produced.

$x + 3y \leq 9$,
$2x + y \leq 8$,
$x \geq 0$ and $y \geq 0$,
max $P = x + 4y$.

Tableau 1

C_B	Vectors in basis	b	1 a_1	4 a_2	0 a_3	0 a_4
0	a_3	9	1	3	1	0
0	a_4	8	2	1	0	1
		0	-1	-4	0	0

Tableau 2

C_B	Vectors in basis	b	1 a_1	4 a_2	0 a_3	0 a_4
4	a_2	3	1/3	1	1/3	0
0	a_4	5	5/3	0	$-1/3$	1
		12	1/3	0	4/3	0

Answer: Produce 0 *Normal* traps,
3 *Big* traps.
Profit $12.

10. Let x = the number of pinball machines to be installed. Let y = the number of hockey games to be installed. The problem then is

$x + 3y \leq 9$,
$2x + y \leq 8$,
$x \geq 0$ and $y \geq 0$,
max $P = x + y$.

Tableau 1

C_B	Vectors in basis	b	1 a_1	1 a_2	0 a_3	0 a_4
0	a_3	9	1	3	1	0
0	a_4	8	2	1	0	1
		0	-1	-1	0	0

Tableau 2

C_B	Vectors in basis	b	1 a_1	1 a_2	0 a_3	0 a_4
0	a_3	5	0	5/2	1	$-1/2$
1	a_1	4	1	1/2	0	1
		4	0	$-1/2$	0	1/2

Tableau 3

C_B	Vectors in basis	b	1 a_1	1 a_2	0 a_3	0 a_4
1	a_2	2	0	1	2/5	$-1/5$
1	a_1	3	1	0	$-1/5$	11/10
		5	0	0	1/5	2/5

Answer: Install 3 pinball machines,
2 hockey games.
Profit $5.

Index